Lecture Notes in Computational Vision and Biomechanics

Volume 23

The research related to the analysis of living structures (Biomechanics) has been a source of recent research in several distinct areas of science, for example, Mathematics, Mechanical Engineering, Physics, Informatics, Medicine and Sport. However, for its successful achievement, numerous research topics should be considered, such as image processing and analysis, geometric and numerical modelling, biomechanics, experimental analysis, mechanobiology and enhanced visualization, and their application to real cases must be developed and more investigation is needed. Additionally, enhanced hardware solutions and less invasive devices are demanded.

On the other hand, Image Analysis (Computational Vision) is used for the extraction of high level information from static images or dynamic image sequences. Examples of applications involving image analysis can be the study of motion of structures from image sequences, shape reconstruction from images, and medical diagnosis. As a multidisciplinary area, Computational Vision considers techniques and methods from other disciplines, such as Artificial Intelligence, Signal Processing, Mathematics, Physics and Informatics. Despite the many research projects in this area, more robust and efficient methods of Computational Imaging are still demanded in many application domains in Medicine, and their validation in real scenarios is matter of urgency.

These two important and predominant branches of Science are increasingly considered to be strongly connected and related. Hence, the main goal of the LNCV&B book series consists of the provision of a comprehensive forum for discussion on the current state-of-the-art in these fields by emphasizing their connection. The book series covers (but is not limited to):

- Applications of Computational Vision and Biomechanics
- Biometrics and Biomedical Pattern Analysis
- Cellular Imaging and Cellular Mechanics
- Clinical Biomechanics
- Computational Bioimaging and Visualization
- Computational Biology in Biomedical Imaging
- Development of Biomechanical Devices
- Device and Technique Development for Biomedical Imaging
- Digital Geometry Algorithms for Computational Vision and Visualization
- Experimental Biomechanics
- Gait & Posture Mechanics
- Multiscale Analysis in Biomechanics
- Neuromuscular Biomechanics
- Numerical Methods for Living Tissues
- Numerical Simulation
- Software Development on Computational Vision and Biomechanics
- Grid and High Performance Computing for Computational Vision and Biomechanics
- Image-based Geometric Modeling and Mesh Generation
- Image Processing and Analysis
- Image Processing and Visualization in Biofluids
- Image Understanding
- Material Models
- Mechanobiology
- Medical Image Analysis
- Molecular Mechanics
- Multi-Modal Image Systems
- Multiscale Biosensors in Biomedical Imaging
- Multiscale Devices and Biomems for Biomedical Imaging
- Musculoskeletal Biomechanics
- Sport Biomechanics
- Virtual Reality in Biomechanics
- Vision Systems

More information about this series at http://www.springer.com/series/8910

Guoyan Zheng · Shuo Li
Editors

Computational Radiology for Orthopaedic Interventions

 Springer

Editors
Guoyan Zheng
Institute for Surgical Technology
  and Biomechanics
University of Bern
Bern
Switzerland

Shuo Li
GE Healthcare and University of Western
  Ontario
London, ON
Canada

ISSN 2212-9391 ·   ISSN 2212-9413   (electronic)
Lecture Notes in Computational Vision and Biomechanics
ISBN 978-3-319-38727-7   ISBN 978-3-319-23482-3   (eBook)
DOI 10.1007/978-3-319-23482-3

Printed on acid-free paper

Springer International Publishing AG Switzerland is part of Springer Science+Business Media
(www.springer.com)

Preface

Due to technology innovations, the applications of medical imaging in orthopaedic interventions are pervasive, ranging from diagnosis and pre-operative surgical planning, intra-operative guidance and post-operative treatment evaluation and follow-up. The rapid adoption of DICOM standard makes the large image databases readily available in orthopaedics for multi-modal, multi-temporal and multi-subject assessment. Consequently, accurate and (semi-) automatic quantitative image computing is indispensable for various orthopaedic inventions, leading to the creation of a new emerging field called computation radiology. The past two decades have witnessed a rapid development and applications of computational radiology.

Responding to the continued and growing demand for computational radiology, this book provides a cohesive overview of the current technological advances in this emerging field, and their applications in orthopaedic interventions. It discusses the technical and clinical aspects of computational radiology and covers intra-operative imaging and computing for orthopaedic procedures. The book is aimed at both the graduate students embarking at a career in computational radiology, and the practicing researchers or clinicians who need an update of the state of the art in both the principles and practice of this emerging discipline.

Contributed by the leading researchers in the field, this book covers not only the basic computational radiology techniques such as statistical shape modelling, CT/MRI segmentation, augmented reality and micro-CT image processing, but also the applications of these techniques to various orthopaedic interventional tasks. Details about following important state-of-the-art development are featured: 3D pre-operative planning and patient-specific instrumentation for surgical treatment of long-bone deformities, computer-assisted diagnosis and planning of periacetabular osteotomy and femoroacetabular impingement, 2D–3D reconstruction-based planning of total hip arthroplasty, image fusion for computer-assisted bone tumour surgery, intra-operative three-dimensional imaging in fracture treatment, augmented reality-based orthopaedic interventions and education, medical robotics for musculoskeletal surgery, inertial sensor-based cost-effective surgical navigation and computer-assisted hip resurfacing using patient-specific instrument guides.

We sincerely thank our colleagues for their hard work in making their contributions and reviewers for providing timely reviews to us. Special thanks to Ms. Alice Ko who coordinated the long and difficult process of editing and review. Finally, we deeply appreciated the intention of the publisher to make this book possible.

Guoyan Zheng
Shuo Li

Contents

Statistical Shape Modeling of Musculoskeletal Structures and Its Applications

Hans Lamecker and Stefan Zachow

Abstract Statistical shape models (SSM) describe the shape variability contained in a given population. They are able to describe large populations of complex shapes with few degrees of freedom. This makes them a useful tool for a variety of tasks that arise in computer-aided medicine. In this chapter we are going to explain the basic methodology of SSMs and present a variety of examples, where SSMs have been successfully applied.

1 Introduction

The morphology of anatomical structures plays an important role in medicine. Not only does the shape of organs, tissues and bones determine the aesthetic appearance of the human body, but it is also strongly intertwined with its physiology. A prominent example is the musculoskeletal system, where the shape of bones is an integral component in understanding the complex biomechanical behavior of the human body. Such understanding is the key to improving therapeutic approaches, e.g. for treating congenital diseases, traumata, degenerative phenomena like osteoporosis, or cancer.

With the advent of modern imaging systems like X-ray computed-tomography (CT), magnetic resonance imaging (MRI), three-dimensional (3D) ultrasound (US) or 3D photogrammetry a variety of methods is available both for capturing the 3D shape of anatomical structures inside or on the surface of the body. This has opened up the opportunity of more detailed diagnosis, planning as well as intervention on a patient-specific basis. In order to transfer such developments into clinical routine and facilitate access for every patient in a cost-effective way, efficient and reliable methods for processing and analyzing shape data are called for.

H. Lamecker (✉) · S. Zachow
Zuse Institute Berlin (ZIB), Takustr. 7, 14195 Berlin, Germany
e-mail: hans.lamecker@1000shapes.com

H. Lamecker
1000shapes GmbH, Wiesenweg 10, 12247 Berlin, Germany

© Springer International Publishing Switzerland 2016
G. Zheng and S. Li (eds.), *Computational Radiology for Orthopaedic Interventions*, Lecture Notes in Computational Vision and Biomechanics 23, DOI 10.1007/978-3-319-23482-3_1

1

In this chapter, we are going to turn the attention on a methodology, which shows great promises for efficient and reliable processing and analysis of 3D shape data in the context of orthopedic applications. We are going to describe the conceptual framework as well as illustrate the potential impact to improving health care with selected examples from different applications.

This chapter is not intended to give an exhaustive overview over the work done in field of statistical shape modeling. Instead, it shall serve as an introduction to the technology and its applications, with the hope that the reader is inspired to convey the presented ideas to his field of work.

2 Statistical Shape Modeling

In this section, the basic conceptual framework of statistical shape modeling is described. There is a large variety of different approaches to many aspects of statistical shape modeling, such as shape representation or comparison techniques, which shall not be covered here. Instead, we are focusing on extracting the essential links and facts in order to understand the power of statistical shape modeling in the context of the applications. The reader interested in more details is referred to [16]. An overview specific to bone anatomy is presented by Sarkalkan et al. [19].

2.1 Representation

3D shape describes the external boundary form (surface) of an object, independent of its location in space. The size of an object hence is part of its shape. For the scope of this chapter, it suffices to know that mathematically, a surface S is represented by a— in general infinite—number of parameters and/or functions x, which describe the embedding of the surface in space, and thus its form. The computerized digital surface representation $S(x)$ in general approximates the shape by reducing the infinite number of parameters x to a finite set. One commonly used representation in computer graphics are triangle meshes, which are point clouds, where the points are connected by triangles, but many other representations like skeletons, splines, etc. are also used.

2.2 Comparison

The fundamental task in analyzing shapes is to compare two shapes S_1 and S_2. This means that for each parameter x_1 for shape S_1 a corresponding parameter x_2 for shape S_2 is identified. One important prerequisite is that such an identification method needs to be independent of the location of the two shapes in 3D space. Such a process is also referred to as matching or registration. For example, for each 3D

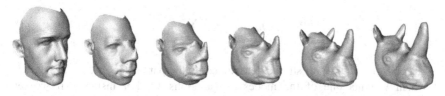

Fig. 1 Transformation of a human into a rhinoceros head is made possible through the representation of shapes in a common shape space

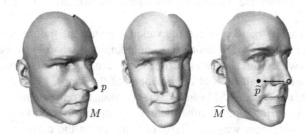

Fig. 2 When the tip of the nose on the *left* head is matched with a point on the cheek on the *right* head, shape interpolation may yield a head with two noses

point on one surface a corresponding 3D point on the other surface may be identified. For other representations, these may not be 3D points but e.g. skeletal parameters, etc. As a consequence one can establish a so-called shape space, where shapes may be treated just like numbers in order to perform calculations on shapes, like e.g. averaging $S_3 = 0.5 \cdot (S_1 + S_2)$ or any other interpolation, see Fig. 1.

Is is obvious that the details of the correspondence identification has a great impact on subsequent analysis, see Fig. 2. Nevertheless, a proper definition for "good" correspondences is difficult to establish in general, and in most cases must be provided in the context of the application. One generic approach to establish correspondence, however, optimizes the resulting statistical shape model built from the correspondences, i.e. its compactness or generalization ability, see Sect. 2.4. Refer to Davies et al. [3] for more details.

2.3 Statistical Analysis

As soon as we are able to perform "shape arithmetic" in a shape space, we can perform any kind of statistical analysis, e.g. like principal component analysis (PCA). This kind of analysis takes as input a set of training shapes $S_1, \ldots, S_n$ and extracts the so-called modes of variations $V_1, \ldots, V_{n-1}$ sorted by their variance in the training set. Together with the mean shape $\overline{S}$ the modes of variations form a statistical shape model (SSM):

$$S(b_1, \ldots, b_{n-1}) = \overline{S} + \sum_{k=1}^{n-1} b_k \cdot V_k \qquad (1)$$

The SSM is a family of shapes determined by the parameters $b_1, \ldots, b_{n-1}$, each of which weights one of the modes of variations V_k. For instance, if we set $b_2 = \cdots = b_{n-1} = 0$ and vary only b_1 we will see the effect of the first mode of variation on the deformations of the shapes within the range of the training population. The PCA may be exchanged with other methods, which will alter the interpretation of both the V_k and their weights b_k. See Sect. 3.3.2 for such an alternative approach.

In the PCA case, the idea is that the variance within the training population is contained in only few modes, hence the whole family of shapes can be described with only a few "essential degrees of freedom", see Fig. 3. Note that the shape variations V_i are generally global deformations of the shape, i.e. they vary every point on the shape. Thus, a SSM can be a highly compact representation of a family of complex geometric shapes. Furthermore, it is straightforward to synthesize new shapes by choosing new weights. These may lie within the range of the training shapes or even extrapolate beyond that range.

2.4 Evaluation

With the SSM we can represent any shape as a linear combination of the input shapes or some kind of transformation of those input shapes, e.g. via PCA. However, up to what accuracy can we represent/reconstruct an unknown shape S^* by an SSM $S(b)$? The idea is to compute the best approximation of the SSM to the unknown shape:

$$b^* = \operatorname*{argmin}_{b} d(S^*, S(b)) \qquad (2)$$

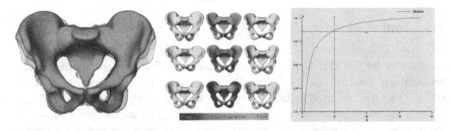

Fig. 3 *Left* Overlay of several training shapes. *Middle* First three modes of variation from *top* to *bottom*. Local deformation strength is color-coded. *Right* 90 % variation lie within 15 shape modes

where $d(\cdot, \cdot)$ is a measure for the distance between two shapes. Then $S(b^*)$ is the best approximation of S^* within the SSM.

If $d(S^*, S(b^*)) = 0$ then the unknown shape is already "contained" in the SSM, otherwise the SSM is not capable of explaining S^*. The smaller $d(S^*, S(b^*))$ for any S^*, the higher the so-called generalizability of the SSM. A better generalizability can be achieved by including more training samples into the model generation process. A good SSM has a high generalizability or reconstruction capability. On the other hand, if $S(b)$ for an arbitrary b is similar to any of the training data sets S_1, ..., S_n, the SSM is said to be of high specificity. In other words, synthesized shapes do indeed resemble real members of the training population. Generalizability and specificity need to be verified experimentally.

3 Applications

3.1 Anatomy Reconstruction

One of the basic challenges in processing medical image data is the automation of segmentation or anatomy reconstruction. Due to noise, artifacts, low contrast, partial field-of-view and other measurement-related issues, the automatic delineation and discrimination of specific structures from other structures or the background—seemingly an easy task for the human brain—is still challenging for the computer.

However, over the last two decades, model-based approaches have shown to be effective to tackle this challenge, at least for well-defined application-specific settings. The basis idea is the use a deformable shape template (such as a SSM) and match it to medical image data like CT or MRI. In this case, the shape S^* from Eq. (2) is not known explicitly. Therefore, such an approach—in addition to the SSM—requires a intensity model, that quantifies how well an instance of the SSM fits to the image data, see Fig. 4. From such a model, S^* can be estimated. Many such intensity models have been proposed in the literature. One generic approach is to "learn" such a model from training images similar to the way SSMs are generated [2]. Shape models are also combined with intensity models in SSIMs or shape and appearance Models. An overview can be found in [10].

The strength of this approach is its robustness stemming from the SSM. Only SSM instances can be reconstructed, thus this method can successfully cope with noisy, partial, low-dimensional or sparse image data.

On the other hand, since the SSM is generally limited in its generalization capability, some additional degrees of freedom are often required in order to get a more accurate reconstruction. Here, finding a trade-off between robustness and accuracy remains an issue. One successful approach for achieving such a trade-off are so called omni-directional displacements [13], see Fig. 5.

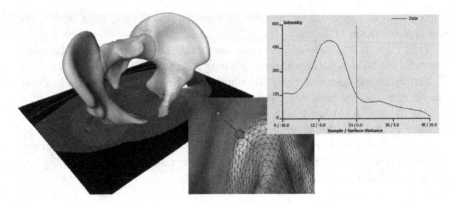

Fig. 4 A SSM is matched to CT data. At each point of the surface an intensity models predicts the desired deformation

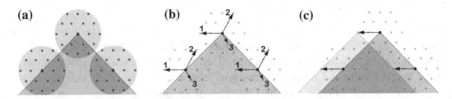

Fig. 5 Instead of just normal displacements as shown in Fig. 4, omnidirectional displacements allow for much more flexibility of local deformations, e.g. in regions of high curvature consistent local translations can be modeled

3.1.1 Example: Knee-Joint Reconstruction from MRI

Osteoarthritis (OA) is a disabling disease affecting more than one third of the population over the age of 60. Monitoring the progression of OA or the response to structure modifying drugs requires exact quantification of the knee cartilage by measuring e.g. the bone interface, the cartilage thickness or the cartilage volume. Manual delineation for detailed assessment of knee bones and cartilage morphology, as it is often performed in clinical routine, is generally irreproducible and labor intensive with reconstruction times up to several hours.

Seim et al. [20] present a method for fully automatic segmentation of the bones and cartilages of the human knee from MRI data. Based on statistical shape models and graph-based optimization, first the femoral and tibial bone surfaces are reconstructed. Starting from the bone surfaces the cartilages are segmented simultaneously with a multi-object technique using prior knowledge on the variation of cartilage thickness.

For evaluation, 40 additional clinical MRI datasets acquired before knee replacement are available. A detailed evaluation is presented in Fig. 7. For tibial and femoral bones the average (AvgD) and the roots mean square (RMS) surface distances were computed. Cartilage segmentation is quantified by volumetric overlap (VOE) and volumetric difference (VD) measures. For all four structures a score was computed indicating the agreement with human inter-observer variability. Reaching the inter-observer variability results in 75 points, while obtaining an exact match to one distinguished manual segmentation results in 100 points. An error twice as high as the human rater's gets 50 points, 3× as high gets 25 points and if 4× as high or more receives 0 points (no negative points). All points are averaged for each image, which results in a total score per image. Details on the evaluation procedure are published in an overview article of the Grand Challenge workshop (www.grand-challenge.org). The average performance of our auto-segmentation system for knee bones and associated cartilage was 64.4 ± 7.5 points. Exemplary results are shown in Fig. 6.

The bone segmentation achieves scores that indicate an error larger than that obtained by human experts. This may be due to relatively large mismatches of the SSM at the proximal and distal end of the MRI data due to missing or weak image features related to intensity inhomogeneities stemming from the MRI sequence (see Fig. 8). A strong artifact of unknown source (see Fig. 8) lead to the worst segmentation result for the tibia. The scores for cartilage segmentation are based on different error measures (volumetric) and are generally better, presumably due to a higher inter-observer variability.

3.1.2 Example: 3D Reconstruction from X-ray

The orientation of the natural acetabulum is useful for total hip arthroplasty (THA) planning and for researching acetabular problems. Currently the same acetabular component orientation goal is generally applied to all THA patients. Creating a patient-specific plan could improve the clinical outcome. The gold standard for surgical planning is from threedimensional (3D) computed tomography (CT) imaging. However, this adds time and cost, and exposes the patient to a substantial radiation dose. If the acetabular orientation could be reliably derived from the standard anteroposterior (AP) radiograph, preoperative planning would be more widespread, and research analyses could be applied to retrospective data, after a postoperative issue is discovered. The reduced dimensionality in 2D X-ray images and its perspective distortion, however, lead to ambiguities that render an accurate assessment of the orientation parameters a difficult task. One goal is to enable robust measurement of the acetabular inclination and version on 2D X-rays using computer-aided techniques.

Based on the idea of Lamecker et al. [17], Ehlke et al. [4] propose a reconstruction method to determine the natural acetabular orientation from a single, preoperative X-ray. The basic idea is to take virtual X-rays of a (extended) SSM and compare them to a clinical X-ray in an optimization framework. The best matching model instance gives an estimated shape and pose (and bone density distribution) of the subject (Fig. 9). The acetabular orientation (Fig. 10) can then be assessed directly from the reconstructed, patient specific anatomy model [5, 6]. The approach

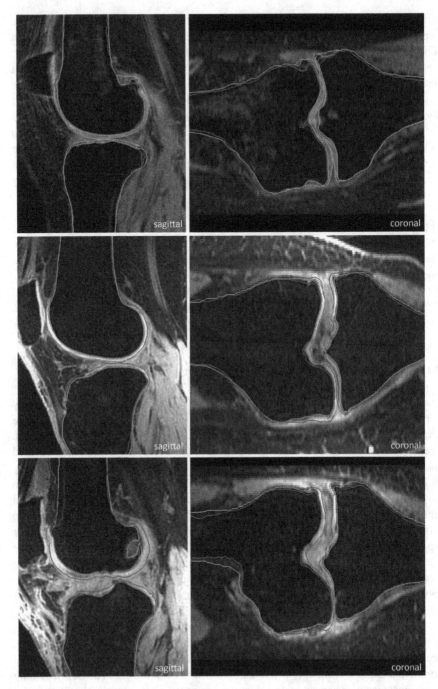

Fig. 6 Selected test cases sorted by quality in terms of achieved score. *Top* to *bottom* bad case, medium case, and good case. *Pink* Automatic segmentation. *Green* Ground truth (Color figure online)

Image	Femur		Tibia		Fem. Cartilage		Tibial Cartilage		Total Score
	AvgD [mm]	RMS [mm]	AvgD [mm]	RMS [mm]	VOE [%]	VD [%]	VOE [%]	VD [%]	
1	0,61	1,15	0,70	1,38	39,3	24,4	45,0	69,3	49
2	0,69	1,17	0,49	0,80	27,3	11,7	12,7	5,5	73
3	0,56	0,87	0,45	0,73	27,3	-14,3	23,5	8,5	71
4	0,73	1,24	0,57	0,83	35,5	31,3	30,7	41,4	54
5	0,67	1,10	0,62	1,07	31,5	16,6	31,1	-12,7	63
6	0,89	1,46	0,57	0,85	22,7	3,9	27,5	-10,5	72
7	1,15	1,64	0,74	1,22	29,5	19,6	35,9	-4,9	59
8	0,76	1,18	0,65	1,04	29,5	19,9	26,7	8,1	63
9	0,54	1,00	0,45	0,74	31,3	21,7	16,5	-3,2	69
10	0,59	1,03	0,42	0,73	25,9	20,5	26,3	-24,1	59
11	0,64	1,02	0,47	0,85	35,8	39,0	31,0	34,6	55
12	0,60	1,05	0,55	1,05	25,4	7,3	30,4	9,4	71
13	1,02	1,98	0,53	0,85	56,6	14,3	34,7	7,2	61
14	0,79	1,22	0,72	1,12	33,0	-20,6	31,7	-8,9	60
15	0,72	1,11	0,50	0,80	28,2	7,1	37,3	-35,3	63
16	0,59	0,97	0,48	0,73	20,5	10,4	25,0	-10,6	72
17	0,61	1,01	0,58	1,02	34,2	20,3	40,2	36,0	54
18	0,76	1,24	0,49	0,84	19,9	11,6	23,5	6,0	72
19	0,57	0,94	0,35	0,60	24,2	8,4	17,4	14,8	73
20	1,43	2,45	0,48	0,76	35,3	16,6	33,9	44,3	51
21	0,76	1,26	0,54	0,89	31,0	6,2	30,5	33,0	63
22	1,06	2,19	0,69	1,28	25,8	7,7	25,1	-4,9	67
23	0,62	1,03	0,54	0,90	22,6	21,6	33,0	-25,9	56
24	0,84	1,44	0,45	0,73	23,3	11,2	30,6	3,7	72
25	0,62	0,97	0,48	0,81	32,6	21,8	23,5	3,1	68
26	0,86	1,48	0,65	1,04	44,0	47,3	28,7	0,3	63
27	0,68	1,20	0,62	1,03	16,6	6,4	20,6	-14,6	70
28	0,84	1,33	0,53	0,81	34,9	30,6	24,6	-9,4	62
29	1,01	1,55	0,54	0,82	44,3	29,4	29,1	-7,4	60
30	0,77	1,23	0,56	0,88	26,9	23,8	44,2	75,6	53
31	0,70	1,05	0,60	0,93	66,2	-33,0	24,6	-5,9	60
32	0,64	1,06	0,72	1,18	24,9	15,3	22,7	5,8	68
33	0,91	1,45	0,78	1,21	28,1	4,7	41,0	-12,1	65
34	0,78	1,21	0,57	0,98	26,4	24,1	21,2	-3,5	66
35	0,86	1,33	0,66	1,21	36,1	24,3	33,1	-16,7	55
36	0,81	1,24	0,60	0,91	23,5	16,6	27,4	-23,4	58
37	0,80	1,32	0,57	0,87	29,4	16,8	21,1	6,9	67
38	0,63	0,98	0,56	0,90	17,3	-1,9	21,3	-6,5	79
39	0,73	1,15	0,45	0,73	25,1	10,4	19,1	-4,1	75
40	0,83	1,29	0,45	0,74	34,3	-8,1	21,3	2,0	75
Avg	0,77	1,26	0,56	0,92	30,7	13,6	28,1	4,3	64
	±0,18	±0,33	±0,10	±0,18	±9,61	±15,00	±7,39	±23,84	±7,46

Fig. 7 Evaluation of automatic segmentations against ground truth for 40 training datasets

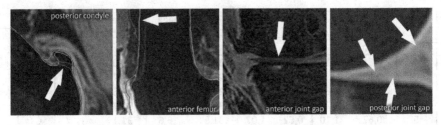

Fig. 8 Challenging regions: large osteophytes with high curvature, low contrast at anterior bone-shaft to soft-tissue interface, strong cartilage wear, low contrast at cartilage to soft-tissue interface, under-segmented bone at cartilage interface

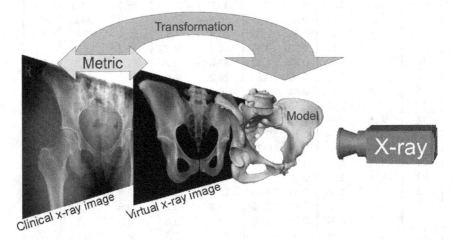

Fig. 9 The 3D shape reconstruction process from X-ray images optimizes the similarity between virtual X-ray images generate from the SSIM with the clinical X-ray images

Fig. 10 AP radiograph with reconstructed ilia, APP and acetabular reference landmarks (*red*), and global acetabular orientation vector (*blue*). Note that the person is standing in front of the X-ray plane, and the projection onto the plane is modeled (Color figure online)

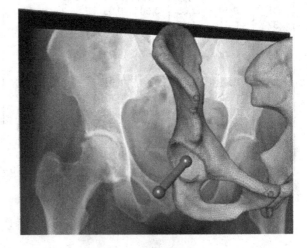

utilizes novel articulated statistical shape and intensity models (ASSIMs) that express the variance in anatomical shape and bone density of the pelvis/proximal femur between individual patients and model the articulation of the hip joints.

Concerning the application of SSMs to the anatomy of joints and their involved bones, one must also consider the variation of joint posture. A straightforward idea is not to care about joints at all and hence to employ multiple, independent models of the individual bones. However, there are two major drawbacks with this approach: First, an objects pose is independent of its adjacent object(s). This allows arbitrary object poses that do not resemble natural joint postures. Second, the shape of two neighbor bones is decoupled, although the adjacent surfaces of contact may correspond with regard to their shape. To eliminate these shortcomings, Bindernagel et al. [1] propose an articulated SSM (ASSM), which considers degrees of freedom that are better suited to such object compounds, (see Fig. 11).

In contrast to previous surface-based methods, the ASSIM-based reconstruction approach not only considers the anatomical shape of the pelvis, but also the bone interior density of both the pelvis and proximal femur. The rationale is to use as much information contained in the X-ray as possible, in order to increase the robustness of the 3D reconstruction (Fig. 12).

A preliminary evaluation on 6 preoperative AP X-ray and matching CT datasets, for the pelvis and femur of 6 different patients, was performed. The patient-specific 3D anatomies were first reconstructed from the 2D images and the inclination and version parameters obtained using the proposed method. These parameters were then compared to gold-standard values assessed independently from the individual patient's CT data. Average acetabular orientation errors, in absolute values, between the 3D reconstructed values and the CT gold standard for the right/left hips

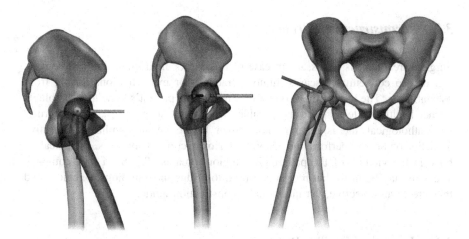

Fig. 11 The spheroidal joint model that is used for the hip realizes three consecutive rotations around the depicted axes. Variation of hip joint posture (default pose is outlined in *gray*): **a** rotation around x-axis (*red*), **b** rotation around z-axis (*blue*), **c** rotation around y-axis (*green*) (Color figure online)

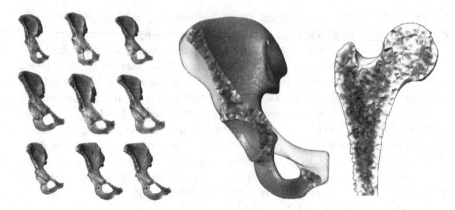

Fig. 12 Statistical shape and intensity models also take into account the intensity distribution inside the volume of the anatomical structure

were 3.86°/2.77° in inclination and 3.73°/4.97° in version. Maximum errors were 8.26°/5.43° in inclination and 6.22°/8.27° in version.

In most cases, the method produced results close to the CT gold standard (e.g. with an error margin of 4°). For the two outliers with an error above 8°, an incorrect fit between the model and reference X-ray was observable in the respective acetabular region, indicating that the statistical model needs to be enhanced further by expanding the training base. Also, currently the anterior pelvic plane is determined from only 4 points. Using a different plane as reference might increase the robustness when computing the inclination and version parameters.

3.2 Reconstructive Surgery

Surgical treatment is necessary in case of missing or malformed bony structures, e.g. due to congenital diseases, tumors, fractures or malformations arising from osteoarthritis. The task of the surgeon is to restore the patient's bone to resemble its former healthy state as good as possible. This is not an easy task as the former non-pathological state is generally not known. Hence the surgeon has no objective guideline on how to perform the restoration. However, in many cases, some healthy bone in the vicinity of the pathological region remains. The SSM of a complete structure may be fit to that region and used to bridge the pathological region, and thus create an objective, yet individual reconstruction guide.

3.2.1 Example: Mandible Dysplasia

Patients with distinct craniofacial deformities or missing bony structures require a surgical reconstruction that in general is a very complex and difficult task. The main

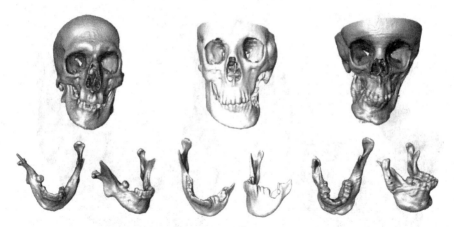

Fig. 13 Three cases of hemifacial microsomia with evidently malformed mandibles

reasons for such malformations, as shown in Fig. 13, are tumor related bone resections or craniofacial microsomia. In cases where the reconstruction cannot be guided by the symmetry of anatomical structures, it becomes particularly challenging. Then a surgeon must compare the individual pathologic situation with a mental image of a regular anatomy to modify the affected structures accordingly. For such a surgical therapy osteotomies are typically performed with either subsequent osteodistraction or osteosynthesis after relocation of bony segments, sometimes even in combination with selective bone and soft tissue augmentation. In many cases of mandibular dysplasia and hemifacial microsomia, any kind of guideline for the perception of a designated objective is highly desired.

Zachow et al. [21] propose a method based on a SSM of the mandible bone for reconstructing missing or malformed bony structures. This is achieved by selecting parts of the mandible that are considered as being regularly shaped and therefore are to be preserved.

A statistical 3D shape model of the human mandible seems to be a valuable planning aid for surgical reconstruction of bone defects. This is particularly useful for severe cases of hemifacial microsomia (Fig. 14). With a best matching candidate of the shape model, regarding the size and the shape of available bone, a surgeon gets a good mental perception of the reconstruction that is to be performed.

3.2.2 Example: Craniosynostosis

Premature ossified cranial sutures of infants (craniosynostosis) often lead to skull deformities in the growth process. This can lead to increased intracranial pressure, vision, hearing, and breathing problems. Since research on the correction of

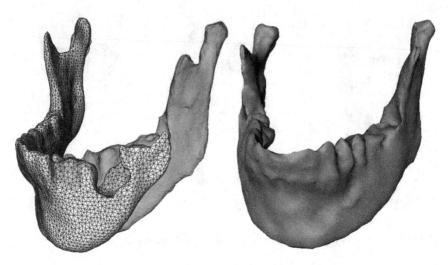

Fig. 14 Template generation for a hypoplastic mandible: **Left:** adaptation of a shape model to the *right* part of the malformed mandible, **Right:** resulting 3D template for mandible reconstruction

underlying disorders on the cellular level is still being carried on patients with craniosynostosis depend on surgical intervention for preventing or reducing functional impairment and improving their appearance. The most commonly used surgical procedure consists of bone fragmentation, deformation (reshaping) and repositioning. A major problem is the evaluation of the aesthetic results of reshaping the cranial vault in small children as the literature does not provide sufficient criteria for assessing skull shape during infancy. A definition of the correct target shape after surgery is missing. The most important and in many cases only indication of the best possible approximation of the skull shape to the unknown healthy shape is left to the subjective aesthetic assessment of the surgeon. This prevents impartial control of therapeutic success and aggravates guidance and instruction of the remodelling process for inexperienced surgeons.

Lamecker et al. [18] and Hochfeld et al. [12] have developed a statistical 3D-shape model of the upper skull to provide an objective, yet patient-specific guidance for the remodelling. To this end, a statistical 3D-shape model of the upper skull is generated from a set of 21 MRI data sets of healthy infants. The 3D cranial model serves as a template for the reshaping process, by finding an optimal fit of any of its variations to a given malformed skull. Usually, no pre-operative MRI scan is available for the infant patients (mostly under the age of one year) in order to avoid unnecessary anesthesia. Hence the matching of the model towards the pathological skull of the patient is performed by non-invasively measuring anthropometric distances that are not affected by the surgical intervention:

- width between both entries of the auditory canals
- distance from nasion to occiput
- height between vertex and the midpoint of the line between the auditory canals

These distances are extrapolated to the skull surface by approximating the skin and skull thickness. The shape model instance that best fits these measurements is selected as a template for the reconstruction process. The resulting shape instance represents an individual interpolation of all shapes contained in the training set.

In a first clinical application, the statistical model was pre-operatively matched to a patient using the method described above. From this computed shape model instance a life-size facsimile of the skull was built and taken to the operating room to guide the reshaping process. Figure 15 illustrate the surgical procedure and the role of the statistical skull model.

3.2.3 Example: Surgical Reconstruction of Complex Fractures in the Midface

Surgical interventions on the craniofacial bone in cases of complex defects spanning the facial mid-plane (e.g. horse kick fractures) require high precision during planning as well intra-operatively (Fig. 16, left). The goal is to restore function but also create aesthetically appealing reconstructions of the fractured parts. Mirroring of the intact side is often not possible. If there no pre-traumatic tomographic data of the anatomical region of the patient is available the re-location of the bone fragments is performed according to the subjective perception and personal skill of the operator.

In order to overcome this situation and possibly restore (or even improve) the pre-traumatic situation an objective guideline for the surgical procedure is desirable. Several central questions need to addressed: (1) What is an objective guideline for re-modeling the bony structures of the mid face? (2) How may the surgeon be supplied with a practical tool to perform the reconstruction based on the planning data? (3) Can the planning guideline be generated fast enough such that it may be used immediately in a first emergency operation?

Zachow et al. [22] propose a way to tackle those challenges based on SSMs. In their work, a SSM representing normal bone anatomy is used to segment a CT data set of a traumatic patient. Fractured regions in the image are masked so they are ignored in the fitting process. The resulting SSM then interpolates those regions thereby creating a "normal" shape guideline in those regions, which matches the patient's anatomy in healthy regions (Fig. 16, right). This computation takes only several minutes on standard computers.

In order to transfer the plan to the intra-operative situation the reconstructed geometry can rasterized to the grid of the CT data and exported in DICOM format directly to the navigation system. There, it can be presented to the surgeon as an overlay on the CT data. The surgeon can mobilize bone fragments towards the targeted position on the SSM, and there perform the osteosynthesis (Fig. 17).

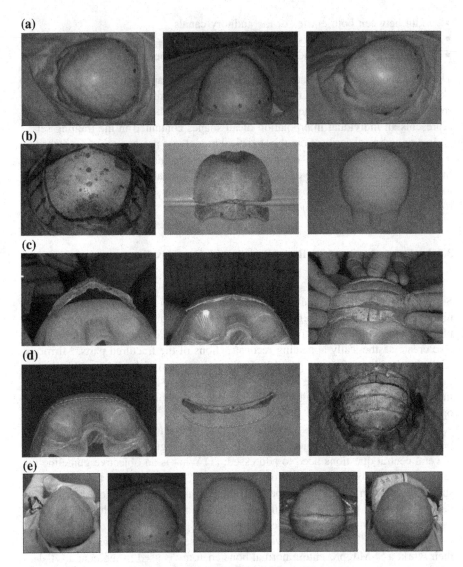

Fig. 15 Craniosynostosis surgery based on a SSM. **a** Three different views of a patient with trigonocephaly (ossification of the suture running down the midline of the forehead)—before surgery. **b** *Cutting lines* indicated on the skull, removed frontal skull region before the reshaping, facsimile of shape model instance on which bone parts are reshaped. **c** Bone stripe before and after reshaping, result of reshaping process on model. **d** Microplates for fixating bone pieces on remaining skull are also shaped on the model, result after fixation of reshaped bone on skull. **e** Comparison between pre- and post-operative situation (from *left* to *right*): patient 2 months before surgery, immediately before surgery, facsimile of the target shape derived from the statistical model, patient immediately after surgery, 3 weeks after surgery

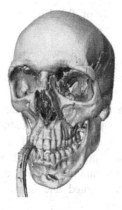
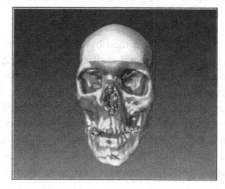

Fig. 16 *Left* Midface trauma visualized by an isosurface of the CT data. *Right* Reconstruction proposal based on a SSM (*red*) (Color figure online)

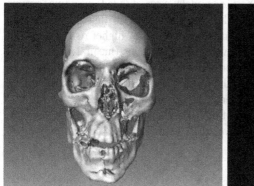
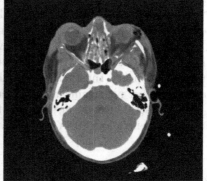

Fig. 17 *Left* 3D visualization of deviation of fractured bone (*white*) to targeted SSM-based planning (*green*). *Right* Overlay of planning result in DICOM data (Color figure online)

3.3 Population-Based Analysis

SSMs represent the shape variability contained in a specific population with few degrees of freedom. Furthermore, they are based on a common parameterization of the shapes. This means that computationally demanding analyses (local, regional or global), such as finite-element studies, across subjects can be performed efficiently. Parameter studies or inverse problems become more tractable due to the compact representation of shape and the small number of shape parameters. This allows to study, e.g. the relationship between shape and socio-demographic or biomechanical parameters (e.g. fracture risk). Instead of performing fully individualized procedures implant design or procedures may be optimized based on population studies, see e.g. [15].

3.3.1 Example: A Large Scale Finite Element Study of Total Knee Replacements

The aim of the study performed by Galloway [7] is to investigate the performance of a cementless osseointegrated tibial tray (P.F.C. Sigmas, Depuy Inc., USA) in a general population using finite element (FE) analysis. Computational testing of total knee replacements (TKRs) typically only use a model of a single patient and assume the results can be extrapolated to the general population. In this study, two SSMs were used; one of the shape and elastic modulus of the tibia, and one of the tibiofemoral joint loads over a gait cycle, to generate a population of FE models. A method was developed to automatically size, position and implant the tibial tray in each tibia, and 328 models were successfully implanted and analyzed, see Fig. 18.

The composite peak strain (CPS) during the complete gait cycle in the bone of the resected surface was examined and the "percentage surface area of bone above yield strain" (PSAY) was used to determine the risk of failure of a model. Using an arbitrary threshold of 10 % PSAY, the models were divided into two groups ("higher risk" and "lower risk") in order to explore factors that may influence potential failure. In this study, 17 % of models were in the "higher risk" group and it was found that these models had a lower elastic modulus (mean 275.7 MPa), a higher weight (mean 85.3 kg), and larger peak loads, of which the axial force was the most significant. This study showed the mean peak strain of the resected surface and PSAY were not significantly different between implant sizes.

It was observed that the distribution of the CPS changed as the PSAY increased. For models with a low PSAY (e.g. "lower risk" case Fig. 19), higher strains were

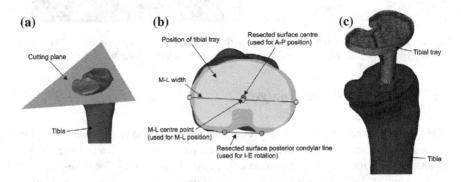

Fig. 18 Selected steps of implanting the tibial tray: **a** shows the position of the cutting plane, **b** shows the landmarks used on the resected surface to position the tibial tray, and **c** is an exploded view of mesh components (Tetra)

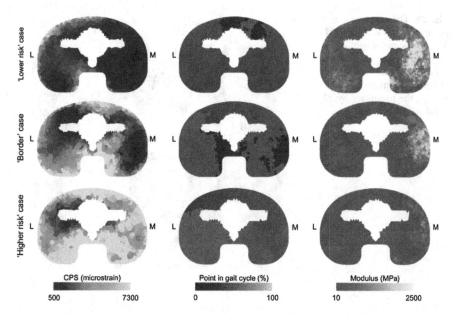

Fig. 19 Three example cases of the resected surface. *Top* is a "lower risk" case (PSAY = 1.45 %), *middle* is a "border" case (PSAY = 9.05 %), and *bottom* is a "higher risk" case (PSAY = 39.75 %). Each is plotted with the CPS (*left*), the point in the gait cycle at which the peak strain occurs (*middle*), and modulus (*right*)

seen around the anterior and posterior edges. As the PSAY increases, models around the 10 % threshold (e.g. "border case" Fig. 19) tended to have bone above yield strain around the periphery. The strains on the lateral side tended to be higher in comparison to the medial side. In the "higher risk" group, bone above yield tended to be distributed over the whole resected surface (e.g. "higher risk" case Fig. 19), although in some cases only the lateral side was above yield. The pattern of strain is most likely due to the difference in modulus between the lateral and medial side of the resection surface.

3.3.2 Example: Digital Morphometry of Facial Anatomy, from Global to Local Shape Variations

Although the following example is not directly related to orthopedic interventions it illustrates the potential of SSMs for population analysis in order to examine clusters or systematic variation related to covariates, such as demographic factors or pathology indicators.

Facial proportions largely depend on maturation (age) and sex. Maturation increases the proportion facial height to head height, the eyes tend to become

Fig. 20 The first two parameters of the face SSM. *Middle column* shape variation color-coded on the average face. *Left/Right column* face instances when varying the parameter by ±2 standard deviations

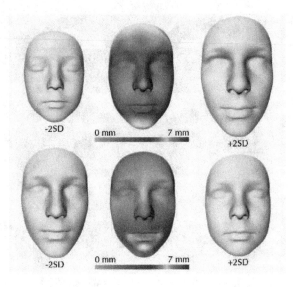

narrower, and the bigonial to bizygonial proportions enlarges. Proportions that associate with sex, involve structures close to nose and cheek. Only 25 % of the variance of facial proportions, mainly those describing variation in the lower and in the mid-face, are independent of sex and age, but are often affected in dysmorphic syndromes [11]. Grewe et al. [9] investigate 52 North German child and adolescent faces aged 2–23 years (26 male, 26 female) acquired with a 3D laser scanner [8]. Based on a mesh with 10,000 points a SSM was created.

Visualization techniques can be employed to explore the nature of the shape model parameters as well as the distribution of the data set. Figure 20 shows that PCA parameters in general have a global effect on the shape, i.e. the shape variation is spread over a large portion of the anatomical region. Sometimes region-specific parameters e.g. for eyes, mouth or nose are desirable. A specifical class of orthogonal transformations, known as VARIMAX rotations [14], lead to parameters with locally concentrated variation, thus yielding meaningful parameters, e.g. for changing the shape of the nose (Fig. 21). This can be used to synthesize new shapes more intuitively.

Shape variation was also regressed on sex and age. Figure 22 shows shape instances computed via the regression function for sex specific shape variation for age. Such information may, for instance, be useful for forensic purposes.

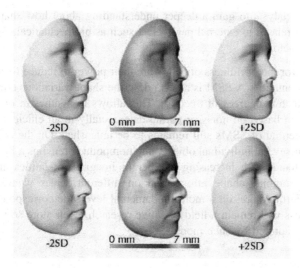

Fig. 21 Varimax transformed SSM with meaningful parameters for changing the shape of the nose. Columns as in Fig. 20

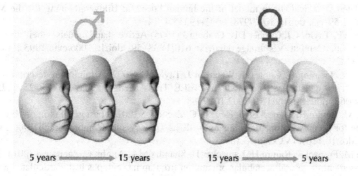

Fig. 22 Sex specific ageing time line produced by the regression function (5, 10, 15 years)

4 Conclusions

In this chapter we have described the concept of statistical shape models. We have illustrated that this technology has a major impact in three different fields:

1. Reconstruction of anatomical structures from medical images, both 3D (CT or MRI) and 2D (X-ray) data.
2. Planning of complex surgical interventions and the use of such plans intra-operatively.

3. Population analysis to gain a deeper understanding about how shape or shape changes are related to external parameters such as biomechanical, demographic or clinical factors.

One reason for the usefulness of SSMs are their power to reduce the complexity of shape representation. A SSM is able to describe shape variations contained in a population with few degrees of freedom. This allows to synthesize or reconstruct new shapes even based on noisy or partial data, usually in an efficient manner.

The full potential of SSMs still remains to be unleashed. On the one hand, the combination of several individual objects into compounds remains a big challenge. On the other hand, with increasing number of imaging modalities and sources, concepts that combine shape information on different scales would offer new possibilities. Furthermore, on a more fundamental level, the correspondence identification process will remain a field of active research, even more so when higher details can be captured by such models.

References

1. Bindernagel M, Kainmueller D, Seim H, Lamecker H, Zachow S, Hege H-C (2011) An articulated statistical shape model of the human knee. In: Bildverarbeitung für die Medizin 2011, pp 59–63. doi:10.1007/978-3-642-19335-4_14
2. Cootes TF, Taylor CJ, Cooper DH, Graham J (1995) Active shape models—their training and application. Comput Vis Image Underst 610(1):38–59. doi:10.1006/cviu.1995.1004 ISSN 1077-3142
3. Davies R, Twining C, Cootes T, Waterton J, Taylor C (2002) A minimum description length approach to statistical shape modeling. IEEE Trans Med Imaging 210(5):525–537. ISSN 0278-0062. doi:10.1109/TMI.2002.1009388
4. Ehlke M, Ramm H, Lamecker H, Hege H-C, Zachow S (2013) Fast generation of virtual x-ray images for reconstruction of 3d anatomy. IEEE Trans Visual Comput Graph 190(12):2673–2682. doi:10.1109/TVCG.2013.159
5. Ehlke M, Frenzel T, Ramm H, Lamecker H, Shandiz MA, Anglin C, Zachow S (2014) Robust measurement of natural acetabular orientation from ap radiographs using articulated 3d shape and intensity models. Technical Report 14–12, ZIB, Takustr.7, 14195 Berlin, 2014
6. Ehlke M, Frenzel T, Ramm H, Shandiz MA, Anglin C, Zachow S (2015) Towards robust measurement of pelvic parameters from ap radiographs using articulated 3d models. In Computer Assisted Radiology and Surgery (CARS), 2015. accepted for publication
7. Galloway F, Kahnt M, Ramm H, Worsley P, Zachow S, Nair P, Taylor M (2013) A large scale finite element study of a cementless osseointegrated tibial tray. J Biomech 460(11):1900–1906. doi:10.1016/j.jbiomech.2013.04.021
8. Grewe CM, Lamecker H, Zachow S (2011) Digital morphometry: the potential of statistical shape models, 2011
9. Grewe CM, Lamecker H, Zachow S (2013) Landmark-based statistical shape analysis. In: Hermanussen M (Ed) Auxology—studying human growth and development url, pp 199–201. Schweizerbart Science Publishers, 2013. URL http://www.schweizerbart.de/publications/detail/isbn/9783510652785
10. Heimann T, Meinzer H-P (2009) Statistical shape models for 3d medical image segmentation: a review. Med Image Anal 130(4):543–563. ISSN 1361-8415. URL http://dx.doi.org/10.1016/j.media.2009.05.004, http://www.sciencedirect.com/science/article/pii/S1361841509000425

11. Hermanussen EM (Ed) Auxology. Schweizerbart Science Publishers, Stuttgart, Germany, 03 2013. ISBN 9783510652785. URL http://www.schweizerbart.de//publications/detail/isbn/ 9783510652785/Hermanussen_Auxology
12. Hochfeld M, Lamecker H, Thomale UW, Schulz M, Zachow S, Haberl H (2014) Frame-based cranial reconstruction. J Neurosurg Pediatr 130(3):319–323. doi:10.3171/2013.11.PEDS1369
13. Kainmüller D, Lamecker H, Heller M, Weber B, Hege H-C, Zachow S (2013) Omnidirectional displacements for deformable surfaces. Med Image Anal 170(4):429–441. doi:10.1016/j. media.2012.11.006
14. Kaiser H (1958) The varimax criterion for analytic rotation in factor analysis. Psychometrika 230(3):187–200. doi:10.1007/BF02289233 ISSN 0033-3123
15. Kozic N, Weber S, Büchler P, Lutz C, Reimers N, Ballester MG, Reyes M (2010) Optimisation of orthopaedic implant design using statistical shape space analysis based on level sets. Med Image Anal 140(3):265–275. ISSN 1361-8415. doi:http://dx.doi.org/10.1016/j.media.2010.02. 008. URL http://www.sciencedirect.com/science/article/pii/S136184151000023X
16. Lamecker H (2009) Variational and statistical shape modeling for 3D geometry reconstruction. PhD thesis, Freie Universität Berlin
17. Lamecker H, Wenckebach T, Hege H-C (2006) Atlas-based 3d-shape reconstruction from x-ray images. In: Proceedings of international conference of pattern recognition (ICPR2006), vol I, pp 371–374. doi:10.1109/ICPR.2006.279
18. Lamecker H, Zachow S, Hege H-C, Zöckler M (2006) Surgical treatment of craniosynostosis based on a statistical 3d-shape model. Int. J. Comput Assist Radiol Surg 1(1):253–254. doi:10. 1007/s11548-006-0024-x
19. Sarkalkan N, Weinans H, Zadpoor AA (2014) Statistical shape and appearance models of bones. Bone 600(0):129–140. ISSN 8756-3282. doi:http://dx.doi.org/10.1016/j.bone.2013.12. 006. URL http://www.sciencedirect.com/science/article/pii/S8756328213004948
20. Seim H, Kainmueller D, Lamecker H, Bindernagel M, Malinowski J, Zachow S (2010) Model-based auto-segmentation of knee bones and cartilage in MRI data. In: Ginneken BV (Ed), Proceedings of MICCAI workshop medical image analysis for the clinic, pp 215–223
21. Zachow S, Lamecker H, Elsholtz B, Stiller M (2005) Reconstruction of mandibular dysplasia using a statistical 3d shape model. In: Proceedings of computer assisted radiology and surgery (CARS), pp 1238–1243. doi:10.1016/j.ics.2005.03.339
22. Zachow S, Kubiack K, Malinowski J, Lamecker H, Essig H, Gellrich N-C (2010) Modellgestützte chirurgische rekonstruktion komplexer mittelgesichtsfrakturen. Proc BMT Biomed Tech 2010 55(1):107–108

Automated 3D Lumbar Intervertebral Disc Segmentation from MRI Data Sets

Xiao Dong and Guoyan Zheng

Abstract This paper proposed an automated three-dimensional (3D) lumbar intervertebral disc (IVD) segmentation strategy from Magnetic Resonance Imaging (MRI) data. Starting from two user supplied landmarks, the geometrical parameters of all lumbar vertebral bodies and intervertebral discs are automatically extracted from a mid-sagittal slice using a graphical model based template matching approach. Based on the estimated two-dimensional (2D) geometrical parameters, a 3D variable-radius soft tube model of the lumbar spine column is built by model fitting to the 3D data volume. Taking the geometrical information from the 3D lumbar spine column as constraints and segmentation initialization, the disc segmentation is achieved by a multi-kernel diffeomorphic registration between a 3D template of the disc and the observed MRI data. Experiments on 15 patient data sets showed the robustness and the accuracy of the proposed algorithm.

1 Introduction

Intervertebral disc (IVD) degeneration is a major cause for chronic back pain and function incapacity [1]. Magnetic Resonance Imaging (MRI) has become one of the key investigative tools in clinical practice to image the spine with IVD degeneration not only because MRI is non invasive and does not use ionizing radiation, but more importantly because it offers good soft tissue contrast which allows visualization of the disc's internal structure [2].

MRI quantification has great potential as a tool for the diagnosis of disc pathology but before quantifying disc information, the IVDs need to be extracted

X. Dong (✉)
Faculty of Computer Science and Engineering, Southeast University, Nanjing, China
e-mail: xiao.dong@seu.edu.cn

G. Zheng
Institute for Surgical Technology and Biomechanics, Bern University, Bern, Switzerland
e-mail: guoyan.zheng@ieee.org

© Springer International Publishing Switzerland 2016
G. Zheng and S. Li (eds.), *Computational Radiology
for Orthopaedic Interventions*, Lecture Notes in Computational
Vision and Biomechanics 23, DOI 10.1007/978-3-319-23482-3_2

from the MRI data. IVD extraction from MRI data comprises two key steps. Firstly, all IVDs have to be detected from the images and secondly, the regions belonging to IVDs have to be segmented. Manual extraction methods [3, 4] as well as automated extraction methods [5–11] have been presented before. Since manual extraction is a tedious and time-consuming process which lacks repeatability, automated methods are preferred.

There are different approaches for automatizing the extraction of IVDs from medical images such as graphical model [5], probabilistic model [6], Random Forest regression and classification [12, 13], watershed algorithm [7], atlas registration [8], statistic shape model [10], graph cuts [9], and anisotropic oriented flux [11]. But stable and accurate IVD segmentation remains a challenge.

In this paper we propose an automated 3D lumbar IVD extraction method with minimal user interaction from MRI data sets. The main contribution of our method is a combination of graphical model-based spine column localization with a multi-kernel diffeomorphic registration-based segmentation. The motivation of the proposed strategy to first identify the spine column structure and then carry out the IVD segmentation stems from the following observation:

> The IVD geometries are highly constrained by the geometry of the spine column. If the geometrical parameters of the spine column and each individual vertebral body can be estimated accurately from the observed images, then they can provide both geometrical and appearance information about the intervertebral discs, which helps to improve the accuracy and robustness of the IVD segmentation.

The work flow of the proposed algorithm consists of the following three steps:

Initialization Two user supplied landmarks on a user selected mid-sagittal slice are required to indicate the centers of L1 and L5 vertebral bodies.

Lumbar spine column identification and modeling Starting from the user initialization, the 3D geometry of the lumbar spine column is automatically extracted from the 3D data sets, which is achieved as a sequential 2D + 3D model fitting procedure. The outputs of the lumbar spine column modeling procedure are the 3D geometric information of each individual vertebral body of L1–L5 and a *soft-tube* model that fits the outer surface of the lumbar spine column.

Lumbar disc segmentation The extracted lumbar spine column can provide reliable prior information for the initialization and constraints of the disc segmentation such as positions, sizes and image appearance of discs. The disc segmentation is finally achieved as a 3D multi-kernel diffeomorphic registration between a disc template and the observed data.

The paper is organized as follows. Section 2 describes details of the method, followed by experimental results in Sect. 3. Discussions and conclusions are presented in Sect. 4.

2 Methods

2.1 Data Sets

All datasets used in this paper were generated from a 1.5 T MRI scanner (Siemens medical solutions, Erlangen, Germany). Dixon protocol was used to reconstruct four aligned high-resolution 3D volumes during one data acquisition: in-phase, opposed-phase, water and fat images, as shown in Fig. 1. Each volume has a resolution of 2 mm × 1.25 mm × 1.25 mm and the data set size is 40 × 512 × 512. The advantage of working with such datasets is that different channels provide complementary information for our disc segmentation task. In our proposed segmentation strategy, we always first extract either intensity or feature information about different tissues on each of the 4 channels and then combine the 4 channel data into a single dataset as explained later.

2.2 Initialization

On the mid-sagittal slice, two landmarks are picked to indicate the centers of L1 and L5 vertebral bodies as shown in Fig. 2a.

2.3 Lumbar Spine Column Identification

Based on the initialization, we first carry out a 2D vertebral body and disc identification to localize vertebrae L1–L5 and the 5 target discs from the mid-sagittal slice. The geometrical information of the 2D identification is then used to guide a further 3D lumbar spine column modeling.

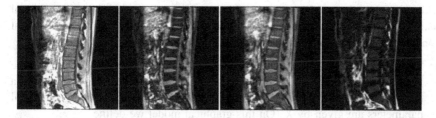

Fig. 1 The four aligned channels of a patient data, *left* to *right* in-phase, opposed-phase, water and fat images (for visualization purpose, we only show the middle sagittal (mid-sagittal) slice of each channel)

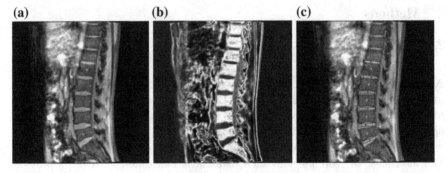

Fig. 2 Initialization and 2D lumbar spine column detection. **a** User initialization by picking *two landmarks* indicating the centers of L1 and L5 in the middle sagittal slice. **b** Probability assignment (displayed as *grey values*) of the bone tissue in the mid-sagittal slice for 2D lumbar spine column detection. **c** 2D lumbar spine column detection result using the graphical model based detection algorithm, *blue* and *green rectangles* representing the vertebral bodies and IVDs respectively (Color figure online)

2.3.1 2D Vertebral Body and Disc Identification

Solutions for spine location and disc labeling include feature-based bottom-up methods, statistical model-based methods and graphical model-based solutions. For a detailed review of the existing methods, we refer to [14, 15]. In this paper, the 2D vertebral body and disc identification is achieved using a graphical model based strategy introduced in [14]. Compared with the graphical models in [5, 6], the advantage of the graphical model in [14] is that both the low level image observation model and the high level vertebra context potentials need not to be learned from training data. Instead they are capable of self-learning from the image data during the inference procedure. The basic idea is to model both the vertebral bodies and discs in the mid-sagittal slice as parameterized rectangles, where the parameters are used to describe the geometries of these rectangles including their centers, orientations and sizes. The graphical model based spine column identification can be understood as matching the parameterized models with the observed images while also considering the geometrical constraints between neighboring vertebral bodies and discs. The exploration of geometrical constraints between vertebral components helps to enhance the identification robustness.

1. *The graphical model*: The graphical model used in this work to represent the lumbar spine column is given in Fig. 3. In this model each node V_i represents a connected disc-vertebra-disc chain of the spine column, whose geometrical parameters are given by X_i. On this graphical model we define

   - *The component observation model* $p(I|X_i)$ of a single component V_i representing the probability that the configuration X_i of the node V_i match the observed images I.

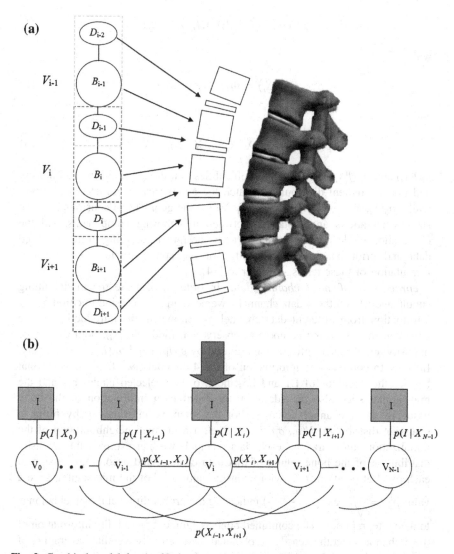

Fig. 3 Graphical model for the 2D lumbar spine column detection. **a** A node V_i represents a disc-vertebra − disc ($D_{i-1} - B_i - D_i$) chain of the lumbar spine and both the discs and vertebrae are modeled as parameterized *rectangles*. **b** The observation model $p(I|X_i)$ of each node V_i and potentials $p(X_i, X_j)$ between nodes V_i, V_j defined on the graphical model

- *The potentials $p(X_i, X_j)$ between neighboring components V_i and V_j encoding the geometrical constraints between components which are defined by the anatomical structure of the spine column.*

The identification of the spine column from the mid-sagittal slice can then be formalized as to find the optimal configurations of $\{V_i\}$, $X = \{X_i\}$ that maximize

$$P(X|I) \propto \Pi_i p(I|X_i) \Pi_{i,j} p(X_i, X_j) \tag{1}$$

with

$$p(I|X_i) = p_I(I|X_i) p_G(I|X_i) \tag{2}$$

and

$$p(X_i, X_j) = p_S(X_i, X_j) p_O(X_i, X_j) p_D(X_i, X_j) \tag{3}$$

$p_I(I|X_i)$ and $p_G(I|X_i)$ stand for the probabilities that the observed image intensity and image gradient distributions match the geometrical parameters X_i respectively. $p_S(X_i, X_j)$, $p_O(X_i, X_j)$ and $p_D(X_i, X_j)$ are the geometrical constraints on the sizes, orientations and distances between neighboring components. All the observation models and constraints can be designed according to the observed data and prior anatomical knowledge of the spine structure. For detailed formulation of these terms, we refer to [14].

2. *Preprocessing of the 4 channel data*: In order to achieve the model fitting simultaneously on the 4 data channels, we need a preprocessing to combine the information from different data channel. As shown in the introduction of the graphical model, in the component observation model $p(I|X_i)$, both the image intensity and gradient information are used by $p_I(I|X_i)$ and $p_G(I|X_i)$ respectively. In order to combine the information of the 4 channels, we firstly observe that besides the positions of L1 and L5, the two user selected landmarks and the mid-sagittal slice also provide intensity distribution information of the bony tissue in the 4 channel volumes. For the intensity information, by fitting a Gaussian distribution $N(\mu_i, \sigma_i)$, $i = 1, 2, 3, 4$ to a small neighbourhood of the initialization landmarks on each data channel, we can estimate the intensity distribution of the bone region of that data channel and accordingly assign to each pixel at position (l, k) with intensity value $x_i^{(l,k)}$ of the mid-sagittal slice a value $p_i^{l,k} = \frac{1}{\sqrt{2\sigma_i^2}} \exp(\frac{-(x_i^{(k,l)} - \mu_i)^2}{2\sigma_i^2})$ indicating the probability that the pixel belongs to the vertebral body. The combined bone assignment probability information of the 4 channels can then easily be obtained by an equally weighted averaging of the 4 channels as $p^{k,l} = \frac{1}{4} \sum_i p_i^{l,k}$. Similarly we can also combine the gradient information of the 4 data channels by simply averaging the gradient amplitude of each channel. The combined intensity and gradient information are then used in the intensity and gradient local observation model components, $p_I(I|X_i)$ and $p_G(I|X_i)$, of the graphical model. Figure 2b shows an example of the bone tissue probability assignment on the mid-sagittal slice computed from the user supplied 2 landmarks (Figs. 2a and 4 channel volume data (see Fig. 1 for an example).

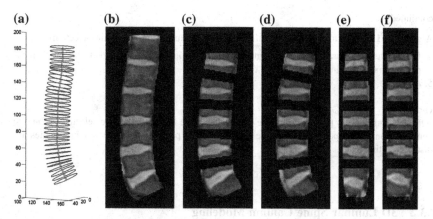

Fig. 4 3D lumbar spine column detection and modeling. **a** The 3D soft-tube model of the lumbar spine column; **b** segmented lumbar spine column image; **c–d** segmented disc candidate regions in sagittal slices; **e–f** segmented disc candidate regions in coronal slices. Although all tasks are conducted in 3D, here we show the results in 2D slices for visualization purpose

3. *Optimization*: The optimization is achieved as an inference on the graphical model, which is essentially a particle based nonparametric belief propagation on the graphical model as described in Algorithm 1. The outputs are then the 2D geometrical parameters of the spine column which best fit the observed mid-sagittal slices of all the 4 data channels.

4. *Detection results*: Fig. 2c gives the 2D lumbar column detection result. It can be observed that the centers, sizes and orientations of the vertebral bodies and IVDs are correctly identified.

Algorithm 1 Graphical model based inference for 2D lumbar spine column detection

---

Input: Bone region assignment map (Fig. 2b) from mid-sagittal slices of the 4 data channel, landmarks from user initialization

Output: 2D geometrical parameters of the lumbar vertebral bodies ($L_1 - L_5$) and discs between $L_1 - S_1$

Initialization: Roughly estimate the possible configuration regions of the positions, sizes and orientations of each vertebral body and disc according to the user initialization and prior anatomical information of the lumbar spine.

Start: $t = 0$, draw N random samples configurations of $\{X_i^n(t), n = 1, ..., N\}$ of each model node V_i from the estimated parameter space.

while not converge **do**

1. Compute the *belief* of each particle $X_i^n(t)$ by the local observation model as $b_i^n(t) \propto p(I|X_i^n(t))$.
2. Run belief propagation till converge on the chain graphical model using the potentials $p(X_i^n, X_j^{n'})$ among nodes to update the *belief* of each particle $X_i^n(t)$ to obtain updated *believes* $\{\bar{b}_i^n(t)\}$, which are the approximations of the marginal probabilities $P(X_i^n|I)$ given in (1) obtained by the belief propagation.
3. Re-sample the particles according to the updated believes $\{\bar{b}_i^n(t)\}$ to obtain new samples $\bar{X}_i^n(t)$ of each node.

(continued)

(continued)

Algorithm 1 Graphical model based inference for 2D lumbar spine column detection

---

4. Update the configuration of each sample $\bar{X}_i^n(t)$ by a Gaussian random perturbation on the parameters to obtain new particles $\{X_i^n(t + 1)\}$.

5. $t = t + 1$.

end while

For each node V_i, the parameters of the particle with the highest belief are selected as the configuration of that node and therefore the geometrical parameters of the vertebral bodies and discs are estimated.

2.3.2   3D Lumbar Spine Column Modeling

The above explained 2D lumbar spine column model only provides an incomplete information of the spine column. Therefore, in order to accurately localize the lumbar column and the geometrical parameters of each vertebral body and disc, a 3D lumbar spine model is needed. To achieve this, we model each lumbar vertebral body as an elliptical cylinder and the lumbar spine column as a variable-radius soft tube. Details of the modeling procedure are described as follows:

- 3D modeling of each vertebral body

  (a) If we approximately model the vertebral body as a cylinder, then from the 2D vertebral body identification results, the position, hight, radius and orientation of each vertebral body and the image intensity distribution of the bone region in each data channel (also modeled as a Gaussian distribution) can be estimated.

  (b) Given the estimated bone tissue intensity distribution of each data channel, then for each voxel in the neighbourhood of the estimated cylinder model of the vertebra body, we can assign the probability if this voxel belongs to the bone tissue. We can also integrate the information of 4 channels in the same way as explained in the 2D model fitting procedure to obtain the combined bone tissue probability assignment $p^{k,l,m}$ for a voxel at position (k, l, m).

  (c) From the bone tissue probability assignment of voxels around each vertebral body, we can further refine the 3D modeling of the vertebral bodies. To achieve this, we further model the vertebral body as an elliptical cylinder. Then a least-squares geometric fitting to the voxels which are assigned with a high enough probability ($p^{k,l,m} > 0.8$) of belonging to the bone tissue can extract the 3D geometry of each vertebral body, including the center, height, orientation and the major radius and minor radius of the elliptical cylinder model.

- *3D modeling of the spine column* Given the 3D model of each vertebral body, we can further construct the 3D lumbar column model. We model the lumbar column as a variable-radius soft tube that best fits the outer surfaces of all the extracted vertebral bodies. Given the 3D models of L1–L5 vertebral bodies, the central axis and the variable-radius of the soft tube can be obtained by a linear interpolation on the centers and radii of the extracted 3D models of vertebral bodies. This results of this 3D variable-radius soft-tube spine column model is shown in Fig. 4a.

Obviously given the 3D soft-tube lumbar spine column model, the spine column region can be extracted from the observed data sets (Fig. 4b). By further eliminating the bony tissue region using the 3D models of vertebral bodies, the candidate region for each target disc can be localized as shown in Fig. 4c–f. The following 3D IVD segmentation is then carried out on the extracted candidate IVD regions.

2.4   3D Disc Segmentation

We solve the 3D disc segmentation as a template based registration between a geometrical disc template and the observed data.

- The IVD template is set as a thin elliptical cylinder. Considering the anatomical structure of the spine column, i.e., each IVD must fall between its neighbouring vertebral bodies, the initial geometries (center, radii, orientation, hight) of the IVD cylinder template can be estimated using the 3D spine column model and the geometries of its neighboring vertebral bodies, which are all available from the previous 3D lumbar spine column modeling procedure.
- For the segmentation of a specific IVD, the correspondent observed data to be matched is just the extracted candidate IVD region as shown in Fig. 4c–f.
- For the registration algorithm we choose the multi-kernel diffeomorphic image matching in the Large Deformation Diffeomorphic Metric Mapping (LDDMM) framework as described in [16] and related literatures [17–20].

2.4.1   Multi-kernel LDDMM Registration

LDDMM framework [19] is one of the two main computational frameworks in computational anatomy [17]. Existing works show that LDDMM is a general solution for nonrigid image registration with a high flexibility and accuracy. In [16, 18, 21] multi-scale LDDMM registration algorithms were investigated. Compared with the LDDMM registration with a single kernel, multi-kernel LDDMM has the capability to optimize the deformation in multiple spatial scales [16].

Following the general idea of LDDMM framework, we formalize the multi-kernel image registration between two images I_0 and I_1 as an optimization problem to find the optimal time dependent velocity field $v(t)$ that minimizes a cost function $\mathcal{E}(\{v^k(t)\})$ as the sum of a deformation energy term and an image similarity term formalized as

$$\mathcal{E}(\{v^i(t)\}) = \frac{1}{2}\sum_i^K w^i \int_0^1 \|v^i(t)\|_{V^i}^2 dt + \|I_0 \circ \phi_v^{-1}(1) - I_1\|_{L^2}^2 \qquad (4a)$$

$$\frac{\partial}{\partial t}\phi_v(t) = v(t) \circ \phi_v(t) \qquad (4b)$$

$$v(t) = \sum_i^K v^i(t) \qquad (4c)$$

$$\phi_v(0) = Id \qquad (4d)$$

where $\|v^i(t)\|_{V^i} = \langle v^i(t), v^i(t)\rangle_{V^i}^{\frac{1}{2}}$ is the norm induced by the inner product $\langle u, v\rangle_{V^i} = \langle L_{V^i}u, L_{V^i}v\rangle_{L^2}$ defined on the ith scale. $\{K_{V^i} = (L_{V^i}^+ L_{V^i})^{-1}\}, i = 1, \ldots, K$ are the K kernels which essentially are used to encode the image deformation energy at different spatial scales and w^i is the weighting factor of the deformation energy of the ith kernel. $\phi_v(t)$ is the time-dependent deformation computed as the integration of the velocity field $v(t)$ and $I_0 \cdot \phi_v^{-1}(t)$ is the transformed image of I_0 by the deformation $\phi_v(t)$.

Using the optimal control based approach introduced in [22, 23], we get the Euler-Poincare equation (EPDiff) (5a)–(5f) of the optimal velocity fields $\{v^k(t)\}$, $k = 1, 2, \ldots, K$ for the multi-kernel LDDMM registration algorithm.

$$\dot{I}(t) = -\nabla I(t) \cdot v(t) \qquad (5a)$$

$$\dot{P}(t) = -\nabla(P(t) \cdot v(t)) \qquad (5b)$$

$$v(t) = \sum_{k=1}^K v^k(t) \qquad (5c)$$

$$v^k(t) = -(w^k)^{-1}K_{V^k}\star(P(t)\nabla I(t)), \quad k = 1, \ldots, K \qquad (5d)$$

$$P(1) = -(I(1) - I_1) \qquad (5e)$$

$$I(0) = I_0 \qquad (5f)$$

The registration algorithm is given as follows:

For more details on the computation routine and the performance of the multi-kernel LDDMM registration algorithm, we refer to [19, 22, 23].

2.4.2 Disc Segmentation by Diffeomorphic Registration

The IVD segmentation is achieved as a template based registration between the thin cylinder IVD template and the correspondent candidate disc region as shown in Fig. 4.

In order to explore both intensity and feature information to enhance the accuracy and robustness of the segmentation, we consider a simultaneous registration of two pairs of images, I_0^I/I_1^I and I_0^E/I_1^E, which stand for the image intensity and edge information template/observation pairs respectively. Accordingly in the cost function of the LDDMM registration (4a), the image similarity term includes two components $\|I_0^I \circ \phi_v^{-1}(1) - I_1^I\|_{L^2}^2 + \beta\|I_0^E \circ \phi_v^{-1}(1) - I_1^E\|_{L^2}^2$ with I_0^I/I_1^I and I_0^E/I_1^E as explained below.

Algorithm 2 Multi-Kernel LDDMM registration

Input: Images to be registered I_0, I_1
Output: Time dependent velocity field $v(t)$, $t \in [0, 1]$ whose integration gives the optimal matching process $I(t)$, $t \in [0, 1]$ which represents a smooth deformation from I_0 to I_1.
Initialization: $I(0) = I_0$, $v^k(t) = 0$, $t \in [0, 1]$, $k = 1, 2, ..., K$
while not converge **do**

1. Compute $I(t)$, $t \in [0, 1]$ by (5a), (5b), (5f).
2. Compute $P(1)$ by (5e).
3. Update $P(t)$, $t \in [0, 1]$ by solving (5c) in the inverse direction.
4. Update $v^i(t)$, $t \in [0, 1]$, $i = 1, 2, ..., K$ by (5d).

end while

Intensity information The template intensity image I_0^I is just the initialized disc template, i.e., a binary 3D image with the interior region of the disc template set as 1. The correspondent target image I_1^I is constructed by a three-step procedure.

1. Based on the 3D spine column model, for each vertebra disc, we can determine a region that belongs to the interior region of the disc with high confidence as explained in Fig. 5.
2. From the extracted high confidence disc voxels, we can estimate the image intensity distribution of the disc tissue in each data set by assuming a Gaussian intensity distribution $N(u_i, \sigma_i)$, $i = 1, 2, 3, 4$, i.e., to estimate the values $\{u_i, \sigma_i\}$, $i = 1, 2, 3, 4$ for each channel volume.
3. Accordingly in each channel volume I_i^{MRI}, $i = 1, 2, 3, 4$, each voxel $v_i^j \in I_i^{MRI}$ in the candidate disc region with an intensity value I_i^j can be assigned a probability

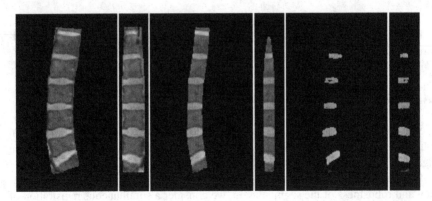

Fig. 5 Determination of the high confidence disc region using the spine column model. *Left* to *right* The spine column region extracted using the spine column model shown in a sagittal and a coronal slice; The central region of the spine column obtained by shrinking the radius of the spine column model by a factor 0.5 shown in the same two slices; The detected high confidence disc regions by further cutting out the bone tissue using the spine column model

$p_i^j = \frac{1}{\sqrt{2\pi\sigma_i^2}} \exp(-\frac{(I_i^j - u_i)^2}{2\sigma_i^2})$ indicating whether voxel v_i^j belongs to the disc region using the correspondent Gaussian distribution model $N(u_i, \sigma_i)$, $i = 1, 2, 3, 4$. The image I_1^I that contains the intensity information of the 4 channel data is then constructed as an average of the probabilities of the 4 data sets $\{P_i = \{p_i^j\}, v_i^j \in I_i^{MRI}, i = 1, 2, 3, 4\}$ as $I_1^I = (\prod P_i)^{\frac{1}{4}}$. Figure 6c–f show an example of the intensity template and the computed correspondent target image.

Edge information For the edge information, the template image I_0^E can be regarded as the outer surface of the disc template as shown in Fig. 6g–h. The target image I_1^E can be obtained by summing up and normalizing the gradient amplitudes of the 4 data set, see Fig. 6i–j.

An example of the template images and the correspondent target images and the time dependent registration procedure is shown in Fig. 6.

By registering the disc template to the observed 3D data volume, the final segmented IVD can then be obtained as the deformed template achieved by the multi-kernel LDDMM registration.

3 Experiments

The proposed algorithms are verified on MRI datasets of 15 patients obtained with the Dixon protocol. In all the data sets, based on the two landmarks obtained from the initialization step, both the 2D lumbar spine column and the 3D spine column

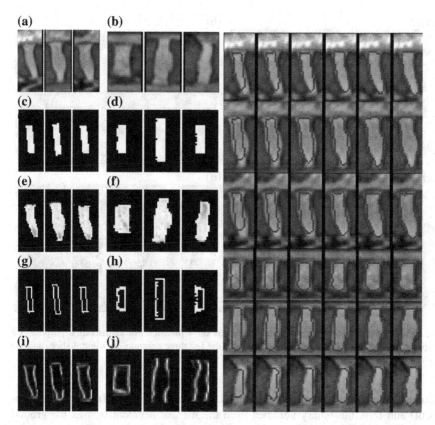

Fig. 6 3D IVD segmentation by multi-kernel LDDMM registration. *Left side* the data used in diffeomorphic registration based 3D lumbar disc segmentation. Although the task was performed in 3D, we show results on 2D slices for visualization purpose. Also be aware that in the target images, the bone tissue regions are extracted using the spine column model. **a, b** 3 sagittal/coronal slices of the candidate disc region (disc L4–L5 in Fig. 2) **c, d** the intensity disc template in 3 sagittal/coronal slices; **e, f** intensity information extracted from MRI data sets in 3 sagittal/coronal slices; **g, h** the edge disc template in 3 sagittal/coronal slices; **i, j** edge information computed from MRI data sets in 3 sagittal/coronal slices; *Right side* the time-dependent deformation of the disc template during the multi-kernel diffeomorphic registration for a L4–L5 disc segmentation. *Left* to *right* the deformations of the template at 6 time slots t = 0, 0.2, 0.4, 0.6, 0.8, 1. t = 0 means the initial template and t = 1 gives the final registration results; from *top row* to *bottom row*: the evolution of the template visualized in 6 different slices

models are correctly extracted. The computational time of each data set varies between 10–15 min depending on the initialization and converge speed with a MATLAB implementation. Examples of the disc segmentation results on 4 patient data sets are shown in Fig. 7.

We also carried out quantitative evaluation of our algorithm. To do this, we manually segmented all datasets (we only need to segment one channel for each

(a)   **(b)**

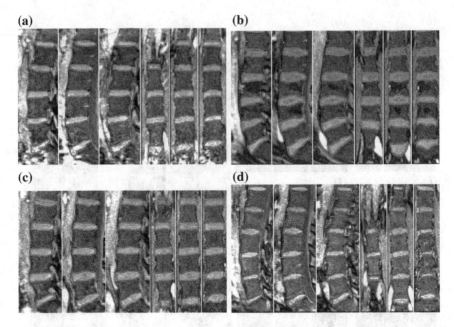

(c)   **(d)**

Fig. 7 3D intervertebral disc segmentation results on 4 patients. For visualization purpose, we display the results on 2D slices. For each image, the *left* three columns are sagittal slices and the *right* three are coronal slices

patient as all four channel volumes are aligned according to Dixon imaging protocol) and took the binary volumes from the manual segmentation as the ground truth to verify the accuracy of the present algorithm. We computed the Dice coefficient D which is usually used to measure the overlap between two binary images:

$$D = \frac{2 \times |A \cap B|}{|A| + |B|} \times 100 \qquad (6)$$

Table 1 shows the average dice coefficients of the 5 discs on all 15 patients when the automated segmentation was compared to the manual segmentation. The highest average dice coefficient was found for patient #8 (87.9 %) and the lowest average dice coefficient was found for patient #9 (80.5 %). We also computed the average dice coefficients for all discs and the results are presented in Table 2. We note that Neubert et al. [10] reported a mean Dice of 76–80 % in their 3D IVD segmentation paper.

Table 1 Average dice coefficients (%) of the 5 discs between the manual segmentation and the proposed algorithm on different patients

Patient	P1	P2	P3	P4	P5	P6	P7	P8	P9	P10	P11	P12	P13	P14	P15
Dice	86.1	81.9	82.6	86.3	86.8	83.6	87.6	87.9	80.5	84.1	86.3	85.4	86.9	87.7	83.1

Table 2 Average dice coefficients (%) between the manual segmentation and the proposed algorithm on different discs on all 15 data sets

Disc	L1–L2	L2–L3	L3–L4	L4–L5	L5–S1
Dice	81.2	87.1	88.2	86.5	82.7

4 Conclusions

In this paper we proposed an automated lumbar intervertebral disc segmentation strategy, whose key components include a graphical model based spine column identification algorithm and a multi-kernel LDDMM registration algorithm to achieve the disc segmentation. By identifying the lumbar spine column structure before carrying out the segmentation, we acquire geometrical and appearance information about the spine column. These information can be used to accurately locate the candidate disc region and provide constraints to enhance the performance of the disc segmentation. By converting the segmentation problem as a template based diffeomorphic registration, we can explore both the intensity and edge information of the observed data while keeping a smooth deformation of the template so that the final segmented discs will possess smooth surfaces. The experiments on 15 patient data sets verified the robustness and accuracy of our method. We also noticed that for abnormal cases, such as with missing/additional vertebrae or the scoliosis case, the automated lumbar column identification may not be reliable although the graphical model can handle the unknown vertebra number as shown in [14]. A possible solution for these extreme cases is to ask the user to indicate the center of each vertebra body during the initialization step. Once the centers are known, the particle filtering based inference can still achieve a reliable 2D lumbar column identification and the following up 3D lumbar column modeling and disc segmentation.

References

1. Modic M, Ross J (2007) Lumbar degenerative disk disease. Radiology 1:43–61
2. Parizel P, Goethem JV, den Hauwe LV, Voormolen M (2007) Degenerative disc disease. In Van Goethem J (ed) Spinal imaging—diagnostic imaging of the spine and spinal cord. Springer, Berlin, pp 122–133
3. Tsai M, Jou J, Hsieh M (2002) A new method for lumbar herniated intervertebral disc diagnosis based on image analysis of transverse sections. Comput Med Imaging Graph 26:369–380
4. Niemelainen R, Videman T, Dhillon S, Battie M (2008) Quantitative measurement of intervertebral disc signal using MRI. Clin Radiol 63:252–255
5. Schmidt S, Kappes JH, Bergtholdt M, Pekar V, Dries S, Bystrov D, Schnorr C (2007) Spine detection and labeling using a parts-based graphical model. In: Karssemeijer N, BL (ed) IPMI 2007. Springer, Berlin, pp 122–133

6. Corso J, Alomari R, Chaudhary V (2008) Lumbar disc localization and labeling with a probabilistic model on both pixel and object features. In: Metaxas D (ed) MICCAI 2008. Springer, Berlin, pp 202–210

7. Chevrefils C, Cheriet F, Aubin C, Grimard G (2009) Texture analysis for automatic segmentation of intervertebral disks of scoliotic spines from MR images. IEEE Trans Inf Technol Biomed 13:608–620

8. Michopoulou S, Costaridou L, Panagiotopoulos E, Speller R, Panayiotakis G, Todd-Pokropek A (2009) Atlas-based segmentation of degenerated lumbar intervertebral discs from MR images of the spine. IEEE Trans Biomed Eng 56:2225–2231

9. Ayed IB, Punithakumar K, Garvin G, Romano W, Li S (2011) Graph cuts with invariant object-interaction priors: application to intervertebral disc segmentation. In: Szekely G, HH (ed) IPMI 2011. Springer, Berlin, pp 221–232

10. Neubert A, Fripp J, Schwarz R, Lauer L, Salvado O, Crozier S (2012) Automated detection, 3d segmentation and analysis of high resolution spine MR images using statistical shape models. Phys Med Biol 57:8357–8376

11. Law M, Tay K, Leung A, Garvin G, Li S (2013) Intervertebral disc segmentation in MR images using anisotropic oriented flux. Med Image Anal 17:43–61

12. Glocker B, Feulner J, Criminisi A, Haynor D, Konukoglu E (2012) Automatic localization and identification of vertebrae in arbitrary field-of-view CT scans. In Ayache N (ed) MICCAI 2012. Springer, Berlin, pp 590–598

13. Glocker B, Zikic D, Konukoglu E, Haynor D, Criminisi A (2013) Vertebrae localization in pathological spine CT via dense classification from sparse annotations. In Mori K (ed) MICCAI 2013. Springer, Berlin, pp 262–270

14. Dong X, Lu H, Sakurai Y, Yamagata H, Zheng G, Reyes M (2010) Automated intervertebral disc detection from low resolution, sparse MRI images for the planning of scan geometries. In: Wang F, Yan P, Suzuki K, Shen D (eds) MLMI2010. Springer, Berlin, pp 10–17

15. Dong X, Zheng G (2015) Automated 3D lumbar intervertebral disc segmentation from MRI data sets. In: Yao J, Glocker B, Klinder T, Li S (eds) Recent advances in computational methods and clinical applications for spine imaging. Lecture notes in computational vision and biomechanics, vol 20. Springer, Berlin, pp 131–142

16. Risser L, Vialard FX, Wolz R, Murgasova M, Holm DD, Rueckert D (2011) Simultaneous multiscale registration using large deformation diffeomorphic metric mapping. IEEE Trans Med Imaging 30:1746–1759

17. Grenander U, Miller MI (1998) Computational anatomy: an emerging discipline. Q Appl Math 4:617–694

18. Sommer S, Nielsen M, Lauze F, Pennec X (2011) A multi-scale kernel bundle for LDDMM: towards sparse deformation description across space and scales. In: Szekely G, HH (ed) IPMI2011. Springer, Berlin, pp 624–635

19. Beg MF, Miller MI, Trouv A, Younes L (2005) Computing large deformation metric mappings via geodesic flow of diffeomorphisms. Int J Comput Vision 61:139–157

20. Miller MI, Trouv A, Younes L (2006) Geodesic shooting for computational anatomy. J Math Imaging Vision 24:209–228

21. Bruveris M, Gay-Balmaz F, Holm DD, Ratiu TS (2011) The momentum map representation of images. J Nonlinear Sci 21:115–150

22. Hart GL, Zach C, Niethammer M (2009) An optimal control approach for deformable registration. In: CVPR2009, pp 9–16

23. Vialard FX, Risser L, Rueckert D, Cotter CJ (2012) Diffeomorphic 3d image registration via geodesic shooting using an efficient adjoint calculation. Int J Comput Vision 97:229–241

Registration for Orthopaedic Interventions

Ziv Yaniv

Abstract Registration is the process of computing the transformation that relates the coordinates of corresponding points viewed in two different coordinate systems. It is one of the key components in orthopaedic navigation guidance and robotic systems. When assessing the appropriateness of a registration method for clinical use one must consider multiple factors. Among others these include, accuracy, robustness, speed, degree of automation, detrimental effects to the patient, effects on interventional workflow, and associated financial costs. In this chapter we give an overview of registration algorithms, both those available commercially and those that have only been evaluated in the laboratory setting. We introduce the models underlying the algorithms, describe the context in which they are used and assess them using the criteria described above. We show that academic research has primarily focused on improving all aspects of registration while ignoring workflow related issues. On the other hand, commercial systems have found ways of obviating the need for registration resulting in streamlined workflows that are clinically more acceptable, albeit at a cost of being sub-optimal on other criteria. While there is no optimal registration method for all settings, we do have a respectable arsenal from which to choose.

1 Introduction

Registration is the process of computing the transformation that relates the coordinates of corresponding points given in two different coordinate systems. It is one of the key components in orthopaedic navigation guidance and robotic systems.

Z. Yaniv (✉)
TAJ Technologies Inc., Mendota Heights, MN 55120, USA
e-mail: zivyaniv@nih.gov

Z. Yaniv
Office of High Performance Computing and Communications,
National Library of Medicine, National Institutes of Health,
Bethesda, MD 20894, USA

© Springer International Publishing Switzerland 2016   41
G. Zheng and S. Li (eds.), *Computational Radiology*
for Orthopaedic Interventions, Lecture Notes in Computational
Vision and Biomechanics 23, DOI 10.1007/978-3-319-23482-3_3

Most commonly, registration is used to transfer preoperative information to the intraoperative setting and for assessment of postoperative results. Slightly less common uses include the use of registration to facilitate automated procedure planning, and computation of population atlases which are later used for intraoperative registration. The standard mathematical formulation of a registration algorithm is as an optimization task, where a problem specific quantity is minimized or maximized. This formulation implies that in most cases registration does not have an analytic solution and therefore we are not guaranteed that the obtained result is the optimal value (global minimum or maximum).

The majority of texts describing registration only focus on algorithmic aspects. In this work we aim to provide a holistic view that we believe benefits both clinical practitioners and system developers. We therefore introduce the concept of a registration framework. That is, we consider both the algorithmic aspects and the integration of the algorithm into the clinical workflow as it pertains to data acquisition and possible effects on clinical outcome.

When evaluating a registration framework it is beneficial to use a theoretical construct, the ideal clinical registration framework. This framework is evaluated using five criteria and has the following characteristics:

1. Accuracy and precision: sub-millimetric target registration error throughout the working space, with sub-millimetric precision.
2. Robustness: not effected by errors and outliers in the input data (e.g. low image quality, user localization errors).
3. Computation speed: takes less than 1 s to complete the computation.
4. Obtrusiveness: no registration specific hardware is introduced into the environment (e.g. US machine used only for registration), and input data is easily acquired in less than 10 s to yield a streamlined workflow.
5. Clinical side effects: has no detrimental side effects (e.g. registration requires additional exposure to ionizing radiation from imaging, additional incisions or surgical procedures).

From an engineering or scientific standpoint the important evaluation criteria of a registration framework are those directly associated with the registration algorithm, the first three criteria listed above. As a result, the later two criteria often do not receive sufficient attention, resulting in sub-optimal clinical registration frameworks. This is readily evident when analyzing the registration framework of one of the earliest robotic systems used in orthopaedics. The ROBODOC system for total hip replacement initially utilized a paired-point registration algorithm to align the robot to the patient. The engineers designing the registration framework placed most of the emphasis on the registration algorithm, choosing to use an optimal algorithm that provides high accuracy and precision, is robust and computes the result in less than 1 s. Unfortunately, while the registration algorithm was optimal, the framework was not. In this case, the system used implanted fiducials to define the points used for registration. This required an additional minor intervention prior to surgery. More importantly, it resulted in significant collateral damage, patients suffered from persistent severe knee pain due to nerve damage caused by the pin implantation [82]. We therefore encourage developers of registration algorithms to consider the whole registration framework and not just the algorithm, as this is what ultimately determines clinical acceptance.

Before we delve into descriptions of specific registration algorithms we will highlight a fundamental characteristic of the errors associated with rigid registration. Spatial errors become larger as one moves farther from the reference frame's origin. This is primarily of importance to operators of such systems, providing the theoretical explanation to observations in the field such as those described in [97]. As we are dealing with rigid registration, we have errors both in the estimated translation and rotation. The translation errors have a uniform effect throughout space and do not depend on the location of the origin. The rotational errors, on the other hand, have a spatially varying effect that is dependent on the distance from the origin, this is sometimes referred to as the "lever effect". Thus, for the same rotational error a point that is close to the origin will exhibit a smaller spatial error as compared to a point that is farther away from the origin.

From a practical standpoint we cannot mitigate the effect of translational errors, but we can mitigate the effect of rotational errors. That is, if we place the origin of our reference frame in the center of the spatial region of interest, we minimize the maximal distance to the origin and thus minimize the maximal spatial error. Figure 1 illustrates these observations in the planar case.

While developers of registration frameworks strive to construct the ideal framework, this is not an easy task. More importantly, it is context dependent. For example, introducing an US machine for the sole purpose of registration makes the framework non-ideal. On the other hand, if the US machine is already available, for instance to determine breaches in pedicle screw hole placement [52], utilizing it for registration purposes does not preclude the framework from being considered ideal.

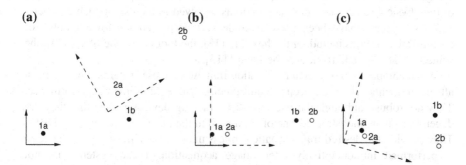

Fig. 1 Characteristics of rigid registration spatial errors: **a** corresponding points (*1a–2a, 1b–2b*) before registration; after registration **b** error only in translation, **c** error only in rotation. In both cases it is clear that registration reduced the spatial error between the points. It is also clear that the effect of the translational error on the spatial errors is uniform throughout space, while the effect of the rotational error depends on the distance from the origin. In theory, one can eliminate the effect of rotational errors for a single point by placing the origin of the reference frame at this point of interest

Given that constructing the ideal framework is hard, it is only natural that developers of guidance systems have devised schemes that bypass this task. We thus start our overview of registration in orthopedics with approaches that perform registration implicitly.

2 Implicit Registration

We identify two distinct approaches to performing implicit registration. The first combines preoperative calibration with intraoperative volumetric image acquisition using cone-beam CT (CBCT). The second uses patient specific templates that are physically mounted onto the patient in an accurate manner. These devices physically constrain the clinician to planned trajectories for drilling or cutting.

2.1 Intraoperative CBCT

This approach relies on the use of a tracked volumetric imaging system to implicitly register the acquired volume to a Dynamic Reference Frame (DRF) rigidly attached to the patient. During construction of the imaging device, it is calibrated such that the location and orientation of the reconstructed volume is known with respect to a DRF that is built into the system. Intraoperatively, a volumetric image is acquired while the patient's DRF and imaging device's DRF are both tracked by a spatial localization system. This allows the navigation system to correctly position and orient the volume in physical space. It should be noted that some manufacturers refer to this registration bypass as automated registration, although strictly speaking the system is not performing intraoperative registration. Figure 2 shows an example of the physical setup and coordinate systems involved in the use of such a system.

These systems have been used to guide various procedures for a number of anatomical structures including the hip [11, 118], the foot [101], the tibia [134], the spine [23, 38, 53, 83, 107], and the hand [115].

When compared to our ideal registration framework these systems satisfy almost all requirements. They are accurate and precise with submillimetric performance. They are robust, although the use of optical tracking devices means that they are dependent on an unobstructed line of sight both to the patient's and device's DRFs. They provide the desired transformation in less than 1 s, as almost no computation is performed intraoperatively after image acquisition. These systems are not obtrusive as they are used to acquire both intraoperative X-ray fluoroscopy and volumetric imaging. The one criterion where they can be judged as less than ideal is that these systems expose the patient to ionizing radiation.

While these systems offer many advantages when evaluated with regard to components of the registration framework, they do have other limitations. The quality of the volumetric images acquired by these devices is lower than that of

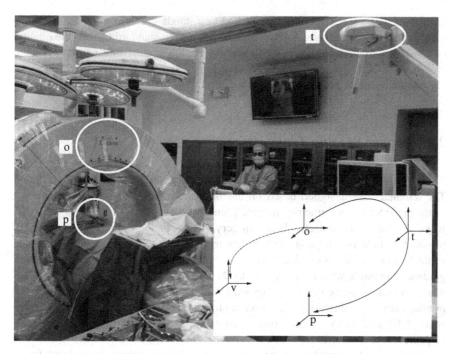

Fig. 2 The O-arm system from Medtronic (Minnisota, MN, USA) facilitates implicit registration via factory calibration. Picture shows the clinical setup (Courtesy of Dr. Matthew Oetgen, Children's National Health System, USA). *Inset* shows corresponding coordinate systems and transformations. Transformations from tracking coordinate system, t, to patient DRF, p, and O-arm DRF, o, vary per procedure and are obtained from the tracking device. The transformation between the O-arm DRF, o, and the volume coordinate system acquired by the O-arm, v, is fixed and obtained via factory calibration. Once the volume is acquired intraoperatively it is implicitly registered to the patient as the transformation from the patient to the volume is readily constructed by applying the three known transformations

diagnostic CT images and this can potentially have an effect on the quality of the visual guidance they provide. Obviously, one can register preoperative diagnostic images (CT, PET, MR) to the intraoperative CBCT, but this changes the registration framework and it is unclear whether such a system retains the advantages associated with the existing one. Another, non technical, disadvantage is the cost associated with these imaging systems. They do increase the overall cost of the navigation system. One potential approach to addressing the cost of high end CBCT systems is to retrofit non-motorized C-arm systems to perform CBCT reconstructions. This requires that the position and orientation of the C-arm be known for each of the acquired X-rays during a manual rotation. Two potential solutions to this task have been described in the literature. The C-Insight system from Mazor Robotics (Caesarea, Israel) [134] uses an intrinsic tracking approach. It both calibrates the X-ray system and estimates the C-arm pose using spherical markers visible in the X-ray images. An extrinsic approach to tracking is describe in [2]. This system utilizes inertial measurement units attached to a standard C-arm to obtain the pose associated with each image. As both approaches retrofit standard C-arms the quality of the reconstructed volumes is expected to be lower than those obtained with motorized CBCT machines, as these also utilize flat panel sensors which yield higher quality input for CBCT reconstruction.

2.2 Patient Specific Templates

This approach relies on the use of physical templates to guide cutting or drilling. The templates are designed based on anatomical structures and plans formulated using preoperative volumetric images, primarily diagnostic CT. This is in many ways similar to stereotactic brain surgery, using patient mounted frames. To physically guide the surgeon the template incorporates two components, the bone contact surface which is obtained via segmentation from the preoperative image and guidance channels which correspond to the physical constraints specified by the plan. Templates can be created either via milling, subtractive fabrication, or via 3D printing, additive fabrication. The later approach has become more common as the costs of 3D printers have come down. Figure 3 illustrates this concept in the context of pedicle screw insertion.

Patient specific templates have been used to guide various procedures for a number of anatomical structures including the spine [15, 68, 69, 98], the hand [60, 70, 85], the hip [41, 98, 119, 148], and the knee [43, 98].

When compared to the ideal registration framework, this approach provides sufficient accuracy, precision and robustness. Registration is obtained implicitly in an intuitive manner, as the template is manually fit onto the bone surface. This approach is not obtrusive as it does not require additional equipment other than the template. In theory there are no clinical side effects associated with the use of a template.

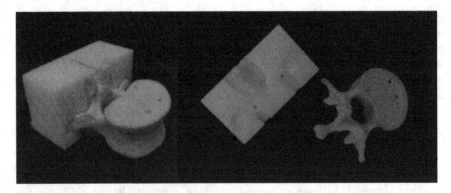

Fig. 3 Patient specific 3D printed template for pedicle screw insertion. Template incorporates drill trajectories planned on preoperative CT (Courtesy Dr. Terry S. Yoo, National Institutes of Health, USA)

Unfortunately, achieving accuracy, precision and robustness requires that the template fit onto the bone in a unique configuration. This potentially requires a larger contact surface, resulting in larger incisions than those used by standard minimally invasive approaches [119]. While a smaller contact surface is desirable it should be noted that this will potentially increase the chance that the operator "fits the square peg into a round hole" (e.g. wrong-level fitting for pedicle screw insertion).

3   3D/3D Registration

In orthopaedics, 3D/3D registration is utilized for alignment of preoperative data to the physical world, spatial alignment of data as part of procedure planning, and for construction of statistical shape and appearance models with the intent of replacing the use of preoperative CT for navigation with a patient specific model derived from the statistical one.

Alignment of preoperative data to the physical world is the most common usage of registration in orthopaedics. It has been described as a component of robotic procedures applied to the knee [7, 67, 78, 79, 96, 103], and hip [73, 81, 109]. In the context of image-guided navigation, it has been described as a component of procedures in the hip [8–10, 54, 61, 94, 110, 118, 119], the femur [8, 16, 94, 99, 154], the knee [54, 99, 116], and the spine [31, 46, 61, 100, 120, 143].

In the context of planning, 3D/3D registration has been used for population studies targeting implant design [58], for the selection and alignment of a patient specific optimal femoral implant in total hip arthroplasty [90], and for planning the alignment of multiple fracture fragments using registration to the contralateral anatomy, thus enabling automated formulation of plans in distal radius osteotomy [27, 28, 102, 111], humerus fracture fixation [14, 35], femur fracture fixation [86] and potentially for scaphoid fracture fixation [63].

In the context of statistical model creation, the first journal publications to describe the use of statistical shape models in orthopedics were [34, 117]. The motivation for this work was to create a patient specific model for guiding ACL reconstruction and total knee arthroplasty without the need of a preoperative 3D scan. A statistical point distribution model of the distal part of the femur was created by digitizing the surfaces of multiple dry bone specimens and creating the required anatomical point correspondence via non-rigid 3D/3D registration. Many others follow a similar approach but with one primary difference, the models used to describe the population are obtained from previously acquired 3D scans, CT or MR, of other patients. This allows modeling of both shape and intensity. For a detailed overview of the different aspects of constructing statistical shape models we refer the interested reader to [44].

In orthopedics, statistical shape models that use registration for establishing anatomical point correspondences have primarily used the free-form deformation

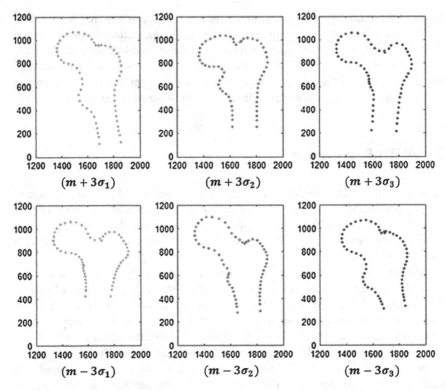

Fig. 4 2D proximal femur statistical point distribution model, three standard deviations from the mean shape, along the first three modes of variation (Courtesy Dr. Guoyan Zheng, University of Bern, Switzerland)

method proposed in [104] and the diffeomorphic-demons method proposed in [131]. These registration methods were utilized for creating models of the femur, pelvis and tibia [9, 16, 58, 108, 137, 147]. This approach was also recently applied in 2D, using 2D/2D registration to establish point correspondences for a proximal femur model [140]. Figure 4 shows the resulting 2D statistical point distribution model.

At this point we would like to highlight two aspects associated with the use of statistical point distribution models which one should always think about: does the input to the statistical model truly reflect the population variability (e.g. using femur data obtained only from females will most likely not reflect the shape and size of male femurs); and if point correspondences were established using registration, how accurate and robust was the registration method.

Having motivated the utility of 3D/3D registration in orthopedics, we now turn our attention to several common algorithms, pointing out their advantages and limitations.

3.1 Point Set to Point Set Registration

In this section we discuss three common algorithm families used to align point sets in orthopedics, paired-point algorithms, surface based algorithms and statistical surface models.

Paired-point algorithms use points with known correspondences to compute the rigid transformation between the two coordinate systems. In this setting points are localized, most often, manually both in image space via mouse clicks, and in physical space via digitizing using a tracked calibrated pointing device such as described in [122]. Both calibration and tracking of pointer devices can be done with sub-millimetric accuracy.

In general, we are given three or more corresponding points in two Cartesian coordinate systems:

$$\mathbf{x}_{li} = \widehat{\mathbf{x}_{li}} + \mathbf{e}_{li}$$
$$\mathbf{x}_{ri} = \widehat{\mathbf{x}_{ri}} + \mathbf{e}_{ri}$$

where $\widehat{\mathbf{x}_{li}}, \widehat{\mathbf{x}_{ri}}$ are the true point coordinates related via a rigid transformation $T(\widehat{\mathbf{x}_{li}}) = \widehat{\mathbf{x}_{ri}}$, $\mathbf{x}_{li}, \mathbf{x}_{ri}$ are the observed coordinates, and $\mathbf{e}_{li}, \mathbf{e}_{ri}$ are the errors in localizing the points.

The most common solution to this problem is based on a least squares formulation:

$$T^* = \operatorname*{argmin}_{T} \sum_{i=1}^{n} \|\mathbf{x}_{ri} - T(\mathbf{x}_{li})\|^2$$

The solution to this formulation is optimal if we assume that there are no outliers and that the Fiducial Localization Errors (FLE)[1] follow an isotropic and homogenous Gaussian distribution. That is, the error distribution is the same in all directions and is the same for all fiducials. Figure 5 visually illustrates the possible FLE categories.

While the formulation is unique, several analytic solutions have been described in the literature, with the main difference between them being the mathematical representation of rotation. These include use of a rotation matrix [3, 128], a unit quaternion [32, 47], and a dual quaternion [132]. All of these algorithms guarantee a correct registration if the assumptions hold. Luckily, empirical evaluation has shown that all choices yield comparable results [30].

In practice, FLE is often anisotropic, such as when using optical tracking systems, where the error along the camera's viewing direction is much larger than the errors perpendicular to it [138]. In this case, the formulation described above does

---

[1]In this context the term fiducial is used to denote a point used to compute the registration, be it an artificial marker or anatomical landmark.

	Homogenous	Inhomogeneous
Isotropic		
Anisotropic		

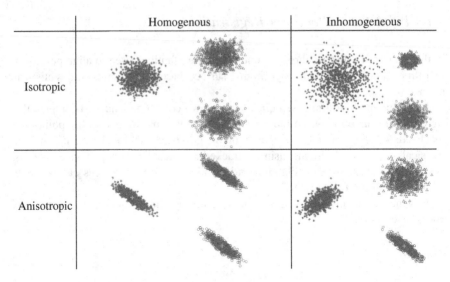

Fig. 5 Categories of fiducial localization error according to their variance. Plus (*red*) indicates fiducial location and *star/circle/triangle* marks (*blue*) denote localization variability (Color figure online)

not lead to an optimal solution. Iterative solutions addressing anisotropic-homogenous noise and anisotropic-inhomogeneous noise models were described in [77, 84] respectively. These methods did not replace the least squares solutions even though they explicitly address the true error distributions. This is possible due to two attributes of the original algorithms, they are analytic, that is they do not require an initial solution as iterative algorithms do, and they are extremely easy to implement. Case in point, Table 1 is a fully functioning implementation of the method described in [128] using MATLAB (The Mathworks Inc., Natick, MA, USA).

One of the issues with paired-point registration methods is that the pairing is explicit, that is the clinician has to indicate which points correspond. This is often performed as part of a manual point localization process which is known to be inaccurate both in the physical and the image spaces [46, 112]. The combination of fixed pairing and localization errors reduces the accuracy of registration, as we are not truly using the same point in both coordinate systems. If on the other hand we allow for some flexibility in matching points then we may improve the registration accuracy. This leads directly to the idea of surface based registration.

Preoperative surfaces are readily obtained from diagnostic CT. Intraoperative surface acquisition is often done using a tracked pointer probe [4]. To ensure registration success one must acquire a sufficiently large region so that the intraoperative surface cannot be ambiguously matched to the preoperatively extracted one. This is a potential issue if it requires increased exposure of the anatomy only for the sake of registration. A possible non-invasive solution is to use calibrated and tracked

Table 1 Source code shows complete implementation of analytic paired point rigid registration in MATLAB

```
function T = absoluteOrientation(pL, pR)
```

```
n = size(pL,2);
meanL = mean(pL,2);
meanR = mean(pR,2);
[U,S,V] = svd(((pL - meanL(:,ones(1,n))) *
                (pR - meanR(:,ones(1,n)))'));
R = V*diag([1,1,det(U*V)])*U';
t = meanR - R*meanL;

T = [R, t; [0, 0, 0, 1]];
```

The analytic nature of the solution and the simplicity of implementation make it extremely attractive for developers, even though this solution assumes noise is isotropic and homogenous, which is most often not the case

ultrasound (US) for surface acquisition. US calibration is still an active area of research with multiple approaches described in the literature [76], most yielding errors on the millimeter scale. More recently published results report sub-millimetric accuracy [80]. Once the tracked calibrated US images are acquired, the bone surface is segmented in the images and its spatial location is computed using the tracking and calibration data. Automated segmentation of the bone surface is not a trivial task. In the works described in [8, 9] the femur and pelvis were manually segmented in the US images prior to registration, this is not practical for clinical use. Automated bone surface segmentation algorithms in US images have been described in [54, 57, 100, 110, 120] for B-mode US and in [79] for A-mode. These algorithms were evaluated as part of registration frameworks which have clinically acceptable errors (on the order of 2 mm).

Surface based registration algorithms use points *without* a known correspondences to compute the rigid transformation between the two coordinate systems. A natural approach for scientists tackling such problems is to decompose them into sub problems with the intent of using existing solutions for each of the sub problems. This general way of thinking is formally known as "computational thinking" and is a common approach in computer science [139].

Given that we have an analytic algorithm for computing the transformation when we have a known point pairing it was only natural for computer scientists to propose a two step approach towards solving this registration task. First match points based on proximity and then estimate the transformation using the existing paired-point algorithm. This process is repeated iteratively with the incremental transformations combined until the two surfaces are in correspondence. This algorithmic approach is now known as the Iterative Closest Point (ICP) algorithm. This algorithm was independently introduced by several groups [13, 20, 149].

While the simplicity of the ICP algorithm makes it attractive, from an implementation standpoint, it has several known deficiencies. The final solution is highly dependent on the initial transformation estimate, and speed is dependent on the computational cost of point pairing. In addition, the use of the analytic least squares algorithm to compute the incremental transformations assumes that there are no outliers and that the error in point localization is isotropic and homogenous. Many methods for improving these deficiencies have been described in the literature, with a comprehensive summary given in [105]. One aspect of the ICP algorithm that was not addressed till recently was that the localization errors are often anisotropic and inhomogeneous. A variant of ICP addressing this issue was recently described in [72].

From a practical standpoint, a combination of paired-point and an ICP variant is often used. The analytic solution most often provides a reasonable initialization for the ICP algorithm which then provides improved accuracy. This was shown empirically in [46]. Unfortunately, this combination still does not guarantee convergence to the correct solution. This is primarily an issue when the intraoperatively digitized surface is small when compared to the preoperative surface. In this situation the surface registration may be trapped by multiple local minima. This is most likely the reason for the poor registration results reported in [4] for registering the femur head in the context of hip arthroscopy.

Statistical surface model based registration [34, 117] are similar to surface based registration as described above, but with one critical difference, they do not use patient specific preoperative data. Instead of a patient specific surface obtained from CT, a surface model is created and aligned to the intraoperative point cloud. The statistical model encodes the variability of multiple example bone surfaces and uses the dominant modes of variation to fit a patient specific model to the intraoperative point cloud. An advantage of using such an approach is that models created from the atlas are limited to the variations observed in the data used to construct it. Thus, these models are plausible. Unfortunately, they often will not provide a good fit to previously unseen pathology. This can be mitigated by allowing the model to locally deform in a smooth manner to better fit the intraoperative point cloud [117].

3.2 Intensity Based Registration

Intensity based registration aligns two images by formulating the task as an optimization problem. The optimized function is dependent on the image intensity values and the transformation parameters. As the intensity values for the images are given at a discrete set of grid locations and the transformation is over a continuous domain, registration algorithms must interpolate intensity values at non grid locations. This means that all intensity based registration algorithms include at least three components: (1) The optimized similarity function which indicates how similar are the two images, subject to the estimated transformation between them; (2) An optimization algorithm; and (3) an interpolation method.

A large number of similarity measures have been described in the literature and are in use. Selecting a similarity measure is task dependent with no "best" choice applicable to all registration tasks. The selection of a similarity measure first and foremost depends on the relationship between the intensity values of the modalities being registered. When registering data from the same modality one may use the sum of squared differences or the sum of absolute differences. For modalities with a linear relationship between them one may use the normalized cross correlation. For modalities with a general functional relationship one can use the correlation ratio. Finally, for more general relationships, such as the probabilistic relationship between CT and PET one may use mutual information or normalized mutual information. These last similarity measures assume the least about the two modalities and are thus widely applicable. This does not mean that they are optimal, as we are ignoring other relevant evaluation criteria: computational complexity, robustness, accuracy, and convergence range. Incorporating domain knowledge when selecting a similarity measure usually improves all aspects of registration performance. More often than not, selecting a similarity measure should be done in an empirical manner, evaluating the selection on all relevant criteria. Case in point, the study of similarity measures described in [92] for 2D/3D registration of X-ray/CT.

Optimization is a mature scientific field with a large number of algorithms available for solving both constrained and unconstrained optimization tasks [33]. The selection of a specific optimization method is tightly coupled to the characteristics of the optimized function. For example, if the similarity measure is discontinuous using gradient based optimization methods is not recommended.

Finally, selecting an interpolation method is dependent on the density of the original data. If the images have a high spatial sampling, we can use simpler interpolation methods as the distance between grid points is smaller. With current imaging protocols linear interpolation often provides sufficiently accurate estimates in a computationally efficient manner. Obviously, other higher order interpolation methods can provide more accurate estimates with a higher computational cost [62].

In orthopedics 3D/3D intensity based non-rigid registration has been used for creating point matches for point distribution based statistical atlases. These models encode the variability of bone shape via statistics on point locations across the population. This in turn assumes that the corresponding points can be identified in all datasets. For sparse anatomically prominent landmarks this can potentially be done manually. For the dense correspondence required to model anatomical variability this is not an option. If on the other hand we non-rigidly align the volumetric data we can propagate a template mesh created from one of the volumes to the others, implicitly establishing the dense correspondence. In [9] this is performed using the free-form deformation registration approach with normalized mutual information as the similarity measure. In [16, 108] registration is performed with a diffeomorphic-demons method with the former using a regularization model which is tailored for improving registration of the femur. It should be noted that the demons set of algorithms assume the intensity values for corresponding points are the same for the two modalities. Finally, in [58] registration is performed using the free-form deformation algorithm and the sum of squared differences similarity measure.

Another setting in which 3D/3D intensity based rigid registration has been utilized in orthopedics is for alignment of preoperative data to the physical world, using intraoperative tracked and calibrated US to align a preoperative CT. In [94] both the US and CT images are converted to probability images based on the likelihood of a voxel being on the bone surface, with the similarity between the probability images evaluated using the normalized cross correlation metric. In [36, 61] simulated US images are created from the CT based on the current estimate of the US probe in the physical world. In both cases simulation of US from CT follows the model described in [135]. Registration is then performed using the Covariance Matrix Adaptation Evolution Strategy, optimizing the correlation ration in the former work and a similarity measure closely related to normalized cross correlation in the later work. In [141, 142] coarse localization of the bone surface is automatically performed both in the CT and US with the intensity values in the respective regions used for registration using the normalized cross correlation similarity measure. To date, US based registration has not become part of clinical practice. This is primarily due to the fact that US machines are not readily available in the operating room as part of current orthopedic procedures. Requiring the availability of additional hardware only for the sake of registration appears to limit adoption of this form of registration.

We now shift our focus to the second form of registration which is relevant for orthopedics, 2D projection (X-ray) to 3D (CT/model/atlas) registration.

4   2D Projection/3D

In orthopaedics 2D X-ray to 3D registration is utilized for alignment of preoperative data to the physical world, and for postoperative evaluation, primarily implant pose estimation. We divide our discussion in two, registration methods that use fiducials or implants and those that use anatomical structures. The former methods are easier to automate and are often more robust and accurate as they use man made structures specifically designed to yield accurate registration results. A broad overview of the literature describing various anatomy based registration approaches, not specific to orthopedics, was recently given in [75]. We refer the reader interested in more detailed algorithm descriptions to that publication.

We start by highlighting two aspects of 2D/3D registration which are often not described in detail in publications but effect the accuracy and success of registration: (1) Calibration of the X-ray device; and (2) in the case of anatomy based registration, how was the registration initialized.

X-ray imaging devices are modeled as a pinhole camera, with distortion when the sensor is an image-intensifier, and without distortion when using a flat panel sensor. By performing calibration we estimate the geometric properties of the imaging system. These are later used by the registration algorithm to simulate the imaging process or to compute geometric information that is dependent on the geometry of the imaging apparatus. We identify two forms of calibration, online which means that calibration is carried out every time an image is acquired [12, 45, 65, 121, 150] and, offline which means calibration is carried out once and we assume the apparatus will return to the same pose whenever an image is acquired [21, 22, 24]. Both forms of calibration image a phantom with known geometry and

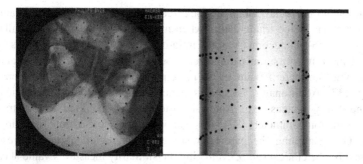

Fig. 6 Calibration images used for (*left*) two tier pattern for online calibration (*right*) helical pattern for offline calibration

compute the geometric properties of the imaging apparatus using the phantom's geometry and its appearance in the image. Figure 6 shows X-ray images of online and offline calibration phantoms.

The online approach is more commonly used when working with image-intensifier based C-arms that are manually manipulated. The offline approach is used when working with systems that also provide cone-beam CT functionality. That is, they are motorized and can be accurately manipulated. In the context of registration, the offline calibration has an advantage over the online approach, the image content is not occluded by the calibration phantom and thus may lead to more accurate registration results. In addition if these systems use flat-panel sensors calibration does not have to include estimation of distortion parameters, with the physical world modeled more closely by the theoretical pinhole model. Finally, if you have worked with a system that utilizes online calibration you may not have noticed the extent of the occlusion introduced by the calibration phantom. This is because many of these systems identify the occluded regions and interpolate the image information so that under visual inspection it appears as if the calibration markers are not there. In practice these images contain less information that is useful for registration than equivalent images without occlusion.

All algorithms that perform anatomy based registration are iterative. That is, they require an initial estimate of the pose of the 3D object being registered. Once the X-ray images of the anatomy are acquired the algorithm is initialized using one of several initialization approaches. The most common methods are:

1. Manual initialization—the operator manipulates the pose of the 3D object using the keyboard and mouse while the X-ray generation process is simulated using the geometric properties of the imaging system. The user manipulates the transformation parameters with the goal of making the simulated image as similar as possible to the real one. A similar, yet clinically more appropriate approach, is to use gestures observed by a depth sensor (e.g. Microsoft Kinect), or a tracking system to manipulate the 3D object [37].
2. Coarse paired point registration—use the paired-point registration algorithm described above with coarsely localized points in the intraoperative setting.

Anatomical or fiducial points are localized either with a tracked and calibrated pointer or via stereo triangulation when multiple X-ray images are available.

3. Clinical setup—the geometry of the intra-operative imaging apparatus is used to bound the transformation parameters. Rough initialization can be obtained by using the intersection point of all principle rays to position the preoperative image [37]. Additionally, the specific patient setup (e.g. supine), and orientation of the X-ray images (e.g. Anterior-Posterior) can be used to constrain the transformation parameters [87]. This can be refined using a brute force approach [88]. Grid sampling the parameter space in the region around the estimate defined by the clinical setup, evaluating the similarity measure's value at each of the grid points and selecting the best one.

Additional less common approaches include an estimate based on the Fourier slice theorem [17] for X-ray/CT registration and the use of a virtual marker [130] to re-register a CT to X-ray with a process that requires an initial paired point registration with wide field of view X-rays with re-registration enabling the use of narrow field of view X-rays.

4.1 Algorithm Classification

We start our overview of 2D/3D registration algorithms by identifying six classes of algorithms, a variation on the classification proposed in [75]. The classification is based on the information utilized by the algorithm, features, intensity, or gradients, and the spatial domain in which optimization is carried out, 2D or 3D.

Feature based algorithms use either anatomical surfaces or markers in 3D and anatomical edges and markers in 2D to formulate the optimization task. This requires segmentation of both 3D and 2D data, something that is not trivial to perform without introducing outliers. This is primarily an issue with the 2D intraoperative X-ray images that often include edges arising from medical equipment associated with the procedure. Using the geometric properties of the X-ray device, either: (1) the features in 3D are projected onto the 2D image and the distance between 2D features and projected features is minimized. We call this approach F2D; (2) the features in 2D are backprojected, defining rays from the camera to the feature location and the distance between the 3D surface points and the backprojected rays is minimized or another option is to identify rays arising from the same 3D feature in multiple images, intersect them to define a 3D point and then minimize the distances between the point clouds. We call this approach F3D.

Intensity based algorithms directly use the intensity values of the images and do not require accurate segmentation, overcoming the main deficiency of feature based algorithms. Using the geometric properties of the X-ray device, either: (1) the 3D image or atlas is used to simulate an X-ray image and the similarity between the actual X-rays and the simulated ones is maximized. This similarity can be between the 2D gradients, edges, or intensity values. This is the most common registration method. In the past, the computational cost of simulating X-rays, also known as Digitally Reconstructed Radiographs (DRRs), was prohibitive. Currently this is no longer an issue, as the use of high performance Graphical Processing Units (GPU) has become commonplace and efficient creation of DRRs on the GPU is both fast and cost effective [29, 126]. We call this approach I2D; (2) the X-ray images are used in a reconstruction framework, similar to cone-beam CT. Either algebraic reconstruction or filtered backprojection methods can be used to perform reconstruction. The reconstructed volume is then rigidly registered to the preoperative volume using 3D/3D intensity based registration algorithms. The main deficiency of this approach is that it requires more X-ray images than any other method, exposing the patient to higher levels of ionizing radiation. This approach has not garnered much acceptance beyond its original proponent [125]. We call this approach I3D.

Gradient based algorithms directly use the gradients computed in 2D and 3D. This is a middle ground between intensity based registration that does not require segmentation and feature based registration which does. While this approach is interesting from an academic standpoint it has not been widely adopted, limited to the original proponents of the approach [66, 74, 124]. Using the geometric properties of the X-ray device, either: (1) the 3D gradients are projected onto the 2D image and the distance and orientation between the 2D gradients and the projected ones is minimized. We call this approach G2D; (2) the 2D gradients are back-projected and combined to form 3D estimates and the distance and orientation between the reconstructed gradients and those arising from the 3D image are minimized. We call this approach G3D.

We start our overview with methods that are based on the alignment of manufactured objects, fiducials or implants.

4.2 Fiducials and Implants

One of the simplest methods for establishing the 3D pose of a manufactured object is to attach fiducial markers to it. Often these are spherical markers that are readily detected in X-ray images. Using a calibrated X-ray device it is straightforward to create a set of backprojected 3D rays, emanating from the location of the X-ray source and going through the marker locations in the X-ray image. By using two or more images the rays corresponding to the same markers are intersected and this intersection point is the 3D location of the marker. Once we have the 3D locations of three or more fiducials we can compute the pose using the paired-point registration algorithm. It should be noted that this approach is not specific to orthopedics and has been described in multiple publications [25, 40, 133, 144, 145]. Given the registration jig used by the robotic spine surgery system described in [113, 114] it is highly likely that this is the registration approach in use, although the publications do not provide the specific details. Figure 7 shows the robot and pose estimation clamp in clinical use. These approaches fall into the F3D category.

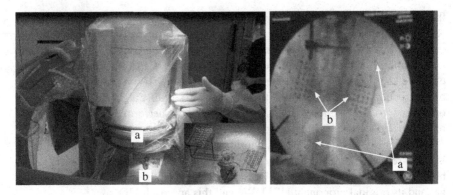

Fig. 7 2D/3D registration setup used by the Mazor Robotics Renaissance system, physical setup and X-ray fluorscopy (Courtesy Ms. Stephani Shipman and Dr. Doron Dinstein, Mazor Robotics, Israel): *a* X-ray calibration phantom, and *b* Pose estimation phantom. Both phantoms consist of metal spheres in known spatial configurations. *Inset* on the *left* shows a detailed view of the pose estimation phantom

Using markers from a single image to estimate the pose of an object is also possible. When we have multiple 2D–3D point correspondences this is the well known perspective-n-point (PnP) camera pose determination problem from computer vision [42]. The problem is solved by minimizing the distances between the 2D points and the projections of the 3D model points. In [146] a drill guide with embedded spherical markers was attached to a robot, enabling the estimation of the robot pose, guiding it to a final pose for performing femoral distal nailing. An evaluation study of the effect of X-ray dose on the accuracy of this form of registration was described in [39]. It was found that increasing X-ray dose increased 2D localization accuracy and the registration accuracy, but only up to a certain point. This approach falls into the F2D category.

More complex marker configurations for estimating the pose of a C-arm have also been described. One such device that has been used in the orthopedic setting is the FTRAC fiducial [49]. This is a complex marker constructed of multiple line segments, ellipses and points. The spatial configuration of these components enables the use of a single image to estimate the marker pose in 3D. In the original work the components of the marker were segmented in the X-ray image and the distance between these points and the projection of the marker components in the given pose was minimized, an F2D approach. This assumes that the segmentation in the image was successful. In subsequent work done in the context of femoral bone augmentation surgery [87], the need for segmentation was eliminated. Instead of using the 2D coordinates of the geometric entities, the CAD model of the marker was used to generate a simulated X-ray. The 3D pose of the marker was then estimated by comparing the simulated X-ray to the actual one, with the correct pose being the one that minimizes the difference between the simulated and actual image. This process was formulated as an optimization task, with similarity between images determined using mutual information and optimization performed with the Nelder-Mead

Downhill Simplex algorithm, an I2D approach. Finally, in [51] only the point markers from the FTRAC were segmented and used in a maximum likelihood framework to estimate the fiducial's pose without requiring an explicit point correspondence between the 3D model and its 2D projection, an F2D approach.

Implants are not specifically designed to facilitate 2D/3D registration, but knowledge of their geometry, their accurate CAD model, is either available from the manufacturer or can be readily measured. This knowledge enables accurate registration as it allows for fast simulation of the X-ray imaging process with highly accurate localization of the implant edges in the simulated image.

In the context of postoperative assessment for total knee arthroplasty, registration was used to assess the relative position of the femoral and tibial implants [71]. The implant CAD models are aligned based on single fluoroscopy image. The optimal alignment is based on minimizing the distance between the 2D edges observed in the X-ray and those created by projecting the CAD model using the known X-ray geometry. A similar approach is described in [56] with the main difference being the use of two images instead of one. In both cases the approach can be classified as an F2D approach. Finally, a method that registers both the implants and femur and tibia to a pair of X-rays using a I2D approach is described in [55]. In this case the edges in the X-ray images are enhanced by a diffusion process so that they extend beyond their actual location. The pose of the implants and boney structures is estimated by generating DRRs, performing edge detection in the DRR and then maximizing the normalized cross correlation between the resulting edge image and the processed X-ray edge image.

In the context of postoperative assessment for total hip replacement, registration was used to assess the cup orientation. In [18, 19] cup and stem orientation are obtained using an I2D approach, aligning the CAD models to a single X-ray. The CAD models of the implants are used in conjunction with a refined DRR generation framework which incorporates both the geometry and material characteristics of the implants to generate the DRR. Comparison between the X-ray and DRR is based on the sum of squared differences between gradient magnitudes, optimized using a Gauss-Newton method. In [50] a similar approach is taken although with a less refined methodology for generating the DRR. Most likely this is why the DRR and X-ray image are compared using mutual information. This is a more forgiving similarity measure which only assume there is a statistical relationship between the two images accommodating less accurate simulations from a physics standpoint.

4.3 Anatomy Based

Registration of anatomical structures using 2D/3D registration is more challenging than registration of fiducials or implants, as the anatomical structures often do not have unique features (e.g. femur shaft) and have higher variability than implants.

In the context of spine surgery, 2D/3D registration has been previously studied extensively [93, 106, 123]. Newer developments have been described in [88, 89] where registration is used to identify vertebral levels in a single X-ray with the intent of reducing wrong site surgery. Preoperatively, the vertebra are identified in the patient's CT. Intraoperatively, DRRs are generated using the GPU and compared with the X-ray using a gradient based similarity measure that is optimized using the Covariance Matrix Adaptation Evolution Strategy (CMA-ES). This method falls into the I2D category. A recent extension of this method to use two images evaluated the effect of angle difference between the images on the accuracy of the registration [129]. Results showed that even with small angular differences of 10–20° registration errors were less than 2 mm. In [64] the vertebra are registered by generating a DRR from the preoperative CT and performing edge detection on it. Edges are also detected in the X-ray and the overlap between the edges in the DRR and X-ray serves as the similarity measure which is maximized.

In the context of femur related interventions, 2D/3D registration was used for kinematic analysis in [127]. The patient's CT was aligned to a single X-ray image by generating DRRs, performing edge detection on the DRR and X-ray image with the goal of maximizing the overlap between the edges, with optimization performed using a genetic algorithm. This method falls into the I2D category. In [87] registration between the preoperative CT and 2–4 X-ray images is performed. DRRs are generated on the GPU and the gradient-information similarity measure is optimized using the CMA-ES algorithm, a classical I2D approach.

While the classical 2D/3D registration problem relies on patient specific data, aligning two datasets from the same patient, a number of groups have investigated the use of a statistical shape model instead of a 3D dataset. This is of interest as it replaces the need for acquiring a preoperative CT, reducing costs, reducing radiation exposure to the patient, and an enabling technology when a CT is not available [5, 6, 48, 152]. In [5, 6, 152] reconstruction of proximal and distal femur surface models is performed using a statistical shape model and two X-ray images. In this framework the patient specific shape is both created and aligned to the X-ray images. The framework uses a two step approach, first project the current model's surface points onto the X-ray image using the known geometric properties of the imaging device and match them with edges detected on X-ray. This defines a matching between the 3D model's surface points and the edges in the X-ray. Then compute the distance between the backprojected rays defined by the edges in the X-ray and their matched 3D surface point. The goal of optimization is to create and align a surface model which minimizes this distance. The differences between the various algorithms are primarily in the 2D matching phase. In addition the approach described in [152] includes one final step, a regularized shape deformation. That is, it allows for modification of the last shape obtained from the statistical model. This accounts for the fact that the shape model reflects the variation of the data used to create it. On the one hand this ensures that the patient specific models created are plausible but on the other hand they are limited to be similar to past observations. By adding this final step the resulting patient specific model is both plausible and accommodates previously unseen minor variations in shape. A related approach that uses a statistical appearance model, encoding both shape and intensities was described in [48]. The femur surface and orientation is estimated using the statistical model and 3–5 images. The process is based on generating DRRs using the current estimated intensity model and pose, then matching edge points between the DRRs and X-ray images via 2D/2D nonrigid registration. Once the matches are established the distance between backprojected rays computed from the X-ray and corresponding 3D model surface points defines the distance between the model and X-ray. This distance is minimized to obtain the final femur surface model and pose.

In the context of total hip replacement 2D/3D registration has been used for postoperative evaluation. In [95] the patient's preoperative CT scan is aligned to a single X-ray image using the intensity based registration approach described in [93]. The CAD model of the cup is manually aligned using a graphical user interface. In [151, 153] cup orientation is estimated from the alignment estimated in the postoperative X-ray and registration of the patient's preoperative CT to the X-ray, without requiring a CAD model. In this case registration is performed in two steps. First anatomical landmarks in the CT and X-ray are manually defined. An initial registration is performed using an iterative solution to the PnP problem, similar to the fiducial based approach described above. This is then followed by an intensity based registration step, which compares the generated DRRs with the X-ray using a similarity measure derived from Markov random field theory. This work was later extended, replacing the second step relying on a patient specific CT with the use of a statistical shape model [155]. This second registration step follows a similar framework to the spastical shape models described above for aligning the femur.

5 Evaluation

From an academic standpoint registration is evaluated for its accuracy and speed. Accuracy is evaluated by establishing a "gold standard" with methods that are clinically not applicable, such as implanting markers to enable the use of methods that are known to be highly accurate (i.e. paired-point fiducial registration) [136]. As we have already noted at the beginning of this chapter this form of evaluation does not address all clinical aspects that determine whether a method will be clinically practical.

To enable comparison of algorithms this "gold standard" needs to be made publicly available. One of the few settings where these gold standards have been made available is in the context of 2D/3D registration [59, 91, 123].

It should be noted that the only registration algorithm with a fully developed theory predicting expected errors and their distribution is the paired-point rigid registration [26]. All other algorithms do not have a solid mathematical error prediction theory. But is this clinically relevant? The answer is, yes! This is a safety issue. If we can theoretically bound the errors then we can ensure the patients safety. On the other hand all of the evaluation methods make assumptions which may be violated in the clinical setting. For instance, most algorithms assume that the tracked reference frame is rigidly attached to the patient during data acquisition. If someone inadvertently moves the frame during data acquisition, no amount of mathematical analysis will be able to bound the registration error introduced by this unexpect act.

We would thus like to enumerate the aspects of registration which all practitioners should keep in mind:

1. Registration errors vary spatially. As long as the expected error at the specific target(s) is sufficiently low one can use the result.
2. Analytical algorithms guarantee an optimal result - as long as their assumptions are met.
3. Iterative algorithms, the majority of registration algorithms, require an initial estimate of the registration parameters. They do not guarantee an optimal result, as they depend on this initial estimate to be sufficiently close to the optimal one.
4. Assumptions made in the lab are sometimes not met in the clinic.
5. A registration result remains valid only if its assumption remain valid too (i.e. reference frame is rigidly attached to the patient).

We therefor recommend that after any registration performed in the clinic, one validate the results, and that this also be done periodically during the intervention to ensure that the registration is still valid. This is a patient safety issue which can have potentially serious repercussions [1].

6   Conclusion

Registration is a key technical technology in navigation and can serve as a tool for preoperative planning and postoperative evaluation. A large number of algorithms have been proposed and have shown clinical utility. In some cases algorithms have not made it into clinical use due to integration issues with existing clinical practices.

For developers of registration algorithms we need to remember that the goal is to integrate our algorithms into clinical practice, a task that requires additional research in terms of workflow analysis in the clinic. Designing our algorithms so that they provide a streamlined workflow and do not require the introduction of additional registration specific hardware.

For practitioners using registration algorithms it is important to understand what are the expectations of the registration algorithm and what are its limitations. Providing the expected environment and input to the algorithm should yield accurate and useful results without the need for repeated data acquisition, something that is not uncommon in the clinic.

In the end the goal of both developers and practitioners is to provide improved healthcare in a safe manner. This goal can only be attained by collaboration and knowledge sharing between the two groups.

References

1. (2014) Class 2 device recall spine & trauma 3D 2.0. URL http://www.accessdata.fda.gov/scripts/cdrh/cfdocs/cfres/res.cfm?id=125729
2. Amiri S, Wilson DR, Masri BA, Anglin C (2014) A low-cost tracked C-arm (TC-arm) upgrade system for versatile quantitative intraoperative imaging. Int J Comput Assist Radiol Surg 9(4):695–711
3. Arun KS, Huang TS, Blostein SD (1987) Least-squares fitting of two 3-D point sets. IEEE Trans Pattern Anal Mach Intell 9(5):698–700
4. Audenaert E, Smet B, Pattyn C, Khanduja V (2012) Imageless versus image-based registration in navigated arthroscopy of the hip: a cadaver-based assessment. J Bone Joint Surg Br 94(5):624–629
5. Baka N, Kaptein BL, de Bruijne M, van Walsum T, Giphart JE, Niessen W, Lelieveldt BPF (2011) 2D–3D shape reconstruction of the distal femur from stereo X-ray imaging using statistical shape models. Med Image Anal 15(6):840–850
6. Baka N, de Bruijne M, van Walsum T, Kaptein BL, Giphart JE, Schaap M, Niessen WJ, Lelieveldt BPF (2012) Statistical shape model-based femur kinematics from biplane fluoroscopy. IEEE Trans Med Imag 31(8):1573–1583
7. Banger M, Rowe PJ, Blyth M (2013) Time analysis of MAKO RIO UKA procedures in comparision with the Oxford UKA. Bone Joint J 95-B(Supp 28):89
8. Barratt DC, Penney GP, Chan CSK, Slomczykowski M, Carter TJ, Edwards PJ, Hawkes DJ (2006) Self-calibrating 3D-ultrasound-based bone registration for minimally invasive orthopedic surgery. IEEE Trans Med Image 25(3):312–323
9. Barratt DC, Chan CSK, Edwards PJ, Penney GP, Slomczykowski M, Carter TJ, Hawkes DJ (2008) Instantiation and registration of statistical shape models of the femur and pelvis using 3D ultrasound imaging. Med Image Anal 12(3):358–374
10. Beaumont E, Beaumont P, Odermat D, Fontaine I, Jansen H, Prince F (2011) Clinical validation of computer-assisted navigation in total hip arthroplasty. Adv Orthop 171783
11. Behrendt D, Mütze M, Steinke H, Koestler M, Josten C, Böhme J (2012) Evaluation of 2D and 3D navigation for iliosacral screw fixation. Int J Comput Assist Radiol Surg 7(2):249–255
12. Bertelsen A, Garin-Muga A, Echeverria M, Gomez E, Borro D (2014) Distortion correction and calibration of intra-operative spine X-ray images using a constrained DLT algorithm. Comput Med Imaging Graph 38(7):558–568
13. Besl PJ, McKay ND (1992) A method for registration of 3D shapes. IEEE Trans Pattern Anal Mach Intell 14(2):239–255
14. Bicknell RT et al (2007) Early experience with computer-assisted shoulder hemiarthroplasty for fractures of the proximal humerus: development of a novel technique and an in vitro comparison with traditional methods. J Shoulder Elbow Surg 16(3 Suppl):S117–S125
15. Birnbaum K, Schkommodau E, Decker N, Prescher A, Klapper U, Radermacher K (2001) Computer-assisted orthopedic surgery with individual templates and comparison to conventional operation method. Spine 26(4):365–370
16. Blanc R, Seiler C, Székely G, Nolte L, Reyes M (2012) Statistical model based shape prediction from a combination of direct observations and various surrogates: application to orthopaedic research. Med Image Anal 16(6):1156–1166
17. van der Bom MJ, Bartels LW, Gounis MJ, Homan R, Timmer J, Viergever MA, Pluim JPW (2010) Robust initialization of 2D–3D image registration using the projection-slice theorem and phase correlation. Med Phys 37(4):1884–1892
18. Burckhardt K, Székely G, Nötzli H, Hodler J, Gerber C (2005) Submillimeter measurement of cup migration in clinical standard radiographs. IEEE Trans Med Imag 24(5):676–688
19. Burckhardt K, Dora C, Gerber C, Hodler J, Székely G (2006) Measuring orthopedic implant wear on standard radiographs with a precision in the 10 μm-range. Med Image Anal 10(4):520–529

20. Chen Y, Medioni G (1992) Object modelling by registration of multiple range images. Image Vis Comput 10(3):145–155
21. Cho Y, Moseley DJ, Siewerdsen JH, Jaffray DA (2005) Accurate technique for complete geometric calibration of cone-beam computed tomography systems. Med Phys 32(4):968–983
22. Claus BEH (2006) Geometry calibration phantom design for 3D imaging. In: Flynn MJ, Hsieh J (eds) SPIE medical imaging: physics of medical imaging, SPIE, p 61422E
23. Costa F et al (2014) Economic study: a cost-effectiveness analysis of an intraoperative compared with a preoperative image-guided system in lumbar pedicle screw fixation in patients with degenerative spondylolisthesis. Spine 14(8):1790–1796
24. Daly MJ, Siewerdsen JH, Cho YB, Jaffray DA, Irish JC (2008) Geometric calibration of a mobile C-arm for intraoperative cone-beam CT. Med Phys 35(5):2124–2136
25. Dang H, Otake Y, Schafer S, Stayman JW, Kleinszig G, Siewerdsen JH (2012) Robust methods for automatic image-to-world registration in cone-beam CT interventional guidance. Med Phys 39(10):6484–6498
26. Danilchenko A, Fitzpatrick JM (2011) General approach to first-order error prediction in rigid point registration. IEEE Trans Med Imag 30(3):679–693
27. Dobbe JGG, Strackee SD, Schreurs AW, Jonges R, Carelsen B, Vroemen JC, Grimbergen CA, Streekstra GJ (2011) Computer-assisted planning and navigation for corrective distal radius osteotomy, based on pre- and intraoperative imaging. IEEE Trans Biomed Eng 58(1):182–190
28. Dobbe JGG, Vroemen JC, Strackee SD, Streekstra GJ (2013) Corrective distal radius osteotomy: including bilateral differences in 3-D planning. Med Biol Eng Comput 51 (7):791–797
29. Dorgham OM, Laycock SD, Fisher MH (2012) GPU accelerated generation of digitally reconstructed radiographs for 2-D/3-D image registration. IEEE Trans Biomed Eng 59 (9):2594–2603
30. Eggert DW, Lorusso A, Fisher RB (1997) Estimating 3-D rigid body transformations: a comparison of four major algorithms. Mach Vis Appl 9(5/6):272–290
31. Ershad M, Ahmadian A, Serej ND, Saberi H, Khoiy KA (2014) Minimization of target registration error for vertebra in image-guided spine surgery. Int J Comput Assist Radiol Surg 9(1):29–38
32. Faugeras OD, Hebert M (1986) The representation, recognition, and locating of 3-D objects. Int J Rob Res 5(3):27–52
33. Fletcher R (1987) Practical methods of optimization, 2nd edn. Wiley, New York
34. Fleute M, Lavallée S, Julliard R (1999) Incorporating a statistically based shape model into a system for computer-assisted anterior cruciate ligament surgery. Med Image Anal 3 (3):209–222
35. Fürnstahl P, Székely G, Gerber C, Hodler J, Snedeker JG, Harders M (2012) Computer assisted reconstruction of complex proximal humerus fractures for preoperative planning. Med Image Anal 16(3):704–720
36. Gill S, Abolmaesumi P, Fichtinger G, Boisvert J, Pichora DR, Borshneck D, Mousavi P (2012) Biomechanically constrained groupwise ultrasound to CT registration of the lumbar spine. Med Image Anal 16(3):662–674
37. Gong RH, Özgür G, Kürklüoglu M, Lovejoy J, Yaniv Z (2013) Interactive initialization of 2D/3D rigid registration. Med Phys 20(12):121911-1–121911-14
38. Gonschorek O, Hauck S, Spiegl U, Weiß T, Pätzold R, Bühren V (2011) O-arm based spinal navigation and intraoperative 3D-imaging: first experiences. Eur J Trauma Emerg Surg 37 (2):99–108
39. Habets DF, Pollmann SI, Yuan X, Peters TM, Holdsworth DW (2009) Error analysis of marker-based object localization using a single-plane XRII. Med Phys 36(1):190–200
40. Hamming NM, Daly MJ, Irish JC, Siewerdsen JH (2009) Automatic image-to-world registration based on X-ray projections in cone-beam CT guided interventions. Med Phys 36 (5):1800–1812

41. Hananouchi T, Saito M, Koyama T, Hagio K, Murase T, Sugano N, Yoshikawa H (2009) Tailor-made surgical guide based on rapid prototyping technique for cup insertion in total hip arthroplasty. Int J Med Robot Comput Assist Surg 5(2):164–169

42. Hartley RI, Zisserman A (2000) Multiple view geometry in computer vision. Cambridge University Press, Cambridge

43. Haselbacher M, Sekyra K, Mayr E, Thaler M, Nogler M (2012) A new concept of a multiple-use screw-based shape-fitting plate in total knee arthroplasty. Bone Joint J 94-B (Supp-XLIV):65

44. Heimann T, Meinzer H (2009) Statistical shape models for 3D medical image segmentation: a review. Med Image Anal 13(4):543–563

45. Hofstetter R, Slomczykowski M, Sati M, Nolte LP (1999) Fluoroscopy as an imaging means for computer-assisted surgical navigation. Comput Aided Surg 4(2):65–76

46. Holly LT, Block O, Johnson JP (2006) Evaluation of registration techniques for spinal image guidance. J Neurosurg Spine 4(4):323–328

47. Horn BKP (1987) Closed-form solution of absolute orientation using unit quaternions. J Opt Soc Am A 4(4):629–642

48. Hurvitz A, Joskowicz L (2008) Registration of a CT-like atlas to fluoroscopic X-ray images using intensity correspondences. Int J Comput Assist Radiol Surg 3(6):493–504

49. Jain AK, Mustafa T, Zhou Y, Burdette C, Chirikjian GS, Fichtinger G (2005) FTRAC-a robust fluoroscope tracking fiducial. Med Phys 32(10):3185–3198

50. Jaramaz B, Eckman K (2006) 2D/3D registration for measurement of implant alignment after total hip replacement. In: Medical image computing and computer-assisted intervention, pp 653–661

51. Kang X, Armand M, Otake Y, Yau WP, Cheung PYS, Hu Y, Taylor RH (2014) Robustness and accuracy of feature-based single image 2-D-3-D registration without correspondences for image-guided intervention. IEEE Trans Biomed Eng 61(1):149–161

52. Kantelhardt SR, Bock HC, Siam L, Larsen J, Burger R, Schillinger W, Bockermann V, Rohde V, Giese A (2010) Intra-osseous ultrasound for pedicle screw positioning in the subaxial cervical spine: an experimental study. Acta Neurochir 152(4):655–661

53. de Kelft EV, Costa F, der Planken DV, Schils F (2012) A prospective multicenter registry on the accuracy of pedicle screw placement in the thoracic, lumbar, and sacral levels with the use of the O-arm imaging system and stealthstation navigation. Spine 37(25):E1580–E1587

54. Kilian P et al (2008) New visualization tools: computer vision and ultrasound for MIS navigation. Int J Med Robot Comput Assist Surg 4(1):23–31

55. Kim Y, Kim KI, hyeok Choi J, Lee K (2011) Novel methods for 3D postoperative analysis of total knee arthroplasty using 2D–3D image registration. Clin Biomech 26(4):384–391

56. Kobayashi K, Sakamoto M, Tanabe Y, Ariumi A, Sato T, Omori G, Koga Y (2009) Automated image registration for assessing three-dimensional alignment of entire lower extremity and implant position using bi-plane radiography. J Biomech 42(16):2818–2822

57. Kowal J, Amstutz C, Langlotz F, Talib H, Ballester MG (2007) Automated bone contour detection in ultrasound B-mode images for minimally invasive registration in computer-assisted surgery—an in vitro evaluation. Int J Med Robot Comput Assist Surg 3 (4):341–348

58. Kozic N, Weber S, Büchler P, Lutz C, Reimers N, Ballester MÁG, Reyes M (2010) Optimisation of orthopaedic implant design using statistical shape space analysis based on level sets. Med Image Anal 14(3):265–275

59. van de Kraats EB, Penney GP, Tomaževič D, van Walsum T, Niessen WJ (2005) Standardized evaluation methodology for 2-D-3-D registration. IEEE Trans Med Imag 24 (9):1177–1189

60. Kunz M, Ma B, Rudan JF, Ellis RE, Pichora DR (2013) Image-guided distal radius osteotomy using patient- specific instrument guides. J Hand Surg Am 38(8):1618–24

61. Lang A, Mousavi P, Gill S, Fichtinger G, Abolmaesumi P (2012) Multi-modal registration of speckle-tracked freehand 3D ultrasound to CT in the lumbar spine. Med Image Anal 16 (3):675–686

62. Lehmann TM, Gönner C, Spitzer K (1999) Survey: interpolation methods in medical image processing. IEEE Trans Med Imag 18(11):1049–1075
63. Letta C, Schweizer A,, Fürnstahl P (2014) Quantification of contralateral differences of the scaphoid: a comparison of bone geometry in three dimensions. Anat Res Int 2014:904275
64. Lin CC et al (2013) Intervertebral anticollision constraints improve out-of-plane translation accuracy of a single-plane fluoroscopy-to-CT registration method for measuring spinal motion. Med Phys 40(3):031–912
65. Livyatan H, Yaniv Z, Joskowicz L (2002) Robust automatic C-arm calibration for fluoroscopy-based navigation: a practical approach. In: Dohi T et al (eds) Medical image computing and computer-assisted intervention, pp 60–68
66. Livyatan H, Yaniv Z, Joskowicz L (2003) Gradient-based 2D/3D rigid registration of fluoroscopic X-ray to CT. IEEE Trans Med Imag 22(11):1395–1406
67. Lonner JH, John TK, Conditt MA (2010) Robotic arm-assisted UKA improves tibial component alignment a pilot study. Clin Orthop Relat Res 468(1):141–146
68. Lu S et al (2009) A novel computer-assisted drill guide template for lumbar pedicle screw placement: a cadaveric and clinical study. Int J Med Robot Comput Assist Surg 5(2):184–191
69. Lu S et al (2009) A novel patient-specific navigational template for cervical pedicle screw placement. Spine 34(26):E959–E964
70. Ma B, Kunz M, Gammon B, Ellis RE, Pichora DR (2014) A laboratory comparison of computer navigation and individualized guides for distal radius osteotomy. Int J Comput Assist Radiol Surg 9(4):713–724
71. Mahfouz MR, Hoff WA, Komistek RD, Dennis DA (2003) A robust method for registration of three-dimensional knee implant models to two-dimensional fluoroscopy images. IEEE Trans Med Imag 22(12):1561–1574
72. Maier-Hein L, Franz AM, dos Santos TR, Schmidt M, Fangerau M, Meinzer H, Fitzpatrick JM (2012) Convergent iterative closest-point algorithm to accomodate anisotropic and inhomogenous localization error. IEEE Trans Pattern Anal Machine Intell 34(8):1520–1532
73. Mantwill F, Schulz AP, Faber A, Hollstein D, Kammal M, Fay A, Jürgens C (2005) Robotic systems in total hip arthroplasty—is the time ripe for a new approach? Int J Med Robot Comput Assist Surg 1(4):8–19
74. Markelj P, Tomaževič D, Pernuš F, Likar B (2008) Robust gradient-based 3-D/2-D registration of CT and MR to x-ray images. IEEE Trans Med Imag 27(12):1704–1714
75. Markelj P, Tomaževič D, Likar B, Pernuš F (2012) A review of 3D/2D registration methods for image-guided interventions. Med Image Anal 16(3):642–661
76. Mercier L, Langø T, Lindseth F, Collins DL (2005) A review of calibration techniques for freehand 3-D ultrasound systems. Ultrasound Med Biol 31(4):449–471
77. Moghari MH, Abolmaesumi P (2007) Point-based rigid-body registration using an unscented Kalman filter. IEEE Trans Med Imag 26(12):1708–1728
78. Momi ED, Cerveri P, Gambaretto E, Marchente M, Effretti O, Barbariga S, Gini G, Ferrigno G (2008) Robotic alignment of femoral cutting mask during total knee arthroplasty. Int J Comput Assist Radiol Surg 3(5):413–419
79. Mozes A, Chang TC, Arata L, Zhao W (2010) Three-dimensional A-mode ultrasound calibration and registration for robotic orthopaedic knee surgery. Int J Med Robot Comput Assist Surg 6(1):91–101
80. Najafi M, Afsham N, Abolmaesumi P, Rohling R (2014) A closed-form differential formulation for ultrasound spatial calibration: multi-wedge phantom. Ultrasound Med Biol 40(9):2231–2243
81. Nakamura N, Sugano N, Nishii T, Miki H, Kakimoto A, Yamamura M (2009) Robot-assisted primary cementless total hip arthroplasty using surface registration techniques: a short-term clinical report. Int J Comput Assist Radiol Surg 4(2):157–162
82. Nogler M, Maurer H, Wimmer C, Gegenhuber C, Bach C, Krismer M (2001) Knee pain caused by a fiducial marker in the medial femoral condyle. Acta Orthop Scand 72(5):477–480

83. Oertel MF, Hobart J, Stein M, Schreiber V, Scharbrodt W (2011) Clinical and methodological precision of spinal navigation assisted by 3D intraoperative O-arm radiographic imaging. J Neurosurg Spine 14(4):532–536

84. Ohta N, Kanatani K (1998) Optimal estimation of three-dimensional rotation and reliability evaluation. In: Computer vision—ECCV'98, LNCS, vol 1406. pp 175–187

85. Oka K, Moritomo H, Goto A, Sugamoto K, Yoshikawa H, Murase T (2008) Corrective osteotomy for malunited intra-articular fracture of the distal radius using a custom-made surgical guide based on three-dimensional computer simulation: case report. J Hand Surg Am 33(6):835–840

86. Okada T, Iwasaki Y, Koyama T, Sugano N, Chen Y, Yonenobu K, Sato Y (2009) Computer-assisted preoperative planning for reduction of proximal femoral fracture using 3-D-CT data. IEEE Trans Biomed Eng 56(3):749–759

87. Otake Y, Armand M, Armiger RS, Kutzer MDM, Basafa E, Kazanzides P, Taylor RH (2012) Intraoperative image-based multiview 2D/3D registration for image-guided orthopaedic surgery: incorporation of fiducial-based C-arm tracking and GPU-acceleration. IEEE Trans Med Imag 31(4):948–962

88. Otake Y, Schafer S, Stayman JW, Zbijewski W, Kleinszig G, Graumann R, Khanna AJ, Siewerdsen JH (2012) Automatic localization of vertebral levels in X-ray fluoroscopy using 3D–2D registration: a tool to reduce wrong-site surgery. Phys Med Biol 57(17):5485–5508

89. Otake Y, Wang AS, Stayman JW, Uneri A, Kleinszig G, Vogt S, Khanna AJ, Gokaslan ZL, Siewerdsen JH (2013) Robust 3D-2D image registration: application to spine interventions and vertebral labeling in the presence of anatomical deformation. Phys Med Biol 58 (23):8535–8553

90. Otomaru I, Nakamoto M, Kagiyama Y, Takao M, Sugano N, Tomiyama N, Tada Y, Sato Y (2012) Automated preoperative planning of femoral stem in total hip arthroplasty from 3D CT data: Atlas-based approach and comparative study. Med Image Anal 16(2):415–426

91. Pawiro SA, Markelj P, Pernus F, Gendrin C, Figl M, Weber C, Kainberger F, Nobauer-Huhmann I, Bergmeister H, Stock M, Georg D, Bergmann H, Birkfellner W (2011) Validation for 2D/3D registration I: a new gold standard data set. Med Phys 38 (3):1481–1490

92. Penney GP, Weese J, Little JA, Desmedt P, Hill DLG, Hawkes DJ (1998) A comparison of similarity measures for use in 2D-3D medical image registration. IEEE Trans Med Imag 17 (4):586–595

93. Penney GP, Batchelor PG, Hill DLG, Hawkes DJ, Weese J (2001) Validation of a two-to three-dimensional registration algorithm for aligning preoperative CT images and intraoperative fluoroscopy images. Med Phys 28(6):1024–1032

94. Penney GP, Barratt DC, Chan CSK, Slomczykowski M, Carter TJ, Edwards PJ, Hawkes DJ (2006) Cadaver validation of intensity-based ultrasound to CT registration. Med Image Anal 10(3):385–395

95. Penney GP, Edwards PJ, Hipwell JH, Slomczykowski M, Revie I, Hawkes DJ (2007) Postoperative calculation of acetabular cup position using 2-D-3-D registration. IEEE Trans Biomed Eng 54(7):1342–1348

96. Petermann J, Kober R, Heinze R, Frölich JJ, Heeckt PF, Gotzen L (2000) Computer-assisted planning and robot-assisted surgery in anterior cruciate ligament reconstruction. Operative Tech Orthop 10(1):50–55

97. Quiñones-Hinojosa A, Kolen ER, Jun P, Rosenberg WS, Weinstein PR (2006) Accuracy over space and time of computer-assisted fluoroscopic navigation in the lumbar spine in vivo. J Spinal Disord Tech 19(2):109–113

98. Radermacher K, Portheine F, Anton M, Zimolong A, Kaspers G, Rau G, Staudte HW (1998) Computer assisted orthopaedic surgery with image based individual templates. Clin Orthop Relat Res Sep 354:28–38

99. Rajamani KT, Styner MA, Talib H, Zheng G, Nolte L, Ballester MÁG (2007) Statistical deformable bone models for robust 3D surface extrapolation from sparse data. Med Image Anal 11(2):99–109

100. Rasoulian A, Abolmaesumi P, Mousavi P (2012) Feature-based multibody rigid registration of CT and ultrasound images of lumbar spine. Med Phys 39(6):3154–3166
101. Richter M, Zech S (2008) 3D imaging (ARCADIS)-based computer assisted surgery (CAS) guided retrograde drilling in osteochondritis dissecans of the talus. Foot Ankle Int 29 (12):1243–1248
102. Rieger M, Gabl M, Gruber H, Jaschke WR, Mallouhi A (2005) CT virtual reality in the preoperative workup of malunited distal radius fractures: preliminary results. Eur Radiol 4 (15):792–797
103. Rodriguez F et al (2005) Robotic clinical trials of uni-condylar arthroplasty. Int J Med Robot Comput Assist Surg 1(4):20–28
104. Rueckert D, Sonoda LI, Hayes C, Hill DLG, Leach MO, Hawkes DJ (1999) Non-rigid registration using free-form deformations: application to breast MR images. IEEE Trans Med Imag 18(8):712–721
105. Rusinkiewicz S, Levoy M (2001) Efficient variants of the ICP algorithm. In: International conference on 3D digital imaging and modeling, pp 145–152
106. Russakoff DB, Rohlfing T, Adler JR Jr, Maurer CR Jr (2005) Intensity-based 2D–3D spine image registration incorporating a single fiducial marker. Acad Radiol 12(1):37–50
107. Schafer S et al (2011) Mobile C-arm cone-beam CT for guidance of spine surgery: Image quality, radiation dose, and integration with interventional guidance. Med Phys 38(8):4563–4574
108. Schuler B, Fritscher KD, Kuhn V, Eckstein F, Link TM, Schubert R (2010) Assessment of the individual fracture risk of the proximal femur by using statistical appearance models. Med Phys 37(6):2560–2571
109. Schulz AP, Seide K, Queitsch C, von Haugwitz A, Meiners J, Kienast B, Tarabolsi M, Kammal M, Jürgens C (2007) Results of total hip replacement using the Robodoc surgical assistant system: clinical outcome and evaluation of complications for 97 procedures. Int J Med Robot Comput Assist Surg 3(4):301–306
110. Schumann S, Nolte LP, Zheng G (2012) Determination of pelvic orientation from sparse ultrasound data for THA operated in the lateral position. Int J Med Robot Comput Assist Surg 8(1):107–113
111. Schweizer A, Fürnstahl P, Harders M, Székely G, Nagy L (2010) Complex radius shaft malunion: osteotomy with computer-assisted planning. HAND 5:171–178
112. Shamir RR, Joskowicz L, Spektor S, Shoshan Y (2009) Localization and registration accuracy in image guided neurosurgery: a clinical study. Int J Comput Assist Radiol Surg 4 (1):45–52
113. Shoham M, Burman M, Zehavi E, Joskowicz L, Batkilin E, Kunicher Y (2003) Bone-mounted miniature robot for surgical procedures: concept and clinical applications. IEEE Trans Robot Automat 19(5):893–901
114. Shoham M et al (2007) Robotic assisted spinal surgery—from concept to clinical practice. Comput Aided Surg 12(2):105–115
115. Smith EJ, Al-Sanawi H, Gammon B, John PS, Pichora DR, Ellis RE (2012) Volume slicing of cone-beam computed tomography images for navigation of percutaneous scaphoid fixation. Int J Comput Assist Radiol Surg 7(3):433–444
116. Smith JR, Riches PE, Rowe PJ (2014) Accuracy of a freehand sculpting tool for unicondylar knee replacement. Int J Med Robot Comput Assist Surg 10(2):162–169
117. Stindel E, Briard JL, Merloz P, Plaweski S, Dubrana F, Lefevre C, Troccaz J (2002) Bone morphing: 3D morphological data for total knee arthroplasty. Comput Aided Surg 7(3):156–168
118. Stöckle U, Schaser K, König B (2007) Image guidance in pelvic and acetabular surgery-expectations, success and limitations. Injury 38(4):450–462
119. Sugano N (2013) Computer-assisted orthopaedic surgery and robotic surgery in total hip arthroplasty. Clin Orthop Surg 5(1):1–9
120. Talib H, Peterhans M, Garća J, Styner M, Ballester MAG (2011) Information filtering for ultrasound-based real-time registration. IEEE Trans Biomed Eng 58(3):531–540

121. Tate PM, Lachine V, Fu L, Croitoru H, Sati M (2001) Performance and robustness of automatic fluoroscopic image calibration in a new computer assisted surgery system. In: Medical image computing and computer-assisted intervention, pp 1130–1136
122. Tensho K, Kodaira H, Yasuda G, Yoshimura Y, Narita N, Morioka S, Kato H, Saito N (2011) Anatomic double-bundle anterior cruciate ligament reconstruction, using CT-based navigation and fiducial markers. Knee Surg Sports Traumatol Arthrosc 19(3):378–383
123. Tomaževič D, Likar B, Pernuš F (2004) "Gold standard" data for evaluation and comparison of 3D/2D registration methods. Comput Aided Surg 9(4):137–144
124. Tomaževič D, Likar B, Slivnik T, Pernuš F (2003) 3-D/2-D registration of CT and MR to X-ray images. IEEE Trans Med Imag 22(11):1407–1416
125. Tomaževič D, Likar B, Pernuš F (2006) 3-D/2-D registration by integrating 2-D information in 3-D. IEEE Trans Med Imag 25(1):17–27
126. Tornai GJ, Pappasa GC (2012) Fast DRR generation for 2D to 3D registration on GPUs. Med Phys 39(8):4795–4799
127. Tsai TY, Lu TW, Chen CM, Kuo MY, Hsu HC (2010) A volumetric model-based 2D to 3D registration method for measuring kinematics of natural knees with single-plane fluoroscopy. Med Phys 37(3):1273–1284
128. Umeyama S (1991) Least-squares estimation of transformation parameters between two point patterns. IEEE Trans Pattern Anal Mach Intell 13(4):376–380
129. Uneri A, Otake Y, Wang AS, Kleinszig G, Vogt S, Khanna AJ, Siewerdsen JH (2014) 3D–2D registration for surgical guidance: effect of projection view angles on registration accuracy. Phys Med Biol 59(2):271–287
130. Varnavas A, Carrell T, Penney GP (2013) Increasing the automation of a 2D–3D registration system. IEEE Trans Med Imag 32(2):387–399
131. Vercauteren T, Pennec X, Perchant A, Ayache N (2009) Diffeomorphic demons: efficient non-parametric image registration. NeuroImage 45(1, Supplement 1):S61–S72
132. Walker MW, Shao L, Volz RA (1991) Estimating 3-D location parameters using dual number quaternions. CVGIP Image Underst 54(3):358–367
133. Wei W, Schön N, Dannenmann T, Petzold R (2011) Determining the position of a patient reference from C-Arm views for image guided navigation. Int J Comput Assist Radiol Surg 6 (2):217–227
134. Weil YA, Liebergall M, Mosheiff R, Singer SB, Joskowicz L, Khoury A (2011) Assessment of two 3-D fluoroscopic systems for articular fracture reduction: a cadaver study. Int J Comput Assist Radiol Surg 6(5):685–692
135. Wein W, Brunke S, Khamene A, Callstrom MR, Navab N (2008) Automatic CT-ultrasound registration for diagnostic imaging and image-guided intervention. Med Image Anal 12:577–585
136. West J et al (1997) Comparison and evaluation of retrospective intermodality brain image registration techniques. J Comput Assist Tomogr 4(4):554–568
137. Whitmarsh T, Humbert L, Craene MD, Barquero LMDR, Frangi AF (2011) Reconstructing the 3D shape and bone mineral density distribution of the proximal femur from dual-energy X-ray absorptiometry. IEEE Trans Med Imag 12(30):2101–2114
138. Wiles AD, Likholyot A, Frantz DD, Peters TM (2008) A statistical model for point-based target registration error with anisotropic fiducial localizer error. IEEE Trans Med Imag 27 (3):378–390
139. Wing JM (2006) Computational thinking. Commun ACM 49(3):33–35
140. Xie W, Franke J, Chen C, Grützner PA, Schumann S, Nolte L, Zheng G (2014) Statistical model-based segmentation of the proximal femur in digital antero-posterior (AP) pelvic radiographs. Int J Comput Assist Radiol Surg 9(2):165–176
141. Yan CXB, Goulet B, Pelletier J, Chen SJS, Tampieri D, Collins DL (2011) Towards accurate, robust and practical ultrasound-CT registration of vertebrae for image-guided spine surgery. Int J Comput Assist Radiol Surg 6(4):523–537
142. Yan CXB, Goulet B, Chen SJ, Tampieri D, Collins DL (2012) Validation of automated ultrasound-ct registration of vertebrae. Int J Comput Assist Radiol Surg 7(4):601–610

143. Yan CXB, Goulet B, Tampieri D, Collins DL (2012) Ultrasound-CT registration of vertebrae without reconstruction. Int J Comput Assist Radiol Surg 7(6):901–909
144. Yaniv Z (2009) Localizing spherical fiducials in C-arm based cone-beam CT. Med Phys 36 (11):4957–4966
145. Yaniv Z (2010) Evaluation of spherical fiducial localization in C-arm cone-beam CT using patient data. Med Phys 37(10):5298–5305
146. Yaniv Z, Joskowicz L (2005) Precise robot-assisted guide positioning for distal locking of intramedullary nails. IEEE Trans Med Imag 24(5):624–635
147. Zeng X, Wang C, Zhou H, Wei S, Chen X (2014) Low-dose three-dimensional reconstruction of the femur with unit free-form deformation. Med Phys 41(8):081–911
148. Zhang YZ, Chen B, Lu S, Yang Y, Zhao JM, Liu R, Li YB, Pei GX (2011) Preliminary application of computer-assisted patient-specific acetabular navigational template for total hip arthroplasty in adult single development dysplasia of the hip. Int J Med Robot Comput Assist Surg 7(4):469–474
149. Zhang Z (1994) Iterative point matching for registration of free-form curves and surfaces. Int J Comput Vision 13(2):119–152
150. Zheng G, Zhang X (2009) Robust automatic detection and removal of fiducial projections in fluoroscopy images: an integrated solution. Med Eng Phys 31(5):571–580
151. Zheng G, Zhang X (2010) Computer assisted determination of acetabular cup orientation using 2D–3D image registration. Int J Comput Assist Radiol Surg 5(5):437–447
152. Zheng G, Gollmer S, Schumann S, Dong X, Feilkas T, Ballester MÁG (2009) A 2D/3D correspondence building method for reconstruction of a patient-specific 3D bone surface model using point distribution models and calibrated X-ray images. Med Image Anal 13 (6):883–899
153. Zheng G, Zhang X, Steppacher SD, Murphy SB, Siebenrock K, Tannast M (2009) Hipmatch: an object-oriented cross-platform program for accurate determination of cup orientation using 2D–3D registration of single standard X-ray radiograph and a CT volume. Comput Methods Programs Biomed 95(3):236–248
154. Zheng G, Schumann S, Ballester MÁG (2010) An integrated approach for reconstructing a surface model of the proximal femur from sparse input data and a multi-resolution point distribution model: an in vitro study. Int J Comput Assist Radiol Surg 5(1):99–107
155. Zheng G, von Recum J, Nolte L, Grützner PA, Steppacher SD, Franke J (2012) Validation of a statistical shape model-based 2D/3D reconstruction method for determination of cup orientation after THA. Int J Comput Assist Radiol Surg 7(2):225–231

3D Augmented Reality Based Orthopaedic Interventions

Xinran Zhang, Zhencheng Fan, Junchen Wang and Hongen Liao

Abstract Augmented reality (AR) techniques, which can merge virtual computer-generated guidance information into real medical interventions, help surgeons obtain dynamic "see-through" scenes during orthopaedic interventions. Among various AR techniques, 3D integral videography (IV) image overlay is a promising solution because of its simplicity in implementation as well as the ability to produce a full parallax augmented natural view for multiple observers and improve surgeons' hand-eye coordination. To obtain a precise fused result, patient-3D image registration is a vital technique in the IV overlay based orthopaedic interventions. Marker or marker-less based registration techniques are alternative depending on a particular clinical application. According to accurate AR information, minimally invasive therapy including cutting, drilling, implantation and other related operations, can be performed more easily and safely. This chapter reviews related augmented reality techniques for image-guided surgery and analyses several examples about clinical applications. Eventually, we discuss the future development of 3D AR based orthopaedic interventions.

1 Introduction

In orthopaedic surgery, complex anatomical structures often block direct observation and operation. With the fast development of medical imaging techniques, such as X-rays, computed tomography (CT) and C-arm fluoroscopy imaging, more

Xinran Zhang and Zhencheng Fan—equally contributed.

X. Zhang · Z. Fan · H. Liao (✉)
Department of Biomedical Engineering, School of Medicine, Tsinghua University,
100084 Beijing, People's Republic of China
e-mail: liao@tsinghua.edu.cn

J. Wang
Department of Bioengineering, Graduate School of Engineering,
The University of Tokyo, Tokyo, Japan

© Springer International Publishing Switzerland 2016
G. Zheng and S. Li (eds.), *Computational Radiology
for Orthopaedic Interventions*, Lecture Notes in Computational
Vision and Biomechanics 23, DOI 10.1007/978-3-319-23482-3_4

detailed anatomy structures are visible. Orthopaedic intervention is a clinical field in which image-guided techniques have been extensively used. Image-guided techniques can help surgeons understand real-time spatial relationships between critical structures, intervention targets and interventional tools [1]. Moreover, image-guided orthopaedic interventions navigate the surgical procedure through real-time guidance and minimally invasive interventions can be achieved effectively and safely.

There are four main technical steps in image-guided surgery (IGS) [2]. Firstly, the patient's specific planning and optimization are a guarantee of a successful operation. According to two-dimensional (2D) cross-sectional images and three-dimensional (3D) reconstructed images, surgeons can decide the entrance and path of the surgical instrument in order to avoid damage to surrounding high-risk structures. The second step is registration. Based on various registration methods, pre-operative medical images are combined with the patient. Thirdly, real-time tracking during surgery is necessary to detect positions of anatomical structures and surgical tools. Nowadays, various types of tracking systems are commercially available. Finally yet importantly, intuitive visualization and friendly mutual interface for users are vital for guiding surgeons.

Image-guided techniques have been applied in orthopaedic procedures from planning, therapy to postoperative evaluation. However, in general, images for guidance are always displayed on a separated screen. In this way, surgeons need to keep switching focus between screen and patient to confirm the entrance of intervention, the position between tool and target, shown in Fig. 1. The operations' efficiency and accuracy are significantly reduced [3]. New techniques in IGS are highly demanded to solve problems mentioned above.

Augmented reality (AR), which can improve accuracy of the decision-making in IGS, is a visualization technique to merge virtual computer-generated images into real surgical scene seamlessly and accurately [4]. Therefore, alternate viewing between images displayed and the actual surgical area is no longer required. Therefore, surgeons' hand-eye coordination is improved [5]. Two crucial requirements for medical AR system are "right time" and "right place". Both of them

Fig. 1 Clinical IGS scene

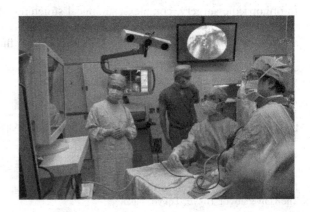

ensure the correspondence between real intervention environments and virtual images dynamically. In order to achieve these goals, a variety of techniques are applied to medical AR systems, such as 3D positioning techniques, intra-operating imaging techniques and 3D visualization techniques [6].

Most studies on AR in surgical environments focus on the usage of AR surgical microscopes, head-mounted displays (HMD) and semi-transparent mirrors overlay systems [7]. They use optical or video-based methods to achieve virtual message injection. Compared with the restriction of the position between the eye and the eyepiece or the screen, image overlay can provide a larger viewing angle and a more comfortable visual perception for multiple observers.

By using image overlay techniques, surgeons look through a semi-transparent mirror and see the reflected virtual images overlaid onto the real scene. Virtual images can be 2D cross-sectional images, 3D surface or volume rendering results and autostereoscopic images. Proper geometrical relationship between the image display and the semi-transparent mirror with accurate patient-image registration ensures that the reflected images appear in a correct position with the appropriate scale. Image overlay guarantees that augmented images can be observed by multiple surgeons, without any assistant tracking devices or special glasses wore by observers [3]. A prototype 2D image overlay system proposed in [8] applies intra-operative CT slices as the intervention guidance, and conducts experiments with phantoms and cadavers. Evaluations prove that this overlay system optimizes spinal needle placement with conventional CT scanners.

Until now, image overlay has been applied in phantom, cadaver or animal studies of knee reconstruction surgery [9], needle insertion [8], dental surgery [10] and some other MIS. The spatial accuracy of virtual images is one of the major determinants of safety and efficiency in surgery, while medical images including cross-sectional, surface or volume rendering image shown in 2D display can't provide adequate geometrical accuracy during precision treatment. This forces surgeons to take extra steps to reconstruct the 3D object with depth information in their mind and match the guidance information. Current 3D visualization techniques in medical AR navigation systems can provide depth perception based on geometry factors such as size, motion parallax and stereo disparity. However, even though all positions are computed correctly, misperceptions of depth may occur when virtual 3D images are merged into real scene [11]. In order to provide a precise depth perception for augmented reality in image-guided intervention, autostereoscopic image overlay is a good solution.

This chapter is organized as follows. Section 2 describes the IV image algorithm, device structure and tracking methods in IV image overlay system. Section 3 introduces the patient-3D image registration and visualization methods used in IV image overlay system. Section 4 shows some typical applications of IV image overlay in orthopaedic interventions. Section 5 summarizes characters of different 3D AR techniques. Section 6 makes a discussion on the current challenges of 3D AR in orthopaedic interventions and concludes the chapter with a brief summary.

2   IV Image Overlay for Orthopaedic Interventions

Autostereoscopic image overlay can superimpose a 3D image onto the surgical area. Autostereoscopic display techniques include lenticular lens, parallax barrier, volumetric display, holographic and light field display. Among all mentioned techniques, integral photography (IP) is a promising approach for creating medical-used autostereoscopic images. The resulting 3D images have arbitrary viewing angles by a series of elemental images covered by a micro convex lens array. The concept of IP was proposed in [12] by Lippmann. IP encodes both horizontal and vertical direction information so that it can generate a full parallax image. The traditional IP technique records elemental images on films, which cannot satisfy real-time IGS applications. IV introduced in [13] uses fast rendering algorithm to accelerate elemental image generation and replaces the film with a high resolution liquid crystal display (LCD) display, which make IV adoptable in clinical AR guidance. Compared with a 2D image or binocular stereoscopic images, it has outstanding features. Firstly, the spatial accuracy of the image is better, especially in depth. Secondly, images can be simultaneously observed by multiple people with less visual fatigue. We will introduce the general system configuration of IV image overlay in the following section.

2.1   IV Image Rendering and Display

The IV overlay system creates 3D images of the anatomical structures around the surgical area and the surgical instruments. Data sources of IV images can be pre-operative/intra-operative medical images or computer-aided design models. Images are fixed in space and observers can see different images from different viewing angles. Therefore, different views of an IV image can be seen in various positions, just like a real 3D object in space.

The calculation of elemental images can be achieved by surface rendering or volume rendering methods. Surface rendering implements reverse mapping from spatial computer-generated (CG) surface model to generate background image. This mapping procedure is conducted from many different predetermined angles. Generation of elemental images is based on the pixel redistribution algorithm [14, 15]. The algorithm separates the coverage display area into $M \times N$ small elements depending on the pre-defined IV image solution. Pixels in different rendered images obtained from the homologous viewport are realignment into elemental image obeying a crossing redistribution rule. Brightness and color of each pixel are determined by the property of the corresponding surface point of the CG model.

Volume rendering simulates that there are rays emitted from each pixel in the LCD display and getting a corresponding point behind each lens at the end. This procedure generates more accurate IV images than surface rendering since every ray is considered. The principle of IV volume rendering is shown in Fig. 2. Each

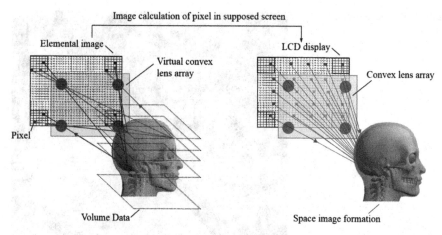

Fig. 2 Principle of volume rendering IV: explaining the method to generate and reconstruct a 3D image by IV. *Left* The principle calculation of elemental image is each point in a voxel data can emit rays pass, which will through the centers of all lenses, and be redisplayed at the back. *Right* When the elemental image is shown on a flat display covered by a lens array, a corresponding 3D image can be reconstructed

light emitted by pixels follows the particular direction that connects the corresponding pixel and lens'center. Each point in space image is reconstructed at the same position by the convergence of rays from the pixel of elemental image on the display.

3D IV images require real-time updating when the surgical scene changes. However, IV rendering is computationally costly, as one rendering needs many conventional surface renderings (for a surface model) or numeric ray-casting procedures (for a volume data). Here we introduce a flexible IV rendering pipeline with graphics processing unit (GPU) acceleration [16]. Modern OpenGL has a programmable rendering pipeline, which can be executed on GPUs. Computer-generated IV method proposed by Wang et al. (2014) uses a consumer-level GPU to realize real-time 3D imaging performance with high image resolution [17].

2.2 IV Image Overlay Devices

General IV image overlay device consists of an IV display, a semi-transparent mirror, a flexible supporting arm and a light-shield frame with a viewing window [18]. The overview of IV image overlay device is shown in Fig. 3. The spatial relationship between the IV image and its corresponding anatomical structure satisfies mirror image relationship. Surgeons can see the IV image reflected in the corresponding location in the surgical area when looking through the semi-transparent mirror. Working distance between the semi-transparent mirror and the surgical area is

Fig. 3 IV image overlay device

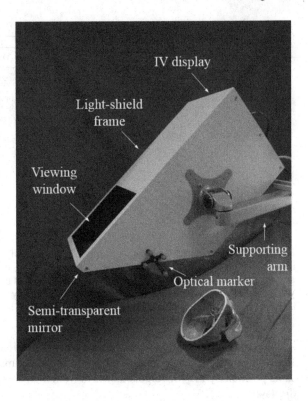

adjustable by changing the distance between the semi-transparent mirror and the IV display. The viewing field and viewing angle are limited by the size of the viewing window.

The spatial relationship between the 3D IV image $p_0(x_0, y_0, z_0)$ and corresponding point p_0' after semi-transparent mirror reflection (Fig. 4) satisfies the following equation [19]:

$$p_0' = p_0 + 2d \frac{\overrightarrow{n}}{|\overrightarrow{n}|} \tag{1}$$

with

$$d = \frac{|a(x_0 - x_1) + b(y_0 - y_1) + c(z_0 - z_1)|}{\sqrt{a^2 + b^2 + c^2}} \tag{2}$$

where $p_1(x_1, y_1, z_1)$ is a point on the plane of the semi-transparent mirror, $\overrightarrow{n}$ is its normal vector and the plane equation is $a(x_0 - x_1) + b(y_0 - y_1) + c(z_0 - z_1) = 0$.

IV overlay-guided surgical operations will be more convenient when the overlay device can be easily manipulated depending on surgeons' postures and viewing angle. AR images can be refreshed automatically when the device is tracked during

Fig. 4 Spatial relationship between 3D IV image and corresponding point after semi-transparent mirror reflection

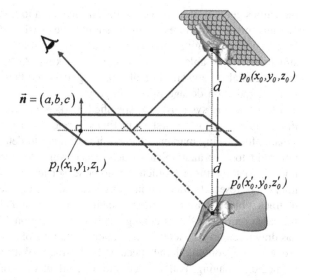

surgery since distance between each point on the target anatomical structures and semi-transparent mirror can be calculated. In this way, correct spatial positions of the IV image are estimated and elemental images will be updated automatically.

2.3 3D Spatial Tracking

A tracking system is necessary in IV image overlay to track surgical instruments, the overlay device and the patient. According to these position data, patient-image registration, intra-operative image registration, IV image rendering and update can be achieved.

The most common clinically used tracking systems are frame-based stereotaxy, optical tracking systems (OTS), electromagnetic tracking systems (EMTS) and intra-operative image-based analysis and tracking method. Frame-based stereotaxy integrates a reference coordinate to access the spatial positions and it is mostly used in oral and maxillofacial surgery. Although this kind of positioning technique proves to have high accuracy and stability in rigid structure tracking, the stereo-tactic frame may obstruct surgical operation. OTS identifies marker patterns and image features in the visible light range or near-illuminated infrared range, and then determine the pose information. OTS can provide a high positioning accuracy and reliability, but the line-of-sight between the marker and the tracking device has to be constantly maintained. EMTS can avoid line-of-sight problem so it is possible to track the instruments inside patient's body. EMTS localizes electromagnetic coils in the generated electromagnetic field by measuring the induced voltage. Nevertheless, its limitations include limited accuracy compared with optical tracking owing to field distortions caused by sensitivity to mental sources. Intra-operative imaging modalities get real-time images of the surgical area to track the pose of instruments or anatomical structures, which should be easily distinguished in images. Intra-operative imaging can easily explain the relationship between instruments and structures and be able to achieve surface tracking. At the same time, image information can also be useful in real-time diagnosis and therapy evaluation, but it may increase radiation dosage during surgery.

Most tracking systems need specified markers or sensors, but there are obvious limitations as follows. First, sometimes surgical space is limited or target is not suitable for placing markers. Second, it is difficult to determine the whole pose of a non-rigid tool or anatomical structure according to positions of some separated points. To solve these problems, marker/sensor-less tracking techniques become a hot area of research. Medical imaging and visible light stereo cameras are available to track objects based on image features and motion detection techniques [10]. Research on marker-less tracking, such as interventional needle, teeth, soft tissues, has drawn remarkable attentions. Major limitations of marker-less tracking in clinic are accuracy, computational speed and robustness. With the technical development of medical imaging, pattern recognition, computer vision and other related fields, marker-less based tracking will have a great potential for clinical applications.

Each tracking systems mentioned above has both advantages and disadvantages, hybrid tracking systems combining two or more tracking techniques are needed in order to achieve more reliable performance and larger working space.

3   Patient-3D Image Registration and Visualization Techniques

3D registration in AR interventions means spatial correlation of the reference position of virtual images and the patient. AR systems calculate the posture of both the virtual and actual objects and align them in a spatial position. Patient-3D image registration is based on a set of distinct external geometrical or natural anatomical features and landmarks, which can be accurately identified in both patient and 3D image spaces [20]. In mathematical terms, registration is represented by a linear coordinate transformation, which transforms coordinates between the image data coordinate system and the patient coordinate system using pair-point or surf-matching techniques. The rigid coordinate transformation, which can be represented by a 4×4 matrix consisting a 3×3 rotation matrix and a 3D translation vector, is solved mainly by minimizing the distance between pair-features using the least-squares fitting algorithm [21]. Following registration and visualization methods can be used in IV image overlay system.

3.1 Marker Based Registration

Accuracy of feature localization and extraction is always the main factor that significantly affects the registration result. Reference frame in stereotactic surgery is one early solution to locate accurate and stable positions of corresponding external points from image and patient in surgical interventions [22]. However, the frame may cause image artifacts during medical image acquisition and reduce the accuracy of registration. Besides, reference frame can bring a lot of uncomfortableness and inconvenience to patients and surgeons.

For the sake of patient's burden and surgical simplicity, reference in positioning becomes much smaller and lighter when using markers instead of frames. Reference markers, which can be mounted on a pointer held by the surgeons, patient and image visualization device, are widely used in orthopaedic interventions. A tracking system is applied to measure the positions of these targets according to the positions of markers.

Two main methods are used to acquire the positions of homologous points. Firstly, specific skin-affixed markers or bone-implanted markers, which can be recognized easily in medical images, are mounted on the patient during imaging procedures. At the same time, the positions of markers can also be determined during the surgery. When markers can be directly tracked, intra-operative patient movements can be tracked and become a real-time feedback to update the image-patient relationship. Bone-implanted markers offer a more stable and precise positioning result than skin-affixed markers since error caused by tissue shifting is almost avoided [21]. The simpler method is to depend on natural anatomical features on the surface such as tip of nose, center of two eyebrows on the surface or the distinct features of bones, which can be identified in images and observation. Positions of these feature points can be determined by putting the tip of a tracked pointer on the targets. Thus, it is not necessary to fix marker on patients, avoiding the invasiveness in imaging and surgery. Figure 5 shows the patient-3D IV image registration of a skull phantom.

3.2 Marker-Less Based Registration

Although marker based methods work well in many applications. However, marker based techniques are limited in some surgical environments. For most anatomical targets, it is difficult to attach marks in advance, because the human body should be protected from damage and infection. Moreover, markers should not be occluded in the effective area of the tracking system. Therefore, marker-less based patient-3D image registration techniques are useful in these situations.

Anatomical landmarks can be utilized to register the 3D data to the patient without fiducial markers [23]. For instance, the patient and the CT data can be correlated based on anatomical landmarks on the skull and the CT data. However,

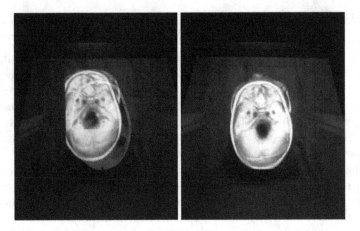

Fig. 5 Patient-3D IV image registration of skull base phantom. *Left* Surgical scene through AR system before registration. *Right* Surgical scene through AR system after registration

landmarks cannot be exactly identified during registration and surgery, therefore the method is not precise enough. The error of the registration with anatomical landmarks is about 2–5 mm [24].

Another approach of marker-less based registration is 3D surface geometry matching. In this technique, one same surface in image and patient space can be described by two large sets of points. Using a matching algorithm, the transformation matrix from points in image space to points in patient space can be calculated. Then, the registration result including the patient's surface and 3D image data can be obtained (Fig. 6). Various solutions have been proposed for the calculation of transformation matrix and the most common algorithm is the iterative closest point (ICP) algorithm [22]. The preoperative image data is from CT, while the patient's data can be acquired by a series of methods. For instance, one surface matching method proposed by Liao et al. (2004) is based on 2D images [25]. In this method, the patient's image can be matched to a 2D rendering surface model, which is extracted from the 3D surface model. After the surface matching, 3D images based on IP can be overlaid on patients and give an AR scene.

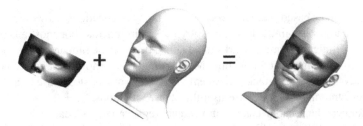

Fig. 6 Concept of surface matching. The fusion result is combined by the patient's surface (*left*) and the data obtained preoperatively by CT (*middle*)

Other methods of surface matching can also be used in 3D AR, like a stereo-camera, laser surface scanning and structured light. Depending on the parameters of the stereo-camera obtained by camera calibration, the surface can be reconstructed from the two images captured by the stereo-camera. Laser surface scanning has no invasiveness and is sufficiently precise in clinical deployment according to previous investigations [26]. The laser scan data can be mapped to the surface data obtained preoperatively by CT using surface matching. The higher the laser scan resolution is, the better the result of marker-less patient registration is. This concept has been applied in commercial systems and used in oral and maxillofacial surgery [23]. Another approach to achieve surface matching is structured light [27, 28]. The structured light is directly projected on patient and an optical device is used to compute and reconstructed the surface of patient. After surface matching, AR images can be seen in the right place.

Another common registration technique is based on anatomical contour. Wang et al. propose an automatic marker-free registration method using stereo tracking and 3D contour matching in AR dental surgery [10]. This method includes four main steps. First, an image template is selected manually in the first left image of the stereo camera, which is used to capture the simulated surgical scene of dental surgery. Then, the corresponding region on the right image can be obtained based on the template matching. Thirdly, within the regions of interest, 2D sharp edges of the teeth are extracted according to the high contrast between the teeth and the oral cavity. The stereo-matched are performed based on epistolary constraint searching. Finally, the 3D contour is reconstructed according to the result of the camera calibration and can be updated in real time. The reconstructed 3D contour is registered with the model derived from CT data which were scanned in advance using the ICP algorithm.

According to the relationship between the reconstructed 3D contour and the teeth model obtained in advance, the model's position and posture can be calculated. The model is merged into the surgical scene using IV image overlay, which was illustrated in Fig. 7. The surgical operation becomes more convenient based on the fusion result and the augmented display of the tool. The accuracy of the dental surgery increases.

Mathematically, patient-3D image registration techniques are similar. The geometrical features, like points, lines and surfaces, are used to correlate the relationship between patient and the data obtained by medical imaging devices. Tracking systems gets the corresponding features from patient in real time. The correlation between the patient and 3D image is calculated by several algorithms while the most well-known one is the ICP algorithm. Moreover, registration of soft tissue is one of the main challenges to solve in surgery currently, where organs and tissues are deformed, cut, dissected, etc. Because of the geometrical features are obtained from the patient in intra-operative situations, non-rigid registration algorithms can be used to optimize the registration. The greater the differences between the reconstructed virtual models and the real organs of the patient in the operating room, the more difficult this challenge is. A flexible AR overlay is required for automatic tracking and compensation of anatomy structural changes requires.

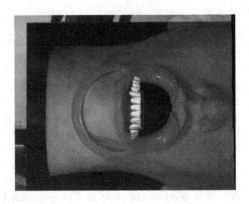

Fig. 7 Real-time
patient-image registration
result using contour tracking
and ICP matching

Accuracy of patient-image rigid registration can be evaluated according to root-mean-square (RMS) distance between corresponding points after registration. In point-based registration, one common measurement is fiducial registration error (FRE), which equals to the RMS distance between registration fiducial points. A more critical evaluation is target registration error (TRE), which represents the distance between corresponding points other than registration fiducial points after registration. Researches show that TRE can reflect the realty more closely than FRE [29]. Accuracy of registration is generally estimated in a phantom experiment, which shall simulate the real surgical scene as far as possible.

3.3 Assisting Visualization Techniques in High-Accurate Theranostics

In order to achieve high-accurate minimally invasive theranostics, assisting visualization techniques are used in the IV overlay system to guide surgical tools.

The traditional visualization technique to locate surgical tools is based on tracking systems. 3D spatial position and posture of surgical tools are recorded by the tracking system and the virtual tools based on IP are overlaid above the actual scene. Surgeons can see the augmented surgical tools, which are hidden in surroundings, which is shown in Fig. 8a [30].

In orthopaedic intervention applications, a laser guidance system can also be integrated into an overlay device to aid accurate needle puncture or implantation (Fig. 8b). The combination of laser guidance and image guidance has advantages in different aspects in MIS. Image guidance can show anatomical relations directly while laser can offer guidance in alignment of surgical instruments. Liao et al. presents a precision-guided navigation system using 3D IV image and laser guidance overlay [30]. The system employs two laser planes to determine a spatial intersection line, which represents the intervention direction of liner instruments. The directions of two laser planes are controlled by orientation of laser shooter

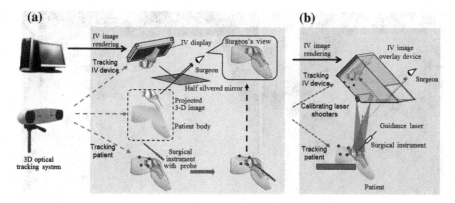

Fig. 8 Assisting visualization techniques. **a** The augmented surgical tools in IV overlay system. **b** Configuration of laser guidance IV image overlay system in knee intervention surgery

modules and mirrors are used to reflect the laser planed to the intervention area. Intervention entrance of instrument is the crossing point of two laser planes formed on the surface of intervention area. After the tip of the instrument is fixed, the correct rotation of the instrument can be distinguished when two intersection lines of laser surfaces and instrument are parallel.

4 Applications of IV Image Overlay in Orthopaedic Interventions

Because of the simplicity in implementation and the ability of full parallax over a wide viewing area, the IV image overlay system using IV is a promising solution in oral surgery, knee surgery, spine surgery and other orthopaedic interventions.

4.1 IV Image Overlay Based Knee Surgery

Knee surgery is a type of minimally invasive surgery to treat the injured bone or the cartilage damage on knee. Anterior cruciate ligament (ACL) is an important internal stabilizer of the knee joint to restrain hyperextension and it is easily injured when the biomechanical limits exceed. ACL injury can cause cartilage lesion, and usually the reconstruction of ACL is performed to treat ACL damage, knee pain and swelling. In such surgery, the surgeon operates through tunnels made on femur and tibia. The positional precision of the tunnel is important, because it is closely related to the patient's rehabilitation after the surgery. In traditional surgery, surgeons use medical images obtained by X-ray or CT to guide operations. Although the common navigation system can guide surgeons by the information shown in the screen,

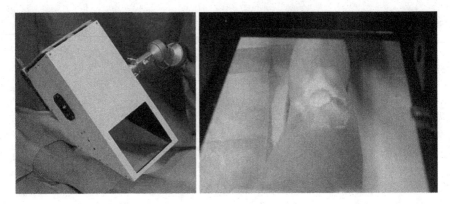

Fig. 9 IV image overlay used knee surgery

it is not vivid and intuitive. Unsatisfying hand-eye coordination as well as the limited visual field requires expert skills of surgeons to grasp the internal structure of knee joint.

Liao et al. [14] presented a surgical navigation system for anterior cruciate ligament reconstruction based on IV image overlay which can display the IV based femur, tibia and surgical tool in real time. This report is the first one that applying IV images with an IV image overlay system in ACL reconstruction surgery.

With the help of the 3D optical tracking system and the semi-transparent mirror, the bone data obtained by CT is shown in the surgical view and the 3D reconstructed surgical instrument can guide the surgeon to see the structure of ACL and surgical tools inside the patient's body based on this system and operate. Experiments testified that the average registration accuracy of the patient-image registration and the IV rendering is about 1.16 mm and frame rate of IV image display is about 3 frames per second for organ while 11 frames per second for surgical tools (Fig. 9) [14].

4.2 IV Image Overlay Based Oral Surgery

Oral surgery is the combination of series surgical processes performed on the teeth and jaw in order to modify dentition. The basic operations in oral surgery include cutting, drilling, fixation, resection, and implantation. Main types are endodontic surgery, oral prosthodontics, and orthodontic treatment. In most cases, operations are limited by the narrow space and surgical targets might be hidden in structures. Therefore, it is difficult to view the surgical targets and the posed of surrounding structures. Surgeons have to do many clinical practices to avoid damaging the surrounding vital structures when accessing the surgical targets. Moreover, dental surgery requires highly precise operations.

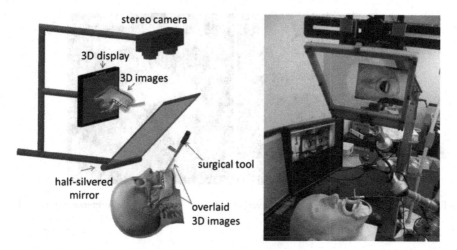

Fig. 10 The overview of the IV image overlay based oral surgery system. *Left* The concept of component parts of the system. *Right* The actual composition of the system

The 3D AR is helpful in computer-assisted oral surgery that has been rapidly evolving since the last decade. Currently, surgical navigation in oral surgery still suffers from the poor hand-eye coordination and the loss of depth perception in visual guidance. Tran et al. and Wang et al. present AR navigation systems with automatic marker-free image registration based on IV image overlay and stereo tracking for dental surgery [9, 10], which overcome the main shortcomings in the currently available technologies [20]. The proposed systems include a stereo camera tracker for tracking patients and instruments, an automatic real-time marker-less based patient-3D image registration method, an accurate IP-camera registration method, and AR visualization using IV overlay (Fig. 10).

As a result, IP based 3D image of the patient's anatomy structure is overlaid on the surgical region by a semi-transparent mirror based on patient-image registration and IP-camera registration, which can guide the surgeon to see the hidden structures. Moreover, the 3D image of the surgical instrument can be also overlaid to show the hidden instrument in the surgical area (Fig. 11). To confirm the feasibility, experiments were performed and the overlay error of the system mentioned above was 0.71 mm [10]. With the help of the 3D AR system, surgeons can obtain the depth perception and operate based on the 3D images with both stereo and motion parallax easily which can guide the operation.

4.3 IV Image Overlay Based Spine Surgery

Spine surgery treats the injured area on spine, which can be injured when it bends, stretch and rotate excessively. Among different types of spine surgery, pedicle

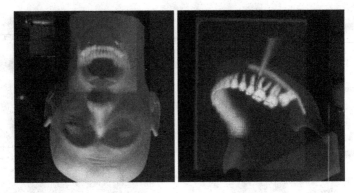

Fig. 11 AR in oral surgery. *Left* Model overlay with surgical scene. *Right* Augmented display of tool (the *pink* one) (Color figure online)

screws and transarticular screws provide strong fixation to the spine. However, a screw in the wrong place can result in neurologic and vascular complication. Exact insertion of a cervical screw is needed, especially when normal anatomic landmarks are difficult to identify during the surgery. To improve the accuracy of screw placement in the spine surgery, microscope, endoscope, the navigation system and series of surgical instruments are used. Among these techniques, the image-guide system is vital to successful access to the target area.

Computer-assisted image guidance system is also used for spine revision cases and recommended a preliminary report including the technique's usefulness and limitations. For instance, an image-guided system was used during the surgery to show the position of the device and surgical planning based on CT data [31]. As a result, all patients recovered well and there were no neurovascular complications or correction loss. All four transarticular screws were exactly placed in the pedicles. Eleven of 47 pedicular screws slightly breached the cortex.

During current spine surgery, the image-guided system still suffers from the problem of hand-eye coordination as well as the lack of depth information while 3D AR can make up for these disadvantages. 3D images can be seen in the surgical region with depth information and guide the operation that surgeon can have a better knowledge of the anatomic structure of the lesion and the position of surgical instruments.

In generally, the lesion as well as surgical instruments might be hidden in other structures during surgery. To solve these problems, 3D AR system is one promising solution that the virtual image in space is 3D with full parallax and surgeons can see through the patient to know the relationship of internal structures. In this way, the accuracy and safety of the surgery will increase. Except from the applications mentioned above, 3D AR has an extensive application orthopaedic surgery. Moreover, combined with the simultaneous fluoroscopy, microscope and other therapy techniques, 3D AR can be used to show clear 3D vision of interested regions over the patient in minimally invasive orthopaedic surgery.

5 Comparison Between IV Overlay Based MIS and Other Image-Guided Surgery Techniques

Several IGS systems are presented to assist surgeons during minimally invasive surgery, mainly containing HMD, AR surgical microscope, 2D overlay display as well as IV overlay.

HMD devices can be worn to see the fusion result of the real scene and the virtual images. In order to display the right images, a tracking system should be used to track the motion of the head and the patient's position. Therefore, surgeons have to wear an additional device, which is not suitable for flexible movement. The overlaid scene is only observed by one surgeon, and surgeons may feel fatigue for long time observing and focusing.

AR surgical microscope is used for minimally invasive surgery. The images shown in two visions are 2D images, which can give surgeons depth information based on stereo vision. However, surgeons easily feel tired because of focusing and the viewing area is limited. With HMD devices and AR surgical microscope, the 3D image is inaccurate based on the stereoscopy and the spatial position lacks accuracy owing to different users.

2D images can be straightly merged into the patient to augment the surgeon's view, while the fusion result of 2D overlay display is not precise enough and lacks depth information. Surgeons can move freely to have a better view of the operative region while the images are overlaid on the patient.

Although IGS techniques mentioned above help surgeons match visual images to the real region without making a hand-eye transformation, there still some problems remain resolved. One vital challenge is that the visualization structures

Table 1 Comparison between IV overlay and other 3D AR systems

	IV overlay	HMD	AR surgical microscope	2D image overlay
Displayed images	3D	2D	2D	2D
Geometrical accuracy of image	Accuracy in theory	Absolute distance impossible	Absolute distance impossible	Accuracy in theory
Supplementary instruments	Needless	Need	Need	Needless
Full-parallax	Possible	Impossible	Impossible	None
Spatial overlay accuracy	Accurate	Inaccurate	Inaccurate	Accurate
View point	Free	Head tracking	Not free	Free
Multiple viewer	Possible	Impossible	Impossible	Possible
Visual fatigue	None	Physiological adjustment and congestion	Physiological adjustment and congestion	None

lack depth information limited by conventional flat displays in current IGS systems. Among different solutions to meet the clinical needs, the IV image overlay system is a promising one because of its flexibility in implementation. IV is a vital autostereoscopic display technique, which has motion parallax in different directions over a wide viewing area. The 3D image calculated by computer-generating algorithms is accurate, and using the tracking system and the semi-transparent mirror, the positional accuracy is enough for clinical application [10]. Moreover, IV images give surgeons an intuitive estimation of the anatomic structures in the target region and it can be updated in real time. Therefore, surgeons can have a better knowledge of high-risk areas and the quality of the surgery can be improved. The comparison between IV overlay and other IGS techniques is shown in Table 1.

6   Discussion and Conclusion

We have described medical 3D AR as an IGS technique of comparative advantage. Among current IGS techniques, 3D AR not only solves the problem about hand-eye coordination, but also helps surgeons observe the hidden anatomical structure from the direct observation. IV image overlay systems provide autostereoscopic visualization of inner anatomical structure and surgical tools with correct depth information directly in the intraoperative scene. Furthermore, the IV images utilized in previous orthopaedic intervention studies can be pre-operative data as well as intraoperative data [32], such as MRI/CT and ultrasound, and it can be rendered in real time based on GPU accelerated algorithm. The geometrical accuracy of IV image is acceptable in surgery. Critical techniques in IV image overlay system include advanced visualization methods, precise patient-3D image registration techniques and reliable tracking methods. Therefore, IV image overlay is a promising solution to achieve clinical demand in orthopaedic interventions and preclinical experiments have revealed a good prospect.

Although it has advantages mentioned above, the IV image overlay still faces the problems about the limitation of related techniques and the reliability. To get a clear autostereoscopic image and a wide viewing angle, a suitable micro lens array as well as a display with high resolution is needed. Except technical limitations, clinical factors including efficiency and usability require further experiments and evaluations. Methods of seamless integration between the AR system and other operation instruments are worth investigating.

We think that 3D AR represented by IV image overlay already becomes a trend in the field of IGS and MIS. Related techniques in orthopaedic interventions are more matured than other areas, which makes orthopaedic a promising field for 3D AR systems to enter clinical practice. Above all, 3D AR techniques are not only suitable for orthopaedic interventions, but also of great help in other complicated MIS situations. 3D AR is a promising means and platform to combine different kinds of advance diagnosis and therapy methods, and has a wide application in precise theranostics systems.

References

1. Cleary K, Peters TM (2010) Image-guided interventions: technology review and clinical applications. Annu Rev Biomed Eng 12:119–142
2. DiGioia AM III, Jaramaz B, Colgan BD (1998) Computer assisted orthopaedic surgery: image guided and robotic assistive technologies. Clin Orthop Relat Res 354:8–16
3. Peters T, Cleary K (2008) Image-guided interventions: technology and applications. Springer, Berlin
4. Liao H, Edwards PJ (2013) Introduction to the special issues of mixed reality guidance of therapy-Towards clinical implementation. Comput Med Imag Gr Off J Comput Med Imaging Soc 37.2: 81
5. Suenaga H, Tran HH, Liao H et al (2013) Real-time in situ three-dimensional integral videography and surgical navigation using augmented reality: a pilot study. Int J Oral Sci 5(2):98–102
6. Lamata P, Ali W, Cano A et al (2010) Augmented reality for minimally invasive surgery: overview and some recent advances. Augment Real 73–98
7. Sauer F, Vogt S, Khamene A (2008) Augmented reality. In: Image-guided interventions. Springer, US, pp 81–119
8. Fichtinger G, Deguet A, Masamune K, et al (2004) Needle insertion in CT scanner with image overlay–cadaver studies. In: Medical image computing and computer-assisted intervention–MICCAI 2004. Springer, Berlin, pp 795–803
9. Tran HH, Suenaga H, Kuwana K et al (2011) Augmented reality system for oral surgery using 3D auto stereoscopic visualization. In: Medical image computing and computer-assisted intervention–MICCAI 2011. Springer, Berlin, pp 81–88
10. Wang J, Suenaga H, Hoshi K et al (2014) Augmented reality navigation with automatic marker-free image registration using 3D image overlay for dental surgery. IEEE Trans Biomed Eng 61(4):1295–1303
11. Liao H, Sakuma I, Dohi T (2007) Development and evaluation of a medical autostereoscopic image integral videography for surgical navigation. In: IEEE/ICME international conference on complex medical engineering, 2007, CME 2007. IEEE, New York
12. Lippmann G (1908) La Photographie integrale. Academie des Sciences, Comtes Rendus, pp 446–451
13. Liao H, Nakajima S, Iwahara M et al (2001) Intra-operative real-time 3-D information display system based on integral videography. In: Medical image computing and computer-assisted intervention–MICCAI 2001. Springer, Berlin
14. Liao H, Nomura K, Dohi T (2005) Autostereoscopic integral photography imaging using pixel distribution of computer graphics generated image. In: ACM SIGGRAPH 2005 Posters. ACM, Texas
15. Liao H, Dohi T, Nomura K (2011) Autostereoscopic 3D display with long visualization depth using referential viewing area-based integral photography. IEEE Trans Vis Comput Gr 17(11):1690–1701
16. Wang J, Sakuma I, Liao H (2013) A hybrid flexible rendering pipeline for real-time 3D medical imaging using GPU-accelerated integral videography. Int J Comput Assist Radiol Surg S287–S288
17. Wang J, Suenaga H, Liao H et al (2014) Real-time computer-generated integral imaging and 3D image calibration for augmented reality surgical navigation. Comput Med Imaging Graph 147:159
18. Liao H, Inomata T, Sakuma I et al (2010) 3-D augmented reality for MRI-guided surgery using integral videography autostereoscopic image overlay. IEEE Trans Biomed Eng 57 (6):1476–1486
19. Liao H, Hata N, Nakajima S et al (2004) Surgical navigation by autostereoscopic image overlay of integral videography. IEEE Trans Inf Technol Biomed 8(2):114–121

20. Lamata P, Ali W, Cano A et al (2010) Augmented reality for minimally invasive surgery: overview and some recent advances. Augment Real 73–98
21. Arun S, Huang S, Blostein D (1987) Least-squares fitting of two 3-D point sets. IEEE Trans Pattern Anal Mach Intell 5:698–700
22. Eggers G, Mühling J, Marmulla R (2006) Image-to-patient registration techniques in head surgery. Int J Oral Maxillofac Surg 35(12):1081–1095
23. Marmulla R, Hassfeld S, Lüth T et al (2003) Laser-scan-based navigation in cranio-maxillofacial surgery. J Cranio-Maxillofac Surg 31(5):267–277
24. Hassfeld S, Mühling J (2001) Computer assisted oral and maxillofacial surgery–a review and an assessment of technology. Int J Oral Maxillofac Surg 30(1):2–13
25. Liao H, Inomata T, Hata N et al (2004) Integral videography overlay navigation system using mutual information-based registration. In: Medical imaging and augmented reality. Springer, Berlin, pp 361–368
26. Luebbers HT, Messmer P, Obwegeser JA et al (2008) Comparison of different registration methods for surgical navigation in cranio-maxillofacial surgery. J Cranio-Maxillofac Surg 36(2):109–116
27. Hostettler A, Nicolau SA, Rémond Y et al (2010) A real-time predictive simulation of abdominal viscera positions during quiet free breathing. Prog Biophys Mol Biol 103 (2):169–184
28. Nicolau S, Soler L, Mutter D et al (2011) Augmented reality in laparoscopic surgical oncology. Surg Oncol 20(3):189–201
29. Fitzpatrick M (2009) Fiducial registration error and target registration error are uncorrelated. In: SPIE Medical Imaging. International Society for Optics and Photonics, Bellingham
30. Liao H, Ishihara H, Tran H et al (2010) Precision-guided surgical navigation system using laser guidance and 3D autostereoscopic image overlay. Comput Med Imaging Graph 34(1):46–54
31. Seichi A, Takeshita K, Nakajima S et al (2005) Revision cervical spine surgery using transarticular or pedicle screws under a computer-assisted image-guidance system. J Orthop Sci 10(4):385–390
32. Herlambang N, Liao H, Matsumiya K et al (2005) Realtime integral videography using intra-operative 3-D ultrasound for minimally invasive heart surgery. J Jpn Soc Comput Aided Surg J JSCAS 7.2:163–166

Fully Automatic Segmentation of Hip CT Images

Chengwen Chu, Junjie Bai, Xiaodong Wu and Guoyan Zheng

Abstract Automatic segmentation of the hip joint with pelvis and proximal femur surfaces from CT images is essential for orthopedic diagnosis and surgery. It remains challenging due to the narrowness of hip joint space, where the adjacent surfaces of acetabulum and femoral head are hardly distinguished from each other. This chapter presents a fully automatic method to segment pelvic and proximal femoral surfaces from hip CT images. A coarse-to-fine strategy was proposed to combine multi-atlas segmentation with graph-based surface detection. The multi-atlas segmentation step seeks to coarsely extract the entire hip joint region. It uses automatically detected anatomical landmarks to initialize and select the atlas and accelerate the segmentation. The graph based surface detection is to refine the coarsely segmented hip joint region. It aims at completely and efficiently separate the adjacent surfaces of the acetabulum and the femoral head while preserving the hip joint structure. The proposed strategy was evaluated on 30 hip CT images and provided an average accuracy of 0.55, 0.54, and 0.50 mm for segmenting the pelvis, the left and right proximal femurs, respectively.

C. Chu · G. Zheng (✉)
Institute for Surgical Technology and Biomechanics, University of Bern,
Bern, Switzerland
e-mail: guoyan.zheng@istb.unibe.ch

C. Chu
e-mail: chengwen.chu@istb.unibe.ch

J. Bai · X. Wu
Department of Electrical and Computer Engineering, The University of Iowa, Iowa City,
IA 52242, USA
e-mail: junjie-bai@uiowa.edu

X. Wu
e-mail: xiaodong-wu@uiowa.edu

© Springer International Publishing Switzerland 2016
G. Zheng and S. Li (eds.), *Computational Radiology*
for Orthopaedic Interventions, Lecture Notes in Computational
Vision and Biomechanics 23, DOI 10.1007/978-3-319-23482-3_5

1   Introduction

Nowadays fully automatic hip joint segmentation from CT data plays an important role in orthopedic diagnostic and surgical procedures, e.g., surgical planning, intra-operative navigation, and postoperative assessment. The hip joint segmentation seeks for completely and accurately separating the pelvis and the proximal femur surfaces from hip CT images. Unfortunately, it is still a big challenge of fully automatic segmentation due to the narrowness of the hip joint and the weak boundary in the resolution-limited CT images.

Various work on hip CT segmentation have been reported and consist of three main categories: intensity-based methods [1–3], statistical shape model (SSM) [4–8], and atlas-based approaches [9–11]. Kang et al. [1] presented an intensity-based segmentation method with four-steps. In the first two steps of this framework, bone regions of the pelvis and the femur were roughly segmented, using local adaptive threshold based region growing and morphological closing based boundary discontinuities. Then in the next two steps, based on the initially extracted surface models of the hip joint, local intensity profiles were obtained for each vertex on the surface models. With local intensity profile analysis, the initially extracted surface models were modified by an oriented boundary adjustment in the combination with periosteal and endosteal surfaces detection. Another similar work which combines multi-fundamental image processing techniques for hip CT segmentation was introduced by zorrofi et al. [2]. Given an unsegmented CT image, this method starts with a definition of region of interest (ROI) for the hip joint. The initial segmentation of the hip joint is achieved by applying histogram based thresholding and morphological operations to the defined ROI. To further improve segmentation of the hip joint on extracted surface models, a Hessian filter and a model-based specific approach were proposed, which followed by a moving disk technique to get refined segmentation. Most recently Cheng et al. [3] improved the method in [2] by incorporating pre-estimated intensity Gaussian model of the hip joint to a Bayes decision rule, for voxel-wised image labeling. Valley-emphasized CT images were used to stand out the contrast between region of the joint space and regions of the femoral head and the acetabulum, which is proved to be useful to preserve the joint space. Similarly, employing intensity Gaussian model guided fussy voxel classification was also applicable to correct hard segmentation errors produced by conventional histogram-based thresholding and morphological operations as described in [2].

Automatic hip CT segmentation with SSM based methods can be divided into two categories: (a) Segmentation of Single anatomical structure by SSM fitting [4, 7, 8], and (b) Simultaneously segmentation of adjacent structures by articulated SSM (aSSM) fitting [5, 6]. The first introduction of the SSM construction and fitting methods to hip CT segmentation was done by Lamecker et al. [4]. This conventional SSM based method starts with a training stage to perform SSM construction from a training population. Given a new image, the constructed SSM is subsequently fitted to this new image to get the segmentation until convergence. An improved work done by Yokota et al. [7] combines inter-organ spatial variations with shape variations to drive the SSM fitting based hip joint segmentation. More specifically, a coarse-to-fine segmentation framework was established, hierarchical SSM of the hip joint was developed by firstly constructing combined SSM for the pelvis-femur regions globally, and secondly constructing a partially divided SSM of the acetabulum and the proximal femur. This method, however, still remains the problem of insufficient segmentation for proximal femur regions which caused by the unbalance in simultaneous optimization of the pelvic and the femoral SSMs [8]. Aiming to improve the segmentation performance for the proximal femur, recent work from Yokota et al. [8] developed a conditional SSM of proximal femur to modify the initial segmentation results obtained by the methods reported in [7]. There exists another way to improve the performance of SSM fitting based methods. Seim et al. [5] addressed this problem by improving the limited ability for shape representation of the SSM due to the small size of the population in SSM construction. More specified, they setup a pipeline by firstly applying adaptation of the initialized mean model to the target image using the similar way as described above, and secondly adding a graph optimization-theory based free-form deformation step. This work has been soon improved and applied to simultaneously segment surface models of the pelvis and the proximal femur with extended a SSM introduced by Kainmueller et al. [6]. The common definition of SSM were extended by approximately modeling joint posture as a parameterized rotation of the femur around the joint center in aSSM construction and model fitting procedures. The fitted aSSM was again improved by a graph optimization-theory based multi-surface detection step [12, 13]. In this method, using a graph optimization theory [12, 13] has shown robustness and achieved promising results in the detection of optimal surfaces for the adjacent structures simultaneously. Therefore, it supples us an elegant solution to address the challenge of joint space preservation in hip CT segmentation.

There also exist atlas-based segmentation methods [9–11], which are recently used for hip CT segmentation. For any given new image, Ehrhardte et al. [9] achieved the segmentation of each anatomical structure in hip CT by implementing free-form based single atlas registration to the target image space. Here we define an atlas as a pair of data consisting of a CT volume and its corresponding ground-truth segmentation. Using the derived deformation filed from registration of CT image of the atlas to the target image, the ground-truth segmentation of the atlas can also be deformed to matching with the target image space. The deformed ground-truth segmentation of the atlas was then defined as the segmentation results of the given target image. In [9], to improve the registration accuracy, a rough segmentation of bony region by threshold based methods and morphological operators was achieved. To speed up the registration process, a multi-resolution strategy was applied to the CT volumes. A very similar framework was presented by Pettersson et al. [10] for segmentation of single pelvic bone from CT images. The only difference between these two methods is Ehrhardte et al. [9] using "demons algorithm" [14] for atlas registration while Pettersson et al. [10] selected "morphon algorithm" [15]. Instead of using only one atlas, recent work by Xia et al. [11] introduced a multi-atlas registration based method for the hip joint segmentation from MRI images. Each individual atlas were firstly transformed to the target image and then fused to produce spatial prior. The spatial prior is then used to guide the SSM fitting for the hip joint segmentation as done by [7]. Either for SSM based methods or atlas-based methods, an important pre-process is needed to initially align generated model (SSM or spatial prior) to the target image before segmentation. The most common way for model initialization is to apply scaled rigid transformation using a set of defined anatomical landmark positions. The last two decades witness the significant success of using machine learning based method to solve the automatic landmark detection problem, especially the random forest regression based methods [16–18]. RF was originally introduced by Breiman [16]. The first introduction of using RF regression based method in the field of high-dimensional medical image processing were reported by Criminisi et al. [17] and Lindner et al. [18]. Since then the RF based method has shown robustness for solving the problems of automatic anatomical landmark detection and object localization. Therefore, we are motivated to use the RF regression based method to automatically detect a set of landmark positions, where the detected landmarks will be further used for model initialization before segmentation step [19]. Basically, RF regression based landmark detection method consists of two stages: training stage and prediction stage (Fig. 1). For each landmark position, during training we first sample some local volumes (patches in case of 2D slice image) in the training image (Fig. 1a, b). Then, the displacements from the sampled volumes to the ground truth of landmark position, as well as the visual features of all the local volumes are calculated. Using the RF regression method, we could then estimate a map between feature space and displacement space. In the prediction stage, given a new image, we sample another set of local volumes and calculate the visual features and centers for them (Fig. 1c). By applying the features of all the sampled volumes

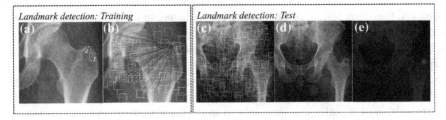

Fig. 1 The example of RF training and landmark detection. Illustration on 2D AP X-ray image for easy understanding. **a** A patch sampled around the true landmark position. **b** Multiple sampled training patches from one training data. **c** A target image. **d** Multiple sampled test patches over target image. **e** Each patch gives a single vote for landmark position. **f** Response image calculated using improved fast Gaussian transform

to the estimated map from feature space to displacement space, we could calculate the predicted displacements of each sampled volume. Then, by adding the calculated centers to the predicted displacements (Fig. 1d), a group of predictions for the unknown landmark position will be calculated. Finally, the landmark position is estimated by a voting scheme considering the individual estimations from all the volumes (Fig. 1e). We have a separated RF landmark detector for each landmark. Subsequently applying all the landmark detectors to the target image will then result a set of landmark positions, which can be used to initialize atlases using landmark based scaled rigid registrations.

In this chapter, we propose a two-stage automatic hip CT segmentation method. In the first stage, we use a multi-atlas based method to segment the regions of the pelvis and the bilateral proximal femurs. An efficient random forest (RF) regression-based landmark detection method is developed to detect landmarks from the target CT images. The detected landmarks allow for not only a robust and accurate initialization of the atlases within the target image space but also an effective selection of a subset of atlases for a fast atlas-based segmentation. In the second stage, we refine the segmentation of the hip joint area using graph optimization theory-based multi-surface detection [12, 13], which guarantees the preservation of the hip joint space and the prevention of the penetration of the extracted surface models with a carefully constructed graph. Different from the method introduced in [6], where the optimal surfaces are detected in the original CT image space, here we propose to first unfold the hip joint area obtained from the multi-atlas-based segmentation stage using a spherical coordinate transform and then detect the surfaces of the acetabulum and the femoral head in the unfolded space. By unfolding the hip joint area using the spherical coordinate transform, we convert the problem of detection of two half-spherically shaped surfaces of the acetabulum and the femoral head in the original image space to a problem of detection of two terrain-like surfaces in the unfolded space, which can be efficiently solved using the methods presented in [12, 13]. Figure 2 presents a schematic overview of the complete workflow of our method.

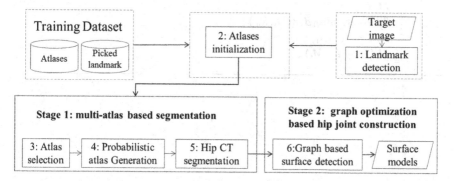

Fig. 2 The flowchart of our proposed segmentation method

2   Multi-atlas Based Hip CT Segmentation

2.1   Landmark Detection Based Atlas Initialization and Selection

In landmark detection step, totally 100 anatomical landmark positions are defined, in which 50 for pelvic model and 25 for each proximal femoral model. We apply landmark detector training and landmark prediction separately for each landmark position. For details, we refer to [19].

In atlas initialization step, using the detected N_l anatomical landmarks, paired-point scaled rigid registrations are performed to align all the N atlases to the target image space. Here, following the Algorithm 1 and 2 we speed up the scaled rigid registrations by aligning all the atlases and the target image to the same reference image, which is randomly selected as one of the training images. Since all the atlases have already been aligned to the reference image prior to the segmentation phase (Algorithm 1), we only need to transform target image to the reference space (Algorithm 2). Based on the scaled rigid registration results, we select N_s atlases with the least paired-point registration errors for the given target image. The selected N_s atlases are then registered to the target image using a discrete optimization based non-rigid registration [20].

Algorithm 1: Atlas alignment to the reference space

> **input** : (A, M, R_m), A stands for the training data set which includs N atlases, M stands for N landmark sets in which one landmark set includs 100 landmark positions for each atlas, R_m stands for the landmark set of the selected reference image;
>
> **output**: (A', M'), A' stands for the set of aligned atlases, and M' stands for the transformed landmark sets of the aligned atlases;
>
> **begin**
>> **for** $i \leftarrow 1$ **to** N **do**
>>> $A_i \leftarrow A(i)$;
>>> $M_i \leftarrow M(i)$;
>>> $T_r^{\star} \leftarrow \operatorname{argmin}_{T_r} dist(M_i, R_m)$;
>>> $A'(i) \leftarrow T_r^{\star}(A_i)$;
>>> $M'(i) \leftarrow T_r^{\star}(M_i)$;

Algorithm 2: Target image alignment

input : (T, T_m, R_m), T stands for the target image, T_m stands for the detected landmark set of the target image, and R_m stands for the landmark set of the reference image;

output: (T', T'_m) T' stands for the aligned target image to the reference space, T'_m stands for the landmark set of the aligned target image;

begin

    $T^\star_r \leftarrow \text{argmin}_{T_r} dist(T_m, R_m)$;

    $T' \leftarrow T^\star_r(T)$;

    $T'_m \leftarrow T^\star_r(T_m)$;

Algorithm 3: Atlas selection

input : (T'_m, A', M'), T'_m stands for the landmark set of the aligned target image, A' stands for the set of aligned atlases, and M' stands for the landmark sets of the aligned atlases;

output: A'_s Selected sub-set of atlases;

begin

    $N_l = Size(T'_m)$;

    Compute paired-point landmark errors between target image and atlases

    for $i \leftarrow 1$ **to** N **do**

        $err_i \leftarrow 0$;

        $M'_i \leftarrow M'(i)$;

        for $j \leftarrow 1$ **to** N_l **do**

            $err_i \leftarrow err_i + dist^2(M'_i(j) - T'_m(j))$;

        $err(i) \leftarrow err_i$;

    Sorting the atlases set with the ascending order of landmark errors

    $A' \leftarrow sort_a scending(A', err)$;

    Get the top N_s atlases

    for $i \leftarrow N - 1$ **to** N **do**

        $A'_s(i) \leftarrow A'(i)$;

2.2 Atlas-Based Segmentation

In this step, we first use the selected N_s atlases after registration to generate probabilistic atlas (PA) both for background and hip joint structures (Fig. 3). We then formulate the multi-atlas based segmentation problem as a Maximum A Posteriori

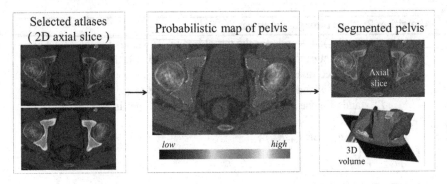

| Selected atlases (2D axial slice) | Probabilistic map of pelvis | Segmented pelvis |

Fig. 3 Example of generating PA for segmentation. *Left* After non-rigidly registered two selected atlases to the target space. *Middle* Generated PA. *Right* Segmented pelvis

(MAP) estimation which can be efficiently solved using a graph cut based optimization method [21]. Using the known label of the selected atlases, PA of the target image $\mathscr{T}$ is computed as:

$$\Lambda_{\mathbf{p}}(l) = \sum_i^{N_s} \omega(A_i, \mathscr{T})\delta(L_i^{\mathbf{p}}, l) / \sum_i^{N_s} \omega(A_i, \mathscr{T}), \qquad (1)$$

where

$$\delta(l, l') = \begin{cases} 1 & \text{if } l = l' \\ 0 & \text{otherwise.} \end{cases} \qquad (2)$$

Here, i indicates the selected atlases, $\mathbf{p}$ denotes the voxel in the CT image, l represents the label of organ regions, $L_i^{\mathbf{p}}$ is the label of the voxel $\mathbf{p}$ in the atlas A_i and $\Lambda_{\mathbf{p}}(l)$ is probability that voxel $\mathbf{p}$ labeling as l. $\omega(A_i, \mathscr{T})$ is the weight of atlas A_i which is evaluated by the similarity between atlases and the target image. We use the normalized cross correlation (NCC) as the similarity between atlases and the target image. The probability $\Lambda_{\mathbf{p}}(l)$ is calculated for every voxels in the target image for background (where $l = 0$), pelvis (where $l = 1$), left femur (where $l = 2$), and right femur (where $l = 3$), respectively. For the given target image, a voxel-wise MAP estimation is defined as

$$L_{\mathbf{p}} = \arg\max_l P(l|I_{\mathbf{p}}) \qquad (3)$$

where $L_{\mathbf{p}}$ is the given label of the voxel $\mathbf{p}$, $I_{\mathbf{p}}$ is the intensity of $\mathbf{p}$, l is the label of each region, and $P(l|I_{\mathbf{p}})$ is the posterior probability. The MAP estimation aims to find a label l which can maximize the posterior probability. In other words, for a voxel $\mathbf{p}$ which has intensity $I_{\mathbf{p}}$, if the label l of any region can maximize the posterior probability, the voxel will be assigned a label of l. The posterior probability is computed according the Bayes theory as $P(l|I_\mathbf{p}) \propto p(I_\mathbf{p}|l)\Lambda_\mathbf{p}(l)$, where $\Lambda_\mathbf{p}(l)$ is the PA describing the prior probability and $p(I_\mathbf{p}|l)$ is the conditional probability that is computed as we did in [22]. The MAP estimation is then solved with the graph cut method [21]. Giving the cost function of

$$E(C) = \sum_{\mathbf{p} \in \mathscr{I}} D_\mathbf{p}(C_\mathbf{p}) + \lambda \sum_{\mathbf{p},\mathbf{q} \in N_\mathbf{p}} V_{\mathbf{p},\mathbf{q}}(C_\mathbf{p}, C_\mathbf{q}) \tag{4}$$

where C is the unseen labelling of target image; $\mathbf{p}$ and $\mathbf{q}$ the voxels in the target image; and $N_\mathbf{p}$ the set of neighbors of voxel $\mathbf{p}$. The date term of $D_\mathbf{p}(C_\mathbf{p})$ is defined based on the estimated posterior probability of $P(l|I_\mathbf{p})$. The relationships between a voxel and its neighborhoods are represented by the smoothness term $V_{(\mathbf{p},\mathbf{q})}(C_\mathbf{p}, C_\mathbf{q})$. Factor λ balances the influence of the two terms. The data term is defined as

$$D_\mathbf{p}(C_\mathbf{p}) = -\gamma \ln \Lambda_\mathbf{p}(l) - (1 - \gamma)\ln p(I_\mathbf{p}|l) \tag{5}$$

which measures the disagreement between a prior probabilistic (PA in our case) and the observed data (conditional probability) with a factor γ. The smoothness term is defined as

$$V_{\mathbf{p},\mathbf{q}}(C_\mathbf{p}, C_\mathbf{q}) = \begin{cases} 0 & \text{if } C_\mathbf{p} = C_\mathbf{q} \\ \frac{1}{(1+|I_\mathbf{p}-I_\mathbf{q}|dist(\mathbf{p},\mathbf{q}))}, & \text{otherwise.} \end{cases} \tag{6}$$

where $dist(\mathbf{p}, \mathbf{q})$ the Euclidean distance between two voxels. From these equations, we can find that $V_{(\mathbf{p},\mathbf{q})}(C_\mathbf{p}, C_\mathbf{q})$ becomes large when the distance between $\mathbf{p}$ and $\mathbf{q}$ is smaller and closer. Hence, two neighborhoods with the large cutting cost is large and they are not separated.

3   Graph Optimization Based Hip Joint Construction

3.1   Problem Formulation

After the multi-atlas segmentation of the hip joint region, we need to refine the segmented region and recover the hip joint structure by separating the surfaces of the acetabulum and the femoral head. In the CT image space, both the acetabulum and the femoral head are ball-like structures and their surfaces can be approximately represented as half-spherically shaped models. To separate these two surfaces, directly applying graph optimization-based surface detection in the CT image space as described in [13] would be an option. However, construction of a graph in the original CT image is not straightforward and requires finding correspondences between two adjacent surfaces obtained from a rough segmentation stage as done in [13], which is challenging.

In our method, instead of performing surface detection in the original CT image space, we first define a hip joint area in the extracted surface models of both the pelvis and the proximal femurs, and then unfold this area using a spherical coordinate transform as shown in Fig. 4. Since the spherical coordinate transform converts a half-spherically shaped surface to a planar surface, the surfaces of the acetabulum and the femoral head can therefore be unfolded to two terrain-like surfaces with a gap (joint space) between them as shown in Fig. 4. We reach this goal with following steps:

1. Detecting rim points of the acetabulum from segmented surface model of the pelvis using the method that we developed before [23] (Fig. 4: 1).
2. Fitting a circle to the detected rim points, determining radius R_c and center of the circle, as well as normal (towards acetabulum) to the plane where the fitted circle is located (Fig. 4: 2).
3. Constructing a spherical coordinate system as shown in Fig. 4: 3, taking the center of the fitted circle as the origin, the normal to the fitted circle as the fixed zenith direction, and one randomly selected direction on the plane where the fitted circle is located as the reference direction on that plane. Now, the position of a point in this coordinate system is specified by three numbers: the radial distance R of that point from the origin, its polar angle Θ measured from the zenith direction and the azimuth angle Φ measured from the reference direction on the plane where the fitted circle is located.
4. Sampling points in the spherical coordinate system from the hip joint area (see Fig. 4: 4) using a radial resolution of 0.25 mm and angular resolutions of 0.03 radians (for both polar and azimuth angles). Furthermore, we require the sampled points satisfying following conditions:

$$\begin{cases} R_c + 20 \leq r \leq R_c/2 \\ 0 \leq \theta \leq \pi/2 \\ 0 \leq \varphi \leq 2\pi \end{cases} \tag{7}$$

5. Getting corresponding intensity values of the sampled points from the CT image, which finally forms an image volume $I(\theta, \varphi, r)$ (Fig. 4: 5), where $0 \leq r \leq (20 + \frac{R_c}{2})/0.25$, $0 \leq \theta \leq \pi/0.06$, $0 \leq \varphi \leq 2\pi/0.03$. The dimension of r depends on the radius of the fitted circle while the dimensions of θ and φ are fixed. To easy the description later, here we define the dimension of r as D_r.

Figure 5 shows an example of generated volume $I(\theta, \varphi, r)$ of a hip joint. With such an unfolded volume, graph construction and optimal multiple-surface detection will be straightforward when the graph optimization-based multiple-surface detection strategy as introduced in the [12, 13] is used.

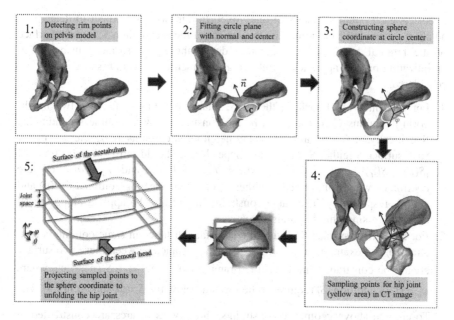

Fig. 4 A schumatic illustration of defining and unfolding a hip joint. Please see text in Sect. 3.1 for a detailed explanation

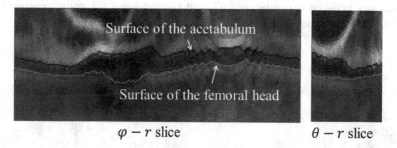

Fig. 5 An example of unfolded volume $I(\theta, \varphi, r)$ of a hip joint, visualized in 2D slices. *Left* A 2-D φ-r slice. *Right* A 2-D θ-r slice. In both slices, the *green line* indicates the surface of the femoral head and the *red line* indicates the surface of the acetabulum. The gap between these two surface corresponds to the joint space of the hip (Color figure online)

3.2 Graph Construction for Multi-surface Detection

For the generated volume $I(\theta, \varphi, r)$ as shown in Fig. 5, we assume that r is implicitly represented by a pair of (θ, φ), e.g. $r = f(\theta, \varphi)$. For a fixed (θ, φ) pair, the voxel subset $\{I(\theta, \varphi, r) | 0 \leq r < R\}$ forms a column along the r-axis and is defined as $Col(p)$. Each column has a set of neighbors and in this paper 4-neighbor system is adopted. The problem is now to find k coupled surfaces such that each surface intersects each column exactly at one voxel. In our case, we expect to detect two adjacent surfaces of a hip joint, i.e., the surface of the acetabulum S_a and the surface of the femoral head S_f. To accurately detect these two surfaces using graph optimization-based approach, following geometric constraints need to be considered:

1. For each individual surface, the shape changes of this surface on two neighboring columns $Col(p)$ and $Col(q)$ are constrained by smoothness conditions. Specifically, if $Col(p)$ and $Col(q)$ are neighbored columns along the θ-axis, for each surface S (either S_a or S_f), the shape change should satisfy the constraint of $|S(p) - S(q)| = |r_p - r_q| \leq \Delta_\theta$, where $r_p = p(\theta_1, \varphi)$ and $r_q = q(\theta_2, \varphi)$ are coordinate values of surface S (either S_a or S_f) intersecting columns $Col(p)$ and $Col(q)$, respectively. The same constraint should also be applied along the φ-axis with a smoothness parameter Δ_φ.

2. For the pair of surface S_a and S_f, their surface distance in same column is constrained. For example, in column $Col(p)$, the distance between these two surface should be constrained in a specified range of $\delta_p^l \leq |S_a(p) - S_f(p)| \leq \delta_p^u$, where $\delta_p^l \geq 0$. In addition, S_f requires to be located below the S_a (as shown in Fig. 5).

To enforce above geometric constraints, three types of arcs are constructed to define a directed graph $G = \{G_a \cup G_s\}$ (see Fig. 6 for details), where G_a and G_s are two subgraphs and each for detecting one surface of S_a and S_f, respectively. For each subgraph, we construct *intra-* and *inter-*column arcs. We also construct *inter-*surface arcs between two subgraphs G_a and G_s, following the graph construction method introduced in [12, 13].

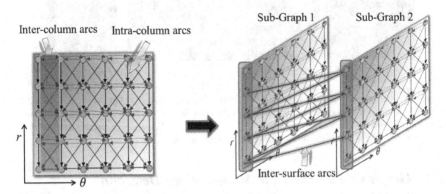

Fig. 6 Graph construction for detecting adjacent two surfaces of a hip joint. An example is presented in 2-D r-θ slice from the unfolded volume $I(\theta, \varphi, r)$. *Left* Intra-column (*black arrows*) and inter-column (*red* and *blue*) arcs for each subgraph; *right* inter-surface arcs to connect two subgraphs. Please note that these two subgraphs share the same nodes as well as the same inter- and intra-column arcs. The inter-surface arcs are constructed between the corresponding two columns which have exactly the same column of voxels in the unfolded volume (Color figure online)

Intra-column arcs: This type of arcs is added to ensure that the target surface intersects each column at exactly one position. In our case, along each column $p(\theta, \varphi)$, every node $V(\theta, \varphi, r)$ has a directed arc to the node immediately below it $V(\theta, \varphi, r - 1)$ with $+\infty$ weight (Fig. 6, left).

Inter-column arcs: This type of arcs is added to constrain the shape changes of each individual surface S on neighboring columns under a 4-neighborhood system. With two pre-defined smoothness parameters Δ_θ and Δ_φ, we construct these arcs with $+\infty$ weight along both the θ-axis and φ-axis (Fig. 6, left). In summary, we have arcs:

$$E = \begin{cases} \{\langle V(\theta, \varphi, r), V(\theta + 1, \varphi, max(0, r - \Delta_\theta))\rangle\} \cup \\ \{\langle V(\theta, \varphi, r), V(\theta - 1, \varphi, max(0, r - \Delta_\theta))\rangle\} \cup \\ \{\langle V(\theta, \varphi, r), V(\theta, \varphi + 1, max(0, r - \Delta_\varphi))\rangle\} \cup \\ \{\langle V(\theta, \varphi, r), V(\theta, \varphi - 1, max(0, r - \Delta_\varphi))\rangle\} \end{cases} \tag{8}$$

To get a smoothed segmentation, we further enforce soft smoothness shape compliance by adding another type of intra-column arcs (Fig. 6, left) [13]:

$$E = \begin{cases} \{\langle V(\theta, \varphi, r), V(\theta + 1, \varphi, r)\rangle | r \geq 1\} \cup \\ \{\langle V(\theta, \varphi, r), V(\theta - 1, \varphi, r)\rangle | r \geq 1\} \cup \\ \{\langle V(\theta, \varphi, r), V(\theta, \varphi + 1, r)\rangle | r \geq 1\} \cup \\ \{\langle V(\theta, \varphi, r), V(\theta, \varphi - 1, r)\rangle | r \geq 1\} \end{cases} \tag{9}$$

Again we construct these arcs along both the θ-axis and φ-axis using a 4-neighbor system. The smoothness penalty that assigned to these arcs are determined by a non-decreasing function $f_{p,q}(|S(p) - S(q)|)$, where $|S(p) - S(q)|$ represent the shape change (determined by the smoothness parameters Δ_θ and Δ_φ) for a surface S on neighbored columns $Col(p)$ and $Col(q)$. We select a linear function $f_{p,q}(|S(p) - S(q)|) = a(|S(p) - S(q)|) + b$ following the method introduced in [13]. Thus, along the θ-axis , we assign a weight a to each arc. Likewise, for the arcs along the φ-axis, we have similar weight to each arc.

Inter-surface arcs: This type of arcs are added to constrain surface distance between S_a and S_f in each column. In our case S_f is required to be below the S_a, assume that distance in column p between surfaces S_a and S_f ranges from δ_p^l and δ_p^u, we add the following arcs (Fig. 6, right):

$$E_s = \begin{cases} \{\langle V_a(\theta, \varphi, r), V_f(\theta, \varphi, r - \delta_p^u)\rangle | r \geq \delta_p^u\} \cup \\ \{\langle V_f(\theta, \varphi, r), V_a(\theta, \varphi, r + \delta_p^l)\rangle | r < R - \delta_p^l\} \cup \\ \{\langle V_a(0, 0, \delta_p^l), V_f(0, 0, 0)\rangle\} \end{cases} \tag{10}$$

where V_a and V_f denote the node in the corresponding column from each subgraph as shown in (Fig. 6, right). For each column $Col(p)$, we have a different distance range (δ_p^l, δ_p^u) which is statistically calculated from a set of training data.

Node cost function: By adding all the arcs as described above, we establish a directed graph $G = (V, E)$, where $V = V_a \cup V_f$ and $E = E_a \cup E_f \cup E_s$. Here, V_a and V_f are node sets from each subgraph, E_a and E_f are intra- and inter-column arcs from each subgraph and E_s is the inter-surface arcs between two subgraphs. In order to detect surfaces based on graph optimization, a new digraph $G_{st}(V \cup \{s, t\}, E \cup E_{st})$ is defined. This is achieved by adding a source node s and a sink node t as well as new edge set E_{st} which includes the edges between nodes in the graph G and the nodes of $\{s, t\}$. Then surface segmentation can be solved using the minimum s-t cuts established by Kolmogorov and Zabih [24]. We add new edges for the edge set E_{st} following the method introduced in [12]. The most important thing here is to assign an appropriate penalty for each edge which is also called *t-links*. As describe in [12], the penalty for each t-link is determined by the pre-computed cost of each node. In our method, an edge-based node cost function is designed by considering both intensity information and a prior information. More specifically, The negative magnitude of the gradient of the volume $I(\theta, \varphi, r)$ is computed at each voxel as $c(\theta, \varphi, r) = -|\nabla I(\theta, \varphi, r)|$. We give each node a weight as:

$$w(\theta, \varphi, r) = \begin{cases} c(\theta, \varphi, r) & if \ z = 0 \\ c(\theta, \varphi, r) - c(\theta, \varphi, r - 1) & otherwise \end{cases} \tag{11}$$

For details, we refer to [12, 13].

4 Experiment and Results

We evaluated the present method on hip CT data of 30 patients after ethical approval. The intra-slice resolutions range from 0.576 mm to 0.744 mm while the inter-slice resolutions are 1.6 mm for all CT data. All CT data are semi-automatically segmented by a trained rater with Amira (www.vsg3d.com/amira). We designed and conducted a validation study to evaluate the performance of the present approach by separating 30 CT data into a training dataset and test dataset. 20 of the hip CT data are selected as the training data both for RF based landmark detection and multi-atlas registration based segmentation. The rest 10 datasets (20 hip joints) are used for evaluation.

To evaluate the performance of the present approach, we first calculate the segmentation accuracy after each stage of the algorithm by comparing the automatic segmentation with the ground-truth segmentation for the pelvis (P), the left femur (LF), and the right femur (RF), respectively. As for performance evaluation, we computed two different metrics. First, average surface distance (ASD) between automatic segmentation and ground-truth segmentation are computed. To compute the ASD, for each vertex on the surface model of automatic segmentation, we found its shortest distance from the surface model derived from the associated ground-truth segmentation. ASD was then computed as the average of all shortest distances. Additionally, Dice overlap coefficient (DOC) between automatic segmentation and

Table 1 Dice overlap coefficient (DOC) (%) and average surface distance (ASD) (mm) between the automatic segmentation and the ground-truth segmentation when evaluated on 10 hip CT data (20 hips) for the pelvis (P), the left femur (LF), and the right femur (RF), respectively

	DOC			SD		
	P	LF	RF	P	LF	RF
MA	93.14 ± 1.15	95.01 ± 1.28	95.48 ± 0.69	0.58 ± 0.11	0.63 ± 0.21	0.58 ± 0.12
GO	93.54 ± 0.99	95.67 ± 1.08	95.94 ± 0.70	0.55 ± 0.09	0.54 ± 0.18	0.50 ± 0.13

Results after stage I (multi-atlas-based segmentation: MA) and after stage II (graph optimization-based surface detection: GO) are shown, where LF stands for the left femur, RF for the right femur and P for the pelvis

ground-truth segmentation are computed. With L_1 being the ground-truth segmentation and L_2 the automatic segmentation, DOC are defined as $DOC = 2|L_1 \cap L_2| /(|L_1| + |L_2|)$. Furthermore, we also looked at the segmentation accuracy around the hip joint local areas which are important for our target clinical applications. For both left acetabulum (LA), right acetabulum (RA), left femoral head (LFH), and right femoral head (RFH), the ASD between automatic segmentation and ground-truth segmentation are computed respectively after each stage of the present method.

Table 1 represents the quantitative evaluation results for ASD and DOC after each stage of the present method, respectively. After the atlas-registration based segmentation, a mean ASD of 0.58 ± 0.11, 0.63 ± 0.21 and 0.58 ± 0.12 mm was found for the pelvis, the left femur and the right femur, respectively, when the extracted surface models was compared to the associated manually segmented surface models. The segmentation results are improved after the second stage of graph optimization based multi-surface detection. The mean ASD of the pelvis, the left femur and the right femur are improved to 0.55 ± 0.09, 0.54 ± 0.18 and 0.50 ± 0.13 mm. Similarly, The present method achieved a mean DOC of 93.54 ± 0.99, 95.67 ± 1.08, and 95.94 ± 0.70 % for pelvis, left femur and right femur, respectively.

Table 2 presents the local quantitative evaluation results of ASD for acetabulum and femoral head regions after each stage of the present method, respectively. It is

Table 2 ASD (mm) between automatic and ground-truth segmentation of the bilateral hip joints from 10 CT data

Bone	Stage	CT 1	CT 2	CT 3	CT 4	CT 5	CT 6	CT 7	CT 8	CT 9	CT 10	Average
LA	MA	0.42	0.30	0.24	0.24	0.26	0.30	0.46	0.40	0.29	0.35	0.33
	GC	0.20	0.13	0.16	0.16	0.19	0.19	0.24	0.17	0.14	0.20	0.18
LFH	MA	0.40	0.56	0.41	0.34	0.51	0.36	0.51	0.81	0.38	0.49	0.48
	GC	0.19	0.15	0.21	0.16	0.20	0.25	0.25	0.19	0.15	0.19	0.19
RA	MA	0.43	0.36	0.22	0.33	0.24	0.30	0.48	0.31	0.25	0.32	0.32
	GC	0.20	0.22	0.15	0.14	0.20	0.16	0.71	0.15	0.15	0.16	0.22
RFH	MA	0.40	0.42	0.54	0.29	0.34	0.39	0.55	0.52	0.43	0.53	0.44
	GC	0.20	0.21	0.20	0.17	0.17	0.17	0.98	0.20	0.22	0.18	0.27

Results after stage I (multi-atlas-based segmentation: MA) and after stage II (graph optimization-based surface detection: GO) are shown, where LA stands for the left acetabulum, LFH for the left femoral head, RA for the right acetabulum and RFH for the right femoral head

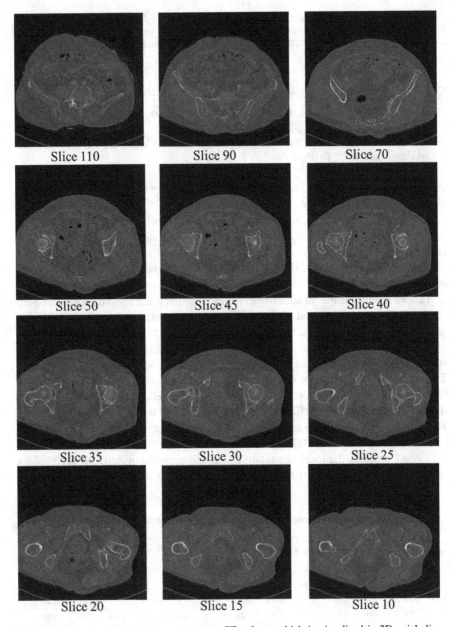

Fig. 7 The segmentation result on one target CT volume which is visualized in 2D axial slice. Please note in the last slice ("Slice 10") there is no segmentation for right femur due to the limited length of selected atlases when comparing to target volume

observed from Table 2 that a mean ASD of 0.33, 0.48, 0.32, and 0.44 mm was achieved for the left acetabulum (LA), right acetabulum (RA), left femoral head (LFH), and right femoral head (RFH) after the atlas-registration based segmentation. The mean ASD are further improved to 0.18, 0.19, 0.22, 0.27 mm after the graph optimization based surface detection, which is the reason for the improvement of the segmentation accuracy for the pelvis and the femurs which can be observed from Table 1.

In Figs. 7 and 8, we visually compare the ground-truth segmentation of a given target image with the result obtained from the present method. Figure 8c, f show the color-coded error distributions of the segmented pelvic model and the right femoral model, respectively. It can be seen that overall the segmentation error is small especially in hip joint areas (see Fig. 8c, f).

We checked whether the present method could preserve the hip joint space and prevent the penetration of the extracted surface models. For all the 20 hip joints that were segmented with the present method, we have consistently found that the hip joint spaces were preserved and that there was no penetration between the extracted adjacent surface models. Figure 9 shows the qualitative results in the defined hip joint area of the present method. It can be observed that penetrated two surfaces of the actabulum and femoral head are successfully recovered using present method. A hip joint structure is accurately reconstructed and there is no penetration between

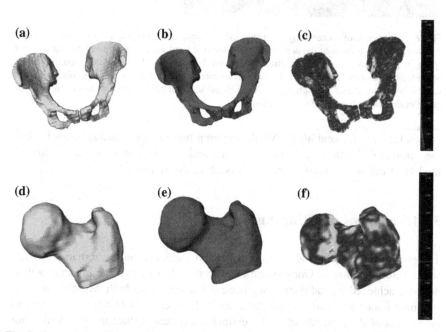

Fig. 8 Comparison of the results obtained by the present method to ground-truth segmentation of a given target image. **a, d** Automatic segmentation: derived pelvic (**a**) and femoral (**d**) models. **b, e** Ground-truth segmentation: derived pelvic (**b**) and femoral (**e**) models. Color-coded error distributions of the automatically segmented pelvic surface model (**c**) and proximal femoral model (**f**) when compared to associated models derived from ground-truth segmentation

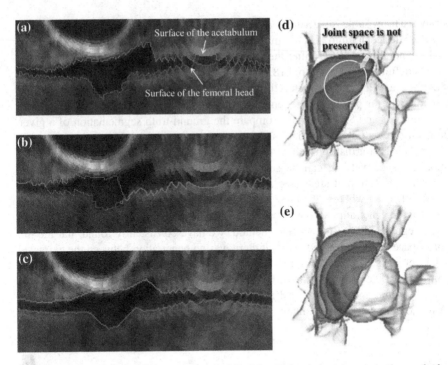

Fig. 9 An example of successfully reconstructed hip joint. Using the graph optimization method, the attached two surfaces of the acetabulum and femoral head are correctly detected and the joint space are preserved to avoid the penetration between the surface of the acetabulum and the surface of the femoral head. **a** Ground truth. **b** Multi-atlas based segmentation. **c** Graph optimization based surface detection. **d** Hip joint after multi-atlas based segmentation. **e** Reconstructed hip joint by graph optimization based surface detection

acetabulum and femoral heads. While mapping the recovered two surfaces back to the original CT image, a decreased segmentation error can be achieved for the models of the pelvis and femur, which is observed in Table 1.

5 Discussions and Conclusion

The goal of the present study is to develop and validate a fully automatic hip joint segmentation approach. Our experimental results showed that the present method not only achieved a good overall segmentation accuracy for both the pelvis and the proximal femur, but also had the advantages of preservation of hip joint space and prevention of the penetration of the extracted adjacent surface models, which are prerequisite conditions to use the segmented models for computer assisted diagnosis and planning for orthopedic surgeries.

Preservation of joint space for the local hip joint region is the main challenge which has been successfully addressed by the present method. Our method achieves

Table 3 Comparison of the results achieved by the present method with those reported in the literature

Method	Preserving hip joint average	SD (mm)	Average DOC (%)
Zoroofi et al. [2]	No	0.91	93.9
Cheng et al. [3]	Yes	0.86	94.4
Lamecker et al. [4]	No		–
Semi et al. [5]	No	0.70	–
Kainmueller et al. [6]	Yes	0.60	–
Yokota et al. [7]	No	1.10	92.7
Yokota et al. [8]	No	0.98	–
The present method	Yes	0.53	95.1

an ASD of 0.20 mm in the local acetabulum region, which is significantly improved compared to the result of 0.80 mm which was reported in [6]. The qualitative results of our study demonstrated that even the hip joint space is very narrow between surfaces of the acetabulum and the femoral head, these two surfaces can be successfully separated by performing optimal surface detection.

The performance of the present method is also compared with those of the state-of-the-art hip CT segmentation methods [2–8]. The comparison results are summarized in Table 3. It is worth to note that due to the fact that different datasets are used in evaluation of different methods, direct comparison of different methods is difficult. Thus, the comparison results in Table 3 should be interpreted cautiously. Nevertheless, as shown in this table, one can see that the performance of the present method is comparable to other state-of-the-art hip CT segmentation methods [2–8].

In conclusion, we presented a fully automatic and accurate method for segmenting CT images of a hip joint. The strength of the present method lies in the combination of a multi-atlas-based hip CT segmentation with a graph optimization-based multi-surface detection. The present method can be extended to segment CT data of other anatomical structures.

References

1. Kang Y, Engelke K, Kalender A (2003) A new accurate and precise 3D segmentation method for skeletal structures in volumetric CT data. IEEE Trans Med Imaging 22(5):586–598
2. Zoroofi RA, Sato Y, Sasama T, Nishii T, Sugano N, Yonenobu K, Yoshikawa H, Ochi T, Tamura S (2003) Automated segmentation of acetabulum and femoral head from 3-D CT images. IEEE Trans Inf Technol Biomed 7(4):329–343
3. Cheng Y, Zhou S, Wang Y, Guo C, Bai J, Tamura S (2013) Automatic segmentation technique for acetabulum and femoral head in CT images. Pattern Recogn 46(11):2969–2984
4. Lamecker H, SeebaÄŸ M, Hege HC, Deuflhard P (2004) A 3D statistical shape model of the pelvic bone for segmentation. In: SPIE 2004, vol. 5370, pp 1341–1351
5. Seim H, Kainmueller D, Heller M, Lamecker H, Zachow S, Hege HC (2008) Automatic segmentation of the pelvic bones from CT data based on a statistical shape model. In: VCBM 2008, pp 93–100

6. Kainmueller D, Lamecker H, Zachow S, Hege HC (2009) An articulated statistical shape model for accurate hip joint segmentation. In: IEEE EMBC, pp 6345–6351

7. Yokota F, Okada T, Takao M, Sugano S, Tada Y, Sato Y (2009) Automated segmentation of the femur and pelvis from 3D CT data of diseased hip using hierarchical statistical shape model of joint structure. In: MICCAI 2009, part II, pp 811–818

8. Yokota F, Okada T, Takao M, Sugano S, Tada Y, Tomiyama N, Sato Y (2013) Automated CT segmentation of diseased hip using hierarchical and conditional statistical shape models. In: MICCAI 2013, part II, pp 190–197

9. Ehrhardte J, Handels H, Plotz W, Poppl SJ (2004) Atlas-based recognition of anatomical structures and landmarks and the automatic computation of orthopedic parameters. Methods Inf Med 43(3):391–397

10. Pettersson J, Knutsson H, Borga M (2006) Automatic hip bone segmentation using non-rigid registration. In: ICPR 2006, pp 946–949

11. Xia Y, Fripp J, Chandra SS, Schwarz R, Engstrom C, Crozier S (2013) Automated bone segmentation from large field of view 3D MR images of the hip joint. Phys Med Biol 58:7375–7390

12. Li K, Wu X, Chen DZ, Sonka M (2006) Optimal surface segmentation in volumetric images A graph-theoretic approach. IEEE Trans Pattern Anal Mach Intell 28(1):119–134

13. Song Q, Wu X, Liu Y, Smith M, Buatti J, Sonka M (2009) Optimal graph search segmentation using arc-weighted graph for simultaneous surface detection of bladder and prostate. In: Proceedings of MICCAI, vol 5762, pp 827–835

14. Thirion JP (1998) Image matching as a diffusion process: an analogy with maxwell™s demons. Med Image Anal 2:243–260

15. Knutsson H, Andersson M (2005) Morphons: segmentation using elastic canvas and paint on priors. In: IEEE international conference on image processing (ICIP05), Genova, Italy, Sep 2005

16. Breiman L (2001) Random forests. Mach Learn 45(1):5–32

17. Criminisi A, Shotton J, Robertson D, Konukoglu E (2010) Regression forests for efficient anatomy detection and localization in CT studies. MCV 2010:106–117

18. Lindner C, Thiagarajah S, Wilkinson JM, arcOGEN Consortium, Wallis G, Cootes TF (2013) Fully automatic segmentation of the proximal femur using random forest regression voting. IEEE Trans Med Imag 32(8):1462–1472

19. Chu C, Chen C, Liu L, Zheng G (2014) FACTS: fully automatic CT segmentation of a hip joint. Ann Biomed Eng. doi:10.1007/s10439-014-1176-4

20. Glocker B, Komodakis N, Tziritas G, Navab N, Paragios N (2008) Dense image registration through MRFs and efficient linear programming. Med Image Anal 12(6):731–741

21. Boykov Y, Jolly MP (2000) Interactive organ segmentation using graph cuts. In: Proceedings of medical image computing and computer-assisted intervention (MICCAI), pp 276–286

22. Chu C, Oda M, Kitasaka T, Misawa K, Fujiwara M, Hayashi Y, Nimura Y, Rueckert D, Mori K (2013) Multi-organ segmentation based on spatially-divided probabilistic atlas from 3D abdominal CT images. In: MICCAI 2013, part II, pp 165–172

23. Liu L, Ecker T, Schumann S, Siebenrock K, Nolte L, Zheng G (2014) Computer assisted planning and navigation of periacetabular osteotomy with range of motion optimization. In: MICCAI 2014, vol 8674, pp 643–650

24. Kolmogorov V, Zabih R (2004) What energy functions can be minimized via graph cuts? IEEE Trans PAMI 26:147–159

Quantification of Implant Osseointegration by Means of a Reconstruction Algorithm on Micro-computed Tomography Images

R. Bieck, C. Zietz, C. Gabler and R. Bader

Abstract One of the most common methods to derive both qualitative and quantitative data to evaluate osseointegration of implants represents histomorphometry. However, this method is time-consuming, destructive, cost-intensive and the two-dimensional (2D) results are only based on one or a few sections of the bone-implant interface. In contrast, micro-computed tomography (μCT) imaging produces three-dimensional (3D) data sets in short time. The present work describes a new image reconstruction algorithm to calculate the effective bone-implant-interface area (eBIIA) by means of μCT. The reconstruction algorithm is based on a series of image processing steps followed by a voxel-boundary-conditioned surface reconstruction. The analysis of the implant-bone interface with μCT is suitable as a non-destructive and accurate method for 3D imaging of the entire bone-implant interface. Despite its limitations in metallic specimen (streak artefacts), μCT imaging is a valuable technique to evaluate the osseointegration of titanium implants.

1 Introduction

For the development of biocompatible bone implants with a high degree of functionality and optimal integrational behaviour, the tissue reactions at the bone-implant-interface (BII) are critical [1]. An established procedure is the quantification of integration into osseous tissue. This osseointegration of an implant is

R. Bieck · C. Zietz · C. Gabler · R. Bader
Department of Orthopaedics, Department of Orthopaedics, Biomechanics and Implant
Technology Research Laboratory, Doberaner Straße 142, 18057 Rostock, Germany

R. Bieck (✉)
Innovation Center Computer Assisted Surgery, Faculty of Medicine, University Leipzig,
Semmelweisstraße 14 D, 04103 Leipzig, Germany
e-mail: Richard.bieck@medizin.uni-leipzig.de

© Springer International Publishing Switzerland 2016
G. Zheng and S. Li (eds.), *Computational Radiology*
for Orthopaedic Interventions, Lecture Notes in Computational
Vision and Biomechanics 23, DOI 10.1007/978-3-319-23482-3_6

conservatively measured by histomorphometric measurements [2, 3]. This long-time gold standard however, is highly invasive, cost-intensive and results in the loss of information involving the unaltered situation of the BII [1, 2]. Furthermore, since the nature of histomorphometry is strictly two-dimensional (2D), complex three-dimensional (3D) bone structures are insufficiently or not at all represented. With the introduction of high resolution imaging technologies such as micro-computed CT (μCT) and powerful reconstruction algorithms the observation of BII surface models in the range of micro millimetres is possible. As a consequence of this development a new definition of osseointegration quality is needed. Furthermore, focus lies on the reconstruction algorithms accuracy, error-proneness, repeatability and significance of the generated images and models [4]. The aim of this work was to introduce an innovate reconstruction algorithm for BII surface models with resolution in the lower μm-realm. For that implant specimen with different surface properties were implanted in rat femura and μCT images acquired. Implants were coated with plasma-polymerized ethylenediamine (PPEDA) and plasma-polymerized allylamine (PPAAm) and compared with a non-coated control group regarding bone-implant-interface-ratio. Pre-processing of images and three-dimensional reconstruction was performed with segmentation software AMIRA (Vers.5.3.3), post-processing with 3D-software GEOMAGIC. With the developed surface reconstruction algorithm the effective contact area of the BII was calculated and then quantitatively compared with histomorphometric slice of the same specimen.

1.1 Micro-CT and Artifacts

Micro-computed computer tomography (μCT) achieves resolutions of 5–50 μm and varies in its build-up from conventional CT setups [4–6]. The main reason is that an improved sub-millimetre resolution is not achievable with standard focus sizes.

Instead μCT systems use either synchrotron sources or micro-focus X-ray tubes with specialized transmission anodes. In these setups the geometry of source and detector defines the highest achievable spatial resolution. In standard μCT systems the specimen are rotated while an X-ray source sends out a beam through them onto a CCD detector (Fig. 1) [4]. For practical reasons the source dimensions should be in the range of the structure to analyse [7]. Another limiting factor is the apparent X-ray beam intensity, which itself is dependent on the tube power and ultimately on the focus size. As a consequence of smaller foci specimen should preferably be in a specified range [7]:

$$D/A = 1000 \tag{1}$$

where in (1) D stands for the specimen diameter and A for the intended spatial resolution.

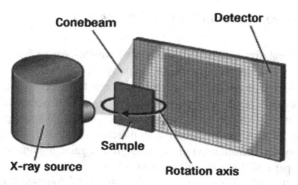

Fig. 1 Standard μCT system components (*source* http://electroiq.com/blog/2011/03/3d-ct-x-ray-imaging-fills-inspection-gaps-says-xradia/)

As in standard CT imaging μCT is prone to a series of artifact. Image acquisition of biological tissue and metal implants generally suffers from so-called "Beam Hardening" [5, 7, 8]. This effect appears when materials with high density difference are imaged with a polychromatic X-ray source. In a polychromatic source the emitted photons have varying energy niveaus. Less energetic photons are easily absorbed in denser materials, while high energy photons travel through the materials. The emerging beam is "hardened" to consist only of high energy photons. If the beam passes a less dense material after hardening, the high energy photons are insignificantly absorbed, resulting in bad resolution along the path of the hardened beam. In images beam hardening results in "cupping"-artifacts such as "streaks" and "dark bands" (Fig. 2). A way to compensate this behaviour is to filter the beam before entering the specimen [8]. Standard filters against beam hardening are aluminium and copper and respectively, combinations of the two, with filter thicknesses between 0.2 and 1.0 mm [7, 8].

Fig. 2 Streaking artifacts (*black lines*) next to the edges of a titanium implant (*white*) integrated into a bone part

1.2   Osseointegration and Surface Properties

For sufficient durability and biocompatibility of weight-bearing bone implants the osseointegration is a determining factor. Initially introduced by *Brânemark* this concept described a procedure to measure the force fit of dental implants in jaw bone [1, 3]. Osseointegration generally describes the direct structural and functional connection between living bone tissue and surface of weight-bearing implant [9]. With growing implant specialization and therefore extending requirements for biocompatibility the measurement of osseointegration gains importance [3, 10, 11]. An implant is per definition osseointegrated when there is no progressive relative movement at the BII. Therefore, from the aspect of long-term compatibility osseointegration stands for an anchoring behavior of implants to be fully functional under system conditions (living tissue) without negative influence on biological reactions [11].

Information about osseointegration using histomorphometric analysis was only attainable by means of trabecular thickness, distance and number. This however, was only usable as a priori knowledge to describe the possible 3D tissue structure. Since the underlying algorithms are based on probabilistic models there is only limited qualitative and even less quantitative data gained from BII models from histomorphometry. The need for a 3D analysis standard is stated in various studies lead to the focus on imaging procedures an 3D tissue reconstruction algorithms and software [10, 12]. As a consequence the evaluation of the BII with modifiable implant surface parameters became an important research area. Customized surface coatings were developed which induce specific biological reactions to support or deny implant integration into tissue. This surface biocompatibility enables implant integration almost independently from the main implant material. The quantitative evaluation of the implant integration by 3D analysis of its BII is an important way to implant optimization and functionalization.

1.3   Medical Image Processing

The processing of clinically relevant medical images is mainly used in diagnose- and treatment-related tasks. In general image processing steps are applied to enhance the visualization of anatomical structures and to detect perceptible pathological symptoms.

Medical image processing technology comprises image registration, segmentation, analysis, pattern detection, 3D-visualization and virtual & augmented reality [13–15].

A crucial part of this study is the processing pipeline for 3D-visualization of medical imaging data (Fig. 3). Depending on the image quality the pipeline has a pre- and post- processing working around segmentation and reconstruction steps.

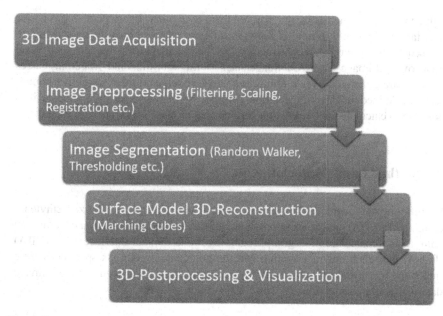

Fig. 3 Image processing pipeline steps

Image preprocessing is used to increase the image information and quality and directly dependent on the study requirements and image acquisition quality. Contents of this step are noise reduction, artifact reduction, contrast enhancement, scaling and resampling. These filtering operations are performed locally pixel- or voxel- wise with so-called "masks". Masks are predefined fields of pixels to be processed on in one step. Depending on the filtering mask dimensions are specified with a "kernel size", where a filtering mask with a kernel of 5 is comprised of a pixel field of 5 pixel height and width. Common filters are used to smooth the grey values of an image with each pixels receiving the mean grey value of all pixels in the specified mask.

1.4 Segmentation and 3D-Reconstruction

The segmentation process is used to label pixels with similar information content and group them in distinct classes of material. This step is used for classification purposes and for pattern recognition. The aim is to raise the image calculation from strictly pixel-related computations to symbol- and object-oriented processing [16, 17]. Segmentation methods range from point-oriented methods (global thresholding) over edge- and contour-oriented methods (random walker, snakes algorithm) to region-based methods (region growing). The global thresholding uses the information stored in the image histograms to separate pixel areas with the same gray value that can be quantitatively separated from each other. Every pixel that is above a certain gray value threshold is added to different segment then the ones below it. Subsequently, a surface reconstruction can be calculated on the separated pixel areas over all images in a μCT image stack. With algorithms like Marching-Cubes surfaces are generated over a finite number of voxels and their respective boundaries as defined in the segmentation process. With two adjacent segments it is possible to calculate the intersecting surface area of these segments with each other.

2 Methods and Material

Sample implants made of TiAlV were coated with plasma-polymerized ethylenediamine (PPEDA) and plasma-polymerized allylamine (PPAAm) and compared with a non-coated control group regarding bone-implant-interface area (BIIA) (Fig. 4). For each polymer coating and the non-coated group six specimens were implanted into distal rat femur. After 5 weeks specimens were removed and image data was acquired individually with a μCT-Setup.

2.1 Image Data Acquisition and Preprocessing

Image data of the specimen were acquired individually with a micro-computer-tomograph (μCT) Nanotom 180nF (phoenix nanotom®, GE Measurement and Control solutions, phoenix|x-ray, Germany). A molybdenum target was used for X-ray beam generation. Voltage and current were set to 70 kV and 135 μA, to reach the optimum contrast. The μCT X-ray source used a cone-beam with vertical specimen alignment. The samples rotated 360° in 0.75° steps. Each step included three 2D images resulting in 1440 2D acquired images for one sample. The region-of-interest (ROI) for the 2D images was set in the range of the implant

Fig. 4 Titanium implant (*left*), reconstructed surface models of implant (*green*) and surrounding bone (*orange*) (*right*) (Color figure online)

surface with approximately 1 mm surrounding bone tissue. Due to anatomical differences of the tibiae, the distance of the samples to the X-ray-tube and the detector, consequently varying the magnification and voxel size. Each specimen 2D data records included 7 GB of data volume and were used for generation of the 3D volume of the sample. The reconstruction of CT data (composing X-ray 2D images to a 3D volume) was performed with datos|x-reconstruction (GE, Germany). Titanium implants are harder to penetrate by X-rays in comparison to bone. Therefore, a beam-hardening correction of 6.7 was used to compensate for the inhomogeneous reconstructed volume of the implant and reach similar grey values inside the entire implant. For further processing, the transformation of the volume data in the DICOM data was required. Due to limited RAM capacity of the PC working station the data volume of each set had to be resampled to reduce the amount of processed data. Datasets were resampled with a Box-filter (2 × 2 × 2-kernel) reducing data volume by factor 8 and voxel resolution by factor 2 while preserving spatial voxel proportions. Subsequently datasets were filtered with an Unsharp-Masking algorithm (x-y-planes, 5×-kernels, Sharpness 0.6) to improve image quality for the following segmentation process. Additionally datasets were cropped to reduce data volume and approximate the volume of interest (VOI).

2.2 Image Data Segmentation and Reconstruction

The BIIA was calculated with a semi-automatically processed segmentation algorithm. Each DICOM dataset had a maximum voxel edge resolution of 8–10 μm after resampling. The segmentation algorithm was executed on a workstation (Intel Quad Core Q9400 2.66 GHz, 2 GB RAM) with the segmentation software AMIRA® (Vers. 5.4.1) and consisted of the steps image data preprocessing, segmentation and surface postprocessing. The segmentation algorithm comprised histogram-oriented threshold detection of voxels with same greyscale intensities followed by inspection and semi-automatic editing of the segmented areas in each slice of a dataset (Fig. 5, left). For all slices corrections at the transition from implant to bone were achieved with morphological opening and island removal. Due to limited image quality separation of less dense bone structures and background areas was difficult. Errors occurring at the transition from bone to background areas where due to image acquisition artifacts (black edges, structure blurring) and where corrected manually.

Following the segmentation surface models were computed with a voxel-conserving algorithm matching the exact voxel boundaries of each segmented material (Fig. 5, right). Since the surface information corresponds to the original number of triangles at the interface of two adjacent materials surfaces, any sort of smoothing algorithm or operation had to be avoided.

From the reconstructed surface models two surface area patches were extracted and the exact number of triangles of each patch computed. The first patch was the

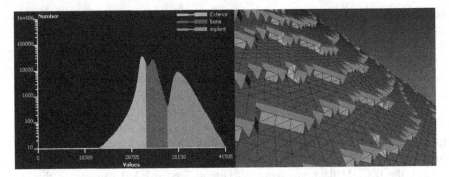

Fig. 5 Image stack histogram with separation of intensities into exterior, bone and implant (*left*), difference between voxel boundary surface reconstruction (*full, blue*) and smoothed surface reconstruction (*triangles, wireframe*) (*right*) (Color figure online)

possible maximum implant-bone-interface area (mBIIA), reduced to a shell area of the implant by subtracting materials at the base and top surface. The second patch was the effective bone-implant-interface area (eBIIA) achieved for each implant (Fig. 6). With the voxel spacing the triangles of each patch were converted into the corresponding real surface in mm^2. As a consequence of the voxel-conserving process the mBIIA and the eBIIA were larger than the real implant shell area (rISA). This was assumed to be caused by Aliasing when a surface curvature is approximated by voxels. Therefor a percental value for the contact area (pVA) was calculated by dividing the eBIIA by the mBIIA. At the moment quantitated results with this algorithm are not possible.

2.3 Surface Models Postprocessing

For some surface models the mBIIA was reduced by adverse implant positioning during implantation. Implant shell areas were partially outside of the bone and had

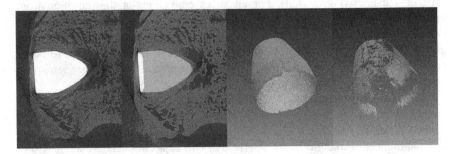

Fig. 6 Image segmentation and reconstruction process (F.l.t.r): resampled µCT image data, bone (*orange*) and implant (*blue*) segmentation, maximum bone-implant-interface area (*Shell Area*) (mBIIA), effective bone-implant-interface area (eBIIA) (Color figure online)

Fig. 7 Cropping of the eBIIA (*left*) and mBIIA (*right*) due to adverse implant positioning during implantation

to be excluded from the segmentation. The exclusion was achieved by cropping the corresponding shell areas in the surface models. The cropping was executed in Geomagic (Geomagic Gmbh). This step automatically resulted in a reduction of the mBIIA and had to be corrected in the calculation of the sVA. Cropping resulted in an improved sVA while cropped implant volume never exceeded 20 % of the original implant volume (Fig. 7).

2.4 Statistical Analysis

The statistical analysis was performed using SPSS Statistics 20 (IBM, Armonk, NY, USA). The statistical data includes mean and standard deviations. All values were proven to have homogeneity of variance (Levene test) before pair-wise comparisons within the independent groups were performed. The Mann-Whitney-U test was used to compare the values of the groups because of the low number of the sample. Values of $p < 0.05$ were considered as statistically significant.

3 Results

For two coatings and one control group six specimen were imaged and the eBIIA reconstructed and normalized in relation to the mBIIA to calculate a standardized value for the bone-implant-interface area (sBIIA) was calculated by dividing the eBIIA by the mBIIA. The calculated sBIIA were compared with histomorphometric measurements for every coating and the control group. At the moment quantitated

Table 1 Results for the bone-implant-interface area for histomorphometry and μCT reconstruction

Surface coating	N	SBIIA ± SD (%) histomorphometry	SBIIA ± SD (%) μCT
Uncoated	5	32.4 ± 27.9	51.3 ± 11.6
PPEDA	6	53.5 ± 19.2	62.0 ± 9.6
PPAAm	4	45.7 ± 22.9	51.8 ± 13.3
Overall	15	44.4 ± 23.5	55.7 ± 11.7

results with this algorithm are not possible. Compared to the uncoated implants, the PPAAm coated implants showed a slight, non-significant increase in sBIIA in μCT ($p = 0.905$) and in histomorphometry ($p = 0.730$). Implants with the PPEDA coating revealed a clear but not significant increase in sBIIA evaluated by μCT ($p = 0.329$) and histomorphometry ($p = 0.126$). Between PPAAm and PPEDA, no significant difference (μCT: $p = 0.257$, histomorphometry: $p = 0.762$) was found. There was an obvious decrease in standard deviation (SD) when sBIIA was evaluated by μCT. Overall, a high correlation coefficient of 0.70 ($p < 0.002$) was found between 3D and 2D quantification of sBIIA (Table 1).

4 Discussion

The quantitative determination of the bone-implant-interface behavior is essential to assess an orthopaedic implants potential for optimization, which aims at improving osseointegration. Although 2D histomorphometry is time consuming, destructive and cost-intensive and 3D incompatible, it is still one of the most commonly conducted methods to qualify bone morphology and to quantify the osseointegration of implants [1, 2]. Compared to histological evaluation, μCT imaging is fast, non-destructive and offers 3D data sets. Therefore, this technology has become of increasing interest in recent years. In the present study, the evaluation of sBIIA by means of μCT was compared with the results evaluated by 2D histomorphometry [2, 12]. The results demonstrate a nearly 11 % higher mean sBIIA with decreased standard deviation calculated by μCT as compared to histomorphometry [6, 7].

Limited picture quality due to beam hardening artefacts at the margin of the implant led to a loss of information and impeded the image reconstruction. Therefore, separation of less dense bone structures and background areas was difficult. Beam hardening artefacts resulted from the use of polychromatic X-ray spectra and the inhomogeneous sample [5]. Hence, beam-hardening correction was used to compensate for the inhomogeneous reconstructed volume of the implants. For further evaluations, beam filtration, e.g. with aluminum or copper, can be used to minimize these artefacts.

Due to limited RAM capacity, the data volume of each set had to be resampled in order to reduce the amount of processed data. The used Box-filter with twofold resampling resulted in a moderate reduction of the storage requirements at sufficiently well recognizable interface details and acceptable resolution. However, even with the use of this filter, the disadvantage factor contributing to the processing time of the data sets was up to 12 h. Nevertheless, reduction of the storage requirements due to resampling and, therefore, a reduced voxel resolution involves the risk of negative influence on the evaluated effective sBIIA. Unsharp- Masking filtering was used to improve data visualization of relevant image structures for the following segmentation process, while achieving sufficient noise suppression and minimal loss of voxel information. The algorithm (5×-kernels, sharpness 0.6) offered a compromise between noise rejection and detail preservation. As a result, the darkest bone constituents were barely visible, but a good sharpening was still achieved. The filter has in large kernels (from 7 × 7) high-pass character, which is associated with the loss of voxel value lower intensities and could potentially lead to a loss of information in the bone area. The implants used in the present study were designed to evaluate osseointegration and implant-associated infections in the tibial metaphysis of rats. Due to the implant design and the implantation procedure, the shell area of the implant had direct bone contact and was therefore, the only analyzed implant area.

In the histomorphometric evaluation, only one slice per implant could be used to determine the sBIIA. The selection of the section plane is an important and influential factor in histomorphometric evaluations. We assume that the low sample size could be one of the main reasons why significant differences between the surface modifications were not found in contrast to previous evaluations.

References

1. Branemark P (1983) Osseointegration and its experimental background. J Prosthet Dent 50 (3):399–410
2. Parfitt AM (1988) Bone histomorphometry: standardization of nomenclature, symbols and units (summary of proposed system). Bone 9:67–69
3. Terheyden H, Lang NP, Bierbaum S, Stadlinger B (2012) Osseointegration—communication of cells. Clin Oral Impl Res 23(10):1127–1135
4. Kuhn JL, Goldstein SA, Feldkamp LA, Goulet RW, Jesion G (1990) Evaluation of a microcomputed tomography system to study trabecular bone structure. J Orthop Res 8:833–842
5. Barrett JF, Keat N (2004) Artifacts in CT: recognition and avoidance 1. Radiographics 24 (6):1679–1691
6. Oest ME, Jones JC, Hatfield C, Prater MR (2008) Micro-CT evaluation of murine fetal skeletal development yields greater morphometric precision over traditional clear-staining methods. Birth Defect Res B 83(6):582–589
7. Engelke K, Karolczak M, Lutz A, Seibert U, Schaller S, Kalender W (1999) Mikro-CT Technologie und Applikationen zur Erfassung von Knochenarchitektur. Der Radiol 39 (3):203–212
8. Meganck JA, Kozloff KM, Thornton MM, Broski SM, Goldstein SA (2009) Beam hardening artifacts in micro-computed tomography scanning can be reduced by X-ray beam filtration and the resulting images can be used to accurately measure BMD. Bone 45(6):1104–1116

9. Bouxsein ML, Boyd SK, Christiansen BA, Guldberg RE, Jepsen KJ, Müller R (2010) Guidelines for assessment of bone microstructure in rodents using micro-computed tomography. J Bone Miner Res 25(7):1468–1486

10. Ko C, Lim D, Choi B, Li J, Kim H (2010) Suggestion of new methodology for evaluation of osseointegration between implant and bone based on μ-CT images. Int J Precis Eng Manuf 11 (5):785–790

11. Mavrogenis AF, Dimitriou R, Parvizi J, Babis G (2009) Biology of implant osseointegration. J Musculoskelet Neuronal Interact 9(2):61–71

12. Akagawa Y, Wadamoto M, Sato Y, Tsuru H (1992) The three-dimensional bone interface of an osseointegrated implant: A method for study. J Prosthet Dent 68(5):813–816

13. Handels H (2009) Medizinische Bildverarbeitung: Bildanalyse, Mustererkennung und Visualisierung für die computergestützte ärztliche Diagnostik und Therapie. ISBN: 978-3-8351-0077-0

14. Humphrey D, Taubman D (2011) A filtering approach to edge preserving MAP estimation of images. IEEE Trans Image Process 20(5):1234–1248

15. Mühlbauer UW (2010) Computergestützte 3D-Visualisierung histologischer Schnittbildserien am Beispiel des bovinen Mesonephros. Dissertation, LMU München: Tierärztliche Fakultät

16. Baril E, Lefebvre LP, Hacking SA (2011) Direct visualization and quantification of bone growth into porous titanium implants using micro computed tomography. J Mater Sci Mater Med 22(5):1321–1332

17. Zachow S, Zilske M, Hege HC (2007) 3D reconstruction of complex anatomy from medical image data: segmentation and geometry processing ZIB-Report 07–41

Surgical Treatment of Long-Bone Deformities: 3D Preoperative Planning and Patient-Specific Instrumentation

Philipp Fürnstahl, Andreas Schweizer, Matthias Graf, Lazaros Vlachopoulos, Sandro Fucentese, Stephan Wirth, Ladislav Nagy, Gabor Szekely and Orcun Goksel

Abstract Congenital or posttraumatic bone deformity may lead to reduced range of motion, joint instability, pain, and osteoarthritis. The conventional joint-preserving therapy for such deformities is corrective *osteotomy*—the anatomical reduction or realignment of bones with fixation. In this procedure, the bone is cut and its fragments are correctly realigned and stabilized with an implant to secure their position during bone healing. Corrective osteotomy is an elective procedure scheduled in advance, providing sufficient time for careful diagnosis and operation planning. Accordingly, computer-based methods have become very popular for its preoperative planning. These methods can improve precision not only by enabling the surgeon to quantify deformities and to simulate the intervention preoperatively in three dimensions, but also by generating a surgical plan of the required correction. However, generation of complex surgical plans is still a major challenge, requiring sophisticated techniques and profound clinical expertise. In addition to preoperative planning, computer-based approaches can also be used to support surgeons during the course of interventions. In particular, since recent advances in additive manufacturing technology have enabled cost-effective production of patient- and intervention-specific osteotomy instruments, customized interventions can thus be planned for and performed using such instruments. In this chapter, state of the art and future perspectives of computer-assisted deformity-correction surgery of the upper and lower extremities are presented. We elaborate on the benefits and pitfalls of different approaches based on our own experience in treating over 150 patients with three-dimensional preoperative planning and patient-specific instrumentation.

P. Fürnstahl · A. Schweizer · M. Graf · L. Vlachopoulos
Computer Assisted Research and Development Group, University Hospital Balgrist,
University of Zurich, Zurich, Switzerland

A. Schweizer · S. Fucentese · S. Wirth · L. Nagy
Department of Orthopedic Surgery, University Hospital Balgrist, University of Zurich,
Zurich, Switzerland

G. Szekely · O. Goksel (✉)
Computer Vision Laboratory, ETH Zurich, Zurich, Switzerland
e-mail: ogoksel@ethz.ch

© Springer International Publishing Switzerland 2016
G. Zheng and S. Li (eds.), *Computational Radiology
for Orthopaedic Interventions*, Lecture Notes in Computational
Vision and Biomechanics 23, DOI 10.1007/978-3-319-23482-3_7

1 Background

Bone and joint disorders are a leading cause of physical disability worldwide and account for 50 % of all chronical diseases in people over 50 years of age [1]. Among these, the degeneration of cartilage (i.e., osteoarthritis) is the most common disease [2], affecting 10 % of men and 18 % of women aged above 60 years [3]. The population is ageing thanks to increased life expectancy; accordingly, osteoarthritis is anticipated to become the fourth leading cause of disability by 2020 [3]. One reason of early arthritis development is abnormal joint loading induced by bone deformity [4, 5]; i.e., the shape of the bone is anatomically malformed. Reasons for bone deformity may be congenital, caused by a birth defect, or posttraumatic if the bone fragments heal in an anatomically incorrect configuration after a fracture (i.e., malunion). Besides the development of osteoarthritis, bone deformity may result in severe pain, loss of function, and aesthetic problems [6]. Particularly, gross deformities can cause quasi-impingement of the bones or increased tension in ligaments, resulting in a limited range of the motion. An unrestricted function is of fundamental importance for performing daily activities such as eating, drinking, personal hygiene, and job-related tasks. If not treated appropriately, salvage procedures such as *arthroplasty* (i.e., restoring the integrity and function of a joint such as by resurfacing the bones or a prosthesis) may be required.

Surgical treatment by corrective osteotomy has become the benchmark procedure for correcting severe deformities of long bones [6–9]. Particularly in young adults, a corrective osteotomy is indicated to avoid salvage procedures such as arthroplasty or fusion [10, 11]. In a corrective osteotomy, the pathological bone is cut and the resulting bone parts are correctly realigned (i.e., anatomical reduction), followed by stable fixation with an osteosynthesis implant to secure their position during bone healing. Treatment goal is the anatomy reconstruction—restoration of length, angular and rotational alignment, and displacement. However, osteotomy is technically challenging to manage operatively and hence it demands careful preoperative planning. It is necessary to assess the deformity as accurate as possible to achieve adequate restoration. Conventionally, the deformity is quantified based on comparison, either to the healthy contralateral side [6, 12] if available or to anatomical standard values otherwise [13].

Three types of deformities are frequently encountered in an isolated or combined form and different types of osteotomies are performed to correct them [13, 14]: *Angular deformity* causes non-anatomical angulation of the distal and proximal bone parts. In general, angulation can occur in any oblique plane between the sagittal and the anterior-posterior planes. Isolated angular deformities are surgically corrected by performing a wedge osteotomy. In a closing wedge osteotomy, the bone is separated with two cuts forming a wedge which, when removed and closed, will correct the deformity. In contrast, in an opening wedge osteotomy one cut is made and the bone parts are aligned, resulting in a wedge-shaped opening. If the resulting gap is too large, which may result in implant failure due to instability, it is filled with structural bone graft to support healing before the parts are fixated [15].

A *rotational deformity* is characterized by an excessive rotation or twist of the bone around its longitudinal axis. That is, the distal segment shows a non-anatomic rotation around its own axis while the proximal part is considered as fixed. Rotational deformities can be corrected by performing a derotation of the fragment, often within a single osteotomy plane. If a *translational deformity* is present, the pathological bone is longer or shorter than it anatomically should have been. Accordingly, a lengthening or shortening osteotomy must be performed.

In practice, most deformities are a combination of the three types of isolated deformities, which makes preoperative planning and the surgery very challenging. A special type of long bone deformities are *intraarticular malunions* [16, 17], i.e., non-anatomical healing of the bone after joint fracture, resulting in gaps or steps on the articular surface. Intra-articular malunions can cause severe damage to the cartilage, making surgical treatment often necessary although the intervention is risky.

Conventionally, deformity assessment relies on plain radiographs or single computed tomography (CT) slices. Angular deformities and length differences are manually measured on plain X-rays in antero-posterior and lateral projections [6, 7]. In CT or magnetic resonance imaging (MRI), rotational deformities are assessed using proximal and distal cross-sectional images to compare torsion between the sides. The problem of the conventional technique is the assumption that a multi-planar deformity can be corrected by subsequently correcting the deformities measured in the anterior-posterior, sagittal, and axial planes separately, which is indeed an invalid assumption [18]. Accordingly, Nagy et al. [6] proposed to calculate the true angle of deformity in 3D space from planar measurements. Although feasible for angular deformities, their approach does not consider the rotational and translational components of a deformity.

Due to the above limitations of traditional approaches, computer assisted planning in 3D has become popular in orthopedic surgery as it permits to quantify a deformity by all 6 degrees of freedom (DoF)—3 translations and 3 rotations. Additionally, in contrast to emergency orthopedic treatments, corrective osteotomies are elective procedures scheduled in advance, therefore providing ample time for computer assisted planning. Since the first 3D osteotomy planning systems for the upper [19, 20] and lower extremities [21] had been described in the 90s, numerous approaches have been proposed for the 3D preoperative planning of long-bone deformities, i.e., osteotomies of the forearm bones, humerus, femur, and tibia. In these approaches, the basis for preoperative planning is a 3D triangular surface model of the patient anatomy generated from X-ray, CT, or MRI data of the patient. Based on the extracted 3D model, a preoperative plan can then be created by quantifying the deformity in 3D, followed by simulating the realignment to the normal anatomy. Dependent on the pathology and anatomy either the healthy contralateral limb, a similar-sized bone template, or anatomical standard values (e.g., axes, distances, and angles) can be used to determine normal anatomy. From a technical point of view, the alignment process to the normal anatomy using registration algorithms received most attention because the optimization target is easy to quantify. Apart from the fact that automatic alignment often achieves poor results, other aspects, such as the position of the osteotomy, are more difficult to define in a standardized way as they do mainly rely on clinical parameters and the experience of the planner. However, all pre- and intra-operative aspects have to be jointly considered for developing a clinically feasible plan. Moreover, each deformity is unique and for this reason the surgical strategy differs from patient to patient, which complicates the development of standardized approaches.

Performing the surgery according to a complex preoperative plan can also be challenging. Missing anatomical reference points due to limited access to and view of the bone makes it almost impossible to perform a 6-DoF reduction without any supporting equipment. Therefore, navigation systems were proposed to execute the preoperative plan intra-operatively. Navigation systems have been used among different disciplines of orthopedic surgeries [22], particularly for bone realignment of the upper [23] and lower extremities [21, 24]. Although they offer superior accuracy [25], navigation systems have been losing popularity since they are costly and laborious to use. Originating from dental surgery [26] and thanks to recent advances in additive manufacturing, the production of patient-specific instruments has instead become a cost-effective and promising alternative to navigation systems [27]. Although the use of patient-specific guides for corrective osteotomies of long bones was described [11, 18, 28–34], the main application of patient-specific guides in orthopedic surgery is currently *arthroplasty* [35]. The key idea behind patient-specific guides is to design targeting devices specific to the anatomy and pathology of the patient for guiding the surgeon intra-operatively. Such patient-specific instruments are extremely versatile and, hence, particularly suited for the treatment of bone deformities because the surgical treatment varies between patients.

Within the last few years, we have successfully treated more than 150 patients by corrective osteotomy using a computer assisted approach. In this chapter, we report on the techniques developed and the experience gained in this field. First, in Sect. 2, the 3D preoperative planning process is described step-by-step, providing a guideline for performing this task in a standardized way. In Sect. 3, current patient-specific instruments required for performing different types of osteotomies are summarized. In Sect. 4, the treatment of intra-articular osteotomies using 3D planning and patient-specific instruments is demonstrated by means of a complex case. Lastly, our results on the accuracy and effort are given in Sect. 5 including a discussion of advantages and limitations of the presented techniques.

2 Pre-operative Planning and Surgical Technique

The most fundamental step of the computer-based planning is the quantification of the deformity in 3D. Dependent on the anatomy and pathology, we either use a template-based approach or an axis-based approach to determine the deformity and, subsequently, the required correction.

2.1 Template-Based Approach

Template-based approaches can be performed if the opposite limb is uninjured. In contrast to the lower limb, bilateral deformities are rarely observed for the upper extremities, making this approach particularly suited for the planning of deformities of the forearm (i.e., radius and ulna) and shoulder bones. As in the conventional method, computer assisted planning relies on the contralateral bone, which serves as a reconstruction template. The input data for the planning tool are 3D models generated from CT scans of the posttraumatic bone as well as the contralateral (healthy) bone. The segmentation is performed in a semi-automatic fashion using thresholding and region-growing algorithms of commercial segmentation software (Mimics, Materialise, Loewen, Belgium). Triangular surface meshes are generated from segmented CT scans using the Marching Cube algorithm [36]. Thereafter, the models are imported in our in-house developed planning-software CASPA (Balgrist CARD AG, Zurich, Switzerland).

In a first step of the planning, the model of the contralateral bone (i.e., the goal model) is mirrored using the sagittal plane as the plane of symmetry (Fig. 1a). Thereafter, a preliminary alignment to the goal model is performed to analyze the underlying deformity (Fig. 1b). Basic principles of comparing a pathological bone with a goal model in 3D are well known [18] and there are also clinically-established surface registration algorithms to facilitate this task. Briefly, the pathological bone is separated into (at least) two parts, both parts are aligned separately to the goal model, and their relative transformation T represents the amount of required reduction (Fig. 1b, c). In posttraumatic cases, a good starting point is to divide the bone into two regions proximal and distal to the former fracture line. Thereafter, point-plane iterative closest point (ICP) registration [37] is applied for aligning the proximal and

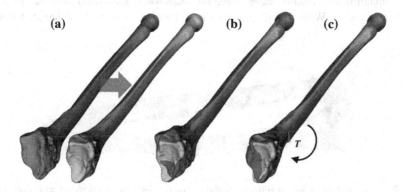

(a) **(b)** **(c)**

Fig. 1 Template-based planning approach demonstrated for the radius bone. **a** Pathological radius (*orange*) and mirrored, contralateral bone (*green*) serving as the reconstruction template. **b** Aligning the proximal part (*orange*) to the goal model reveals the deformity (*violet*). **c** The reduction is simulated by subsequently aligning the distal part. The relative transformation T between the proximal and distal bone parts quantifies the amount of reduction (Color figure online)

distal bone parts with the goal model in an automatic fashion. In this ICP variant, the distance between each point of the source surface to the surface formed by the closest triangle on the target surface is iteratively minimized in a least squares sense. The required closest point queries can be efficiently performed with a KD-Tree [38]. In each ICP iteration, a distance threshold is computed for outlier identification, i.e., all closest point pairs which are above the threshold are discarded.

During this planning stage, potential side-to-side differences between the bones must also be identified and compensated because several studies showed that asymmetry between the left and the right limbs exist for long bones [39–41] such as in length and torsion. Asymmetry in size can be compensated easier for anatomy where the joint is formed by two bones (forearm: radius and ulna; lower leg: tibia and fibula). In such cases, the normal bones of both sides are compared to determine the size differences as demonstrated in Fig. 2. Thereafter, the goal model is first scaled by the same amount before its registration with the pathological bone.

A computer-planned corrective osteotomy does not involve only determining the optimal reduction, but also additional clinical factors crucial for a successful outcome shall be considered and optimized. After preliminary alignment, the surgeon is able to specify these parameters and constraints; e.g., possible access to the bone, osteotomy site, and type of implant. Based on such specifications, the optimal surgical parameter values are determined in the planning tool: Besides the correction amount, the optimal selection of the osteotomy type, the position/orientation of the cutting plane(s), and the implant/screws position are important for a successful outcome and subsequent bone healing process.

A fundamental rule of conventional orthopedic planning is that a malunited bone is ideally cut at the apex of the deformity, which is the center of rotational angulation (CORA) [13] or the point of maximal deformity [7]. However, in practice a malunion has to be always regarded as a 3D deformity with 6 DoF to correct, hence restoring the normal anatomy requires one rotation—around a 3D axis often oblique to the anatomical standard planes—and one additional translation along a 3D displacement vector. While the direction r of the rotation axis can be calculated from

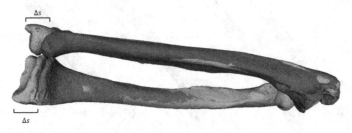

Fig. 2 Compensation of length differences shown for the forearm bones radius and ulna. The normal ulna from the pathological side (pathological radius not shown) is denoted by *orange color*. The contralateral forearm bones are shown in *yellow*. The ulnae of both sides are compared to assess the length differences Δs between the *left* and *right* forearm. The contralateral radius is scaled by Δs, resulting in the goal model (*green bone*) used as the reconstruction template (Color figure online)

the rotation matrix $R_{3\times3}$ of T [42], the center of rotation, C, can be chosen arbitrarily if no additional constraints are introduced. From the clinical point of view, it is desirable to choose the osteotomy site such that the fragments are minimally displaced after reduction to avoid creating gaps or steps at the osteotomy site. This can be achieved if C is close to the osteotomy site, on the one hand, and if C is only minimally translated after applying T, on the other hand. The latter holds true if $\|T \cdot C - C\|$ is minimal. Decomposition of T yields the ill-posed minimization problem $\min\|(I - R_{3\times3}) \cdot C - t\|$, where t denotes the translational component of T. Further constraints can be introduced since any point on the desired axis would be a valid solution of the equation system $(I - R) \cdot x = t$. Additionally defining C as the point on the axis being closest to a reference point p_r (e.g., the center of the bone model) yields the equation with a unique solution

$$\begin{pmatrix} I - R \\ r \end{pmatrix} \cdot x = \begin{pmatrix} t \\ <p_r, r> \end{pmatrix}$$

where the $<\ >$ operator denotes the inner product. The resulting axis (r, C) can be visualized in the planning application indicating the osteotomy position as demonstrated in Fig. 3.

The type of osteotomy and the orientation of the osteotomy plane directly influence how the bone surfaces will be in contact after reduction. Achieving maximum contact between the cortical bone layers of the fragments is strongly

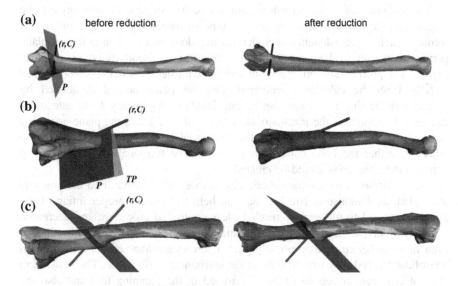

Fig. 3 Different types of osteotomies planned in 3D. The rotation axis (r, C) and the osteotomy plane(s) are shown in *red* and *grey*, respectively. The pathological (*orange*) and contralateral bone (*green*) are given before (*first column*) and after (*second column*) simulated reduction. The reduced bone fragment is depicted in *violet*. **a** Distal radius opening wedge osteotomy. **b** Closing wedge osteotomy of the radius. **c** Single cut osteotomy of the tibia (Color figure online)

desired to promote healing. Again, the previously calculated axis (r, C) can provide a starting point to improve such contact.

As demonstrated in Fig. 3, a closing or opening wedge osteotomy is indicated if angular deformity is predominant; i.e., if the direction r is considerable different from the direction of the bone long axis. Adding or removing a bone wedge is also indicated if shortening or lengthening of the bone is required, i.e., a translation along the bone length axis between the aligned bone fragments was assessed. Given the osteotomy plane P and transformation T, the wedge is defined by P and $T \cdot P$, where the line of intersection between the planes represents the hinge of the wedge. As demonstrated in Fig. 3b, so-called incomplete closing wedge osteotomies can be preoperatively planned if P is oriented such that the line of intersection coincides with (r, C) Incomplete osteotomies are often preferred for bones of the lower limb, promoting bone consolidation and early loading. In this procedure, the bone is not entirely cut but a small cortical bone bridge remains, almost constraining the reduction to a rotation around the bone bridge (i.e., the hinge of the wedge). Vice-versa, the preoperatively planned reduction can be only achieved if the hinge corresponds to (r, C). More examples of this osteotomy type will be given later in Sect. 2.2. Dependent on the position of the hinge, a situation may arise where the wedge type is not clearly defined; i.e., part of the bones overlap while also a gap is created. For such situations, CASPA permits the visualization of the wedge in real time by simultaneously rendering P and $T \cdot P$ on display. By doing so, the osteotomy plane can be interactively translated until the desired wedge osteotomy can be achieved.

If a rotational deformity is predominant, a so-called single cut osteotomy may be preferable to a wedge osteotomy. In this type of osteotomy, a single cut is performed which is pre-calculated such that sliding along and rotation in the cut plane permits reduction as planned [43]. An example of this osteotomy type is given in Fig. 3c. The position and orientation of a single cut osteotomy plane can be derived directly from the calculated reduction, i.e., the plane normal is defined by r. Although the single cut plane can be calculated exactly for any T, the osteotomy can be only applied if the translational component of T along the plane normal is negligible (i.e., below 1 mm) and the tangential translation is small, because otherwise either the bone contact surface would be too small or, worse, a gap between the bone parts would be created.

The inclusion of the osteosynthesis plate model, if available, into the preoperative plan, as demonstrated in Fig. 8a, can help to ensure a proper fitting of the plate. A poor plate fitting may result in less controlled procedure and decreased stability, which may lead to inaccurate reduction, delayed or stalled bone healing, or even implant failure. Inappropriately placed screws are another cause of delayed consolidation, and they may also harm the surrounding soft tissue. The best-fitting implant type can indeed be easily determined in the planning tool and also the optimal position of such implant can already be defined preoperatively. Optimally, the implant also contains the position and direction of the screws. The screw models are used by the surgeon to verify that all screws will be placed sufficiently inside the bone but without penetrating the osteotomy plane or anatomical regions at risk.

In severe deformities, one osteotomy may not be sufficient to achieve a satis-factory result. In such cases, multiple mutually-dependent osteotomies must be performed, resulting in a considerable increase in the complexity of the preopera-tive plan. As demonstrated in the example case given in Fig. 4, we propose a step-wise approach in such cases. The 14-years-old patient suffered from a post-traumatic malunion of both forearm bones, resulting from a fall four years ago. The deformity was assessed by aligning the distal third of each pathological bone to the contralateral model using ICP (Fig. 4a). Next, mid-shaft osteotomies at the apex of the deformity were performed for both radius and ulna, and the reduction was simulated by registering the proximal parts to the goal models using ICP. As shown in Fig. 4b, the resulting reduction of the radius fragment was not acceptable from a clinical point of view, due to the large gap and cortical surface mismatch on the radius. Therefore, an additional osteotomy was defined in the distal third of the radius and the resulting fragment was manually aligned to the goal model. In a second iteration, the transformation between the bone fragments and the osteotomy sites were fine-tuned in a manual fashion in order to achieve correct aligment by applying only single-cut osteotomies. The final result, given in Fig. 4c, was obtained after recalculating the surgical plan.

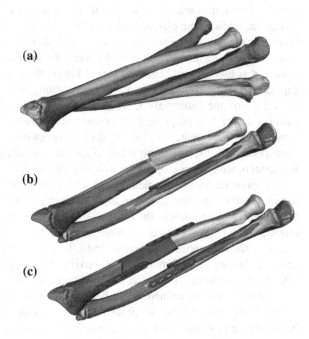

Fig. 4 Stepwise preoperative planning of a triple osteotomy for treating a complex deformity affecting both bones of the forearm. The reconstruction template is shown in *green*. **a** First step: radius and ulna were distally aligned to the template. **b** Second step: for both bones mid-shaft osteotomies were performed. **c** Third step: final result after an additional radius osteotomy (Color figure online)

2.2 Axis Realignment

Realignment of the lower limb is primarily indicated in patients having a varus-(out-knee) or valgus-deformity (in-knee), resulting in a pathologically deviated weight-bearing line of the leg (i.e., the mechanical axis) [13]. In many cases, the deformity is caused by a congenital defect affecting both limbs and, therefore, the contralateral bone cannot be used as a reconstruction template. Instead, the required correction is determined by realigning the mechanical axis to its anatomically normal position. According to Paley et al. [13], the mechanical leg axis passes not exactly through the knee joint center, but is located 8 ± 7 mm medial to the knee joint center. This deviation is called mechanical axis deviation (MAD) and is measured as the perpendicular distance from the mechanical leg axis to the knee joint center. A malalignment can be considered as pathological if the MAD lies outside of this normal range.

Before describing the 3D method for the preoperative planning of axis realignment procedures, the conventional approach is next briefly summarized. In the conventional preoperative assessment, long-leg standing X-rays are acquired for determining pathological and normal mechanical leg axes, as depicted in Fig. 5. Based on these axes, the position C and opening angle φ of a corrective osteotomy is calculated using a method described by Miniaci et al. [44]. In a first step, the preoperative mechanical leg axis l is determined as the straight line from the hip joint center A to the upper ankle joint center B. The postoperative axis l' can be calculated by considering the fact that it should pass through a point F on the tibia plateau, at 62 % of the plateau width measured from medial (for varus-deformity). Point F is the so-called Fujisawa point, named after the author of the study [45], in which the optimal intersection point between the mechanical leg axis and the tibial plateau was investigated. The position of the rotation center of the osteotomy C is dependent on the pathology (e.g., closing/opening tibial/femur osteotomy). After the center of the osteotomy C is defined by the surgeon, the angle φ can be calculated. For this purpose, C is connected by the line r with the preoperative upper ankle joint center B. Lastly, the postoperative ankle joint center B' and, correspondingly, the osteotomy angle φ are determined by rotating r around C until it intersects with the postoperative axis l' at point B'.

We have developed a method for quantifying the required 3D osteotomy parameters (i.e., position, rotation axis, and angle). Our approach requires having triangular surface models of the hip, knee, and ankle joints. We apply a CT protocol that scans the regions of interests while skipping the irrelevant mid-shaft regions to reduce radiation exposure. Based on the CT scan, a limb model consisting of the proximal femur, distal femur, patella, proximal tibia, distal tibia, distal fibula and talus, is reconstructed. Thereafter the hip, knee, and ankle joint centers of the limb model have to be determined in 3D before the mechanical axes can be calculated.

The joint center of the proximal femur A is defined by the center of the best fitting sphere [46], minimizing the distance to a user-selected region of the femoral head points, as demonstrated in Fig. 6a. The bone model of the proximal tibia is

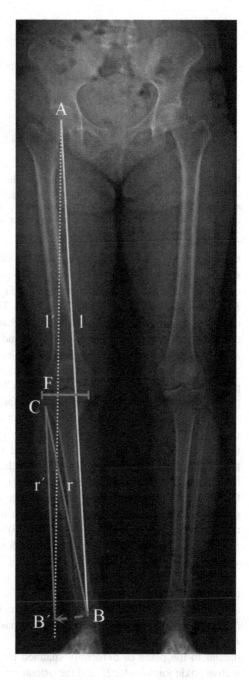

Fig. 5 Measurement of the pathological mechanical leg axis l and calculation of the realigned normal axis l' in 2D

used to determine the center of the knee joint and an analytical description of the tibia plateau plane P_{tib} (Fig. 6b), both required for computing the Fujisawa point. We follow Moreland [47] who defined the knee joint center as the midpoint

(a) **(b)** **(c)**

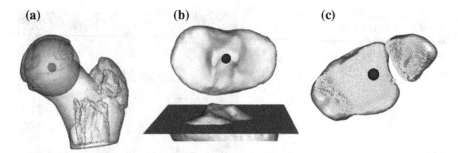

Fig. 6 Determination of the relevant joint parameters. **a** The center of the femoral head is calculated by the center (*dark blue*) of the best fitting sphere (*cyan*). **b** Knee joint center (*dark blue*) and tibia plateau plane P_{tib} (*dark grey*). **c** The ankle joint center (*dark blue*) is defined as the center of mass of the articular surface (*cyan points*) (Color figure online)

between the intercondylar eminences of the tibial plateau. As for the femur, a least squares approach is applied for finding the plane P_{tib}, minimizing the distance to user-selected plateau regions as depicted in Fig. 6b. Lastly, the center of ankle joint B is calculated as demonstrated in Fig. 6c. The ankle center can be anatomically described as the midpoint between the medial and lateral malleolus [47], i.e., the center of the articular surface of both tibia and fibula with respect to the talus bone. Accordingly, we propose a method for calculating the 3D ankle joint center by analyzing the opposing articular surfaces between these bones. First, tibia and fibula points having a small distance to the talus bone are identified, because they are potential candidates for being articular surface points. These candidate points are efficiently found by calculating the closest-point distance to the talus bone model using a KD-tree [38] and considering only points below a user-defined distance threshold. A second criterion is introduced to eliminate false positive candidate points: All candidates, for which the angle between its surface normal vector and the direction vector to its closest point on the talus surface is above a user-defined threshold, are eliminated. Both thresholds are visually determined per case until the entire articular surface is detected as shown in Fig. 6c. Typically the distance and angle thresholds are between 4–8 mm and 40°–70°, respectively. The joint center is finally computed as the center of mass of all selected points on the tibia and fibula surface.

After the joint centers are calculated, the pathological and corrected mechanical axes l and l' can be calculated in 3D (Fig. 7a–c). As depicted in Fig. 7b, axis l is defined as a straight line from the hip joint center A to the ankle joint center B. Axis l' is defined by the direction vector from A, passing through the Fujisawa point F on P_{tib}. For performing the desired correction, the osteotomy axis r must be perpendicular to the plane of deformity spanned by l and l'; i.e., $r = l \times l'$. The postoperative ankle joint center B' and the osteotomy angle φ can be calculated by solving the line/sphere intersection problem shown in Fig. 7c. That is, since the length of the leg axis should not change, B' is located on the sphere S with the center C and radius $|CB|$, i.e. at the intersection of axis l' with S. Once B' is determined, the

Fig. 7 Measurement of the pathological mechanical leg axis l and calculation of the realigned normal axis l' in 3D. **a** Mechanical leg axis before (*red*) and after (*green*) realignment. **b** Calculation of the Fujisawa point F for determining the direction of the postoperative leg axis l'. **c** Calculation of the postoperative ankle joint center B' (Color figure online)

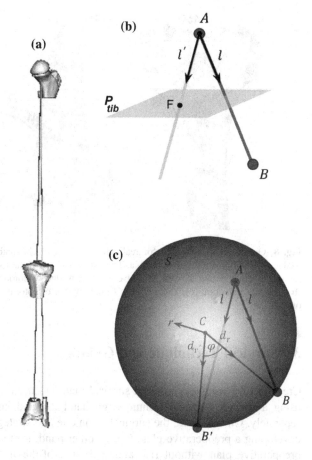

osteotomy angle φ is computed from the angle between the direction vectors $d_{r'}$ and d_r (Fig. 7c).

One benefit of the 3D approach is that the axis realignment does not only encode the deformity correction in the anterior-posterior plane, but it also indicates the correction in the sagittal plane, if desired (e.g., for the correction of the tibial slope). Once the quantitative planning parameters are defined, an opening or closing wedge osteotomy can be simulated as in the template based approach. Examples of an opening wedge proximal tibia osteotomy and a closing wedge distal femur osteotomy are given in Fig. 8. The transformation from B to B' can be expressed by a matrix T, representing the amount of the required reduction. The osteotomy plane P is then uniquely defined by the axis (r, C) forming the hinge of the wedge (see Fig. 8b). The planning tool shall allow for interactive positioning and fine-tuning of the osteotomy plane, because the optimal position is dependent also on the soft-tissue anatomy and the particular pathology of the patient. Note that φ must be recalculated after moving the plane and the axis to a new axis position, because the opening angle is dependent on the position of the axis.

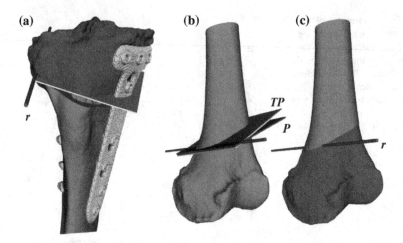

Fig. 8 Osteotomies performed for realignment of the mechanical axis of the leg. **a** Opening wedge osteotomy of the proximal tibia. The exact position of the osteosynthesis plate (*yellow*) is preoperatively planned as well. **b** and **c** Closing wedge osteotomy of the distal femur before (**b**) and after reduction (**c**). The wedge is formed by the osteotomy plane *P* and *TP* (Color figure online)

3   Surgical Technique and Guides

Creation of a computer assisted surgical plan and the implementation of the plan using a specific surgical technique go hand in hand and cannot be considered separately. On one hand, the surgical technique and strategy must be known before developing a preoperative plan. On the other hand, it must be possible to realize a preoperative plan without risk and with state-of-the-art surgical techniques. For multi-planar deformities, the reduction task is often too complex to be performed conventionally without additional tools supporting the surgeon. We use patient-specific instruments in the surgery, enabling the surgeon to reduce the bone exactly as planned on the computer. These so-called *surgical guides* have proved successful over years in different orthopedic interventions. The basic principle is that the body of a guide is shaped such that it can be uniquely placed on its planned position on the bone by using characteristics of the irregular shaped bone surface and it helps intra-operatively to define the osteotomy plane and correction.

Each malunion and, consequently, each surgery is different which makes not only preoperative planning but also patient-specific instrumentation challenging. The goal is to develop guides that are sufficiently flexible to be applied to various types of osteotomies, but nevertheless enabling generation in a standardized and time-efficient way. In our experience, patient-specific guides have to be considered as an integral part of the surgical plan; i.e., they should be designed in the planning tool to define the osteotomy position, the direction of the cut, the position of the implant, and the relative transformation of the fragment after reduction. The CASPA tool enables the generation of the guides in a semi-automatic fashion. In a first step, the guide body is created based on a 2D outline drawn around the osteotomy location on the bone surface. It is crucial that this step is performed by a surgeon, because the guide surface must not cover soft-tissue structures (e.g., ligaments) that cannot be removed from the bone for guide placement. To generate a 3D guide body, the outline is extruded normal to the bone surface by a user defined height, followed by boolean subtraction of the bone surface [48]. To achieve a unique fit, the shape of the guide body is designed to contain irregular convex and concave parts covering the bone from different directions.

As demonstrated in Fig. 9, we have developed different cutting guides to support the surgeon in performing osteotomy. One way of defining the cut position and direction is to use a metallic inlay (Fig. 9a) that contains a cutting slit to guide the saw blade. The inlay is inserted into a dedicated frame in the guide body for alignment according to the planned osteotomy plane. Although very accurate, the technique is limited to certain saw blade types and it requires the bone to be sufficiently exposed. Alternatively, a cutting slit can be directly integrated into the plastic guide or the edge of the guide body can be used for guiding the saw blade (Fig. 9b). These types of cutting guides are particularly helpful for performing closing wedge osteotomies, where the distance between the two cuts may be very small. Lastly, K-wires set by a drilling guide can be used to approximate the osteotomy plane. In this case, the direction of the saw blade can be also controlled inside of the bone.

In our experience, two different approaches have been proven to be successful for supporting correct realignment of the bone fragments with the help of patient-specific reduction guides. In one method, separate reduction guides are used

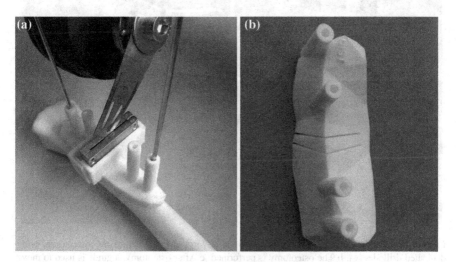

Fig. 9 Different types of cutting guides for guiding the saw blade. **a** Metallic inlay with a cutting slit. **b** Cutting slit directly integrated in the plastic guide body

to predefine the reduction by their shape. To do so, two guides combined with K-wires are required in the surgery (see Fig. 10). First, a reduction guide is generated in CASPA based on four parallel K-wires that are positioned on the fragments in their reduced positions. This guide uniquely defines the relative transformation between the fragments after reduction. Next, the K-wires are transformed back to the pathological bone by applying the transformation T^{-1}, and, accordingly, a guide is constructed. In the surgery, the latter guide is applied first to set the K-wires before the osteotomy, as demonstrated in Fig. 10a. After the osteotomy (Fig. 10b), the reduction guide is inserted over the K-wires to move the fragments to their planned positions (Fig. 10c), followed by plate fixation (Fig. 10d).

Note that, if K-wires are considerably divergent, a multi-part guide is required to allow for the removal of the guide. For this reason, we have developed multi-part guides that can be stably connected but removed separately. As shown in Fig. 10a, the first variant results in a very stable connection between the parts due to cylinders that are plugged into corresponding holes in the opposite guide parts. The cylinders must point in the same direction as the K-wires to permit removal of the guide part after K-wire insertion. The system can be used only for guides which are sufficiently high (e.g., 1 cm). Alternatively, a V-shaped connector as shown in Fig. 10b can be applied if the space for the guide is limited (Fig. 11).

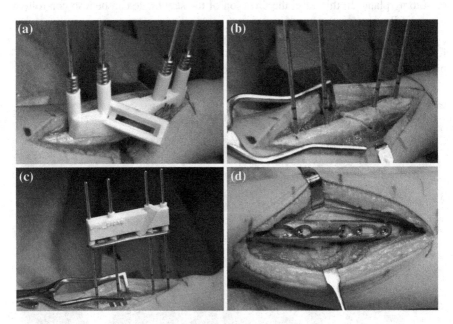

Fig. 10 Reduction guides based on K-wires. **a** K-wires are set before the osteotomy using dedicated drill sleeves. **b** The osteotomy is performed. **c** After osteotomy, a guide is used to move the K-wires and, consequently, the mobilized fragments into their planned position. **d** Fixation with an implant

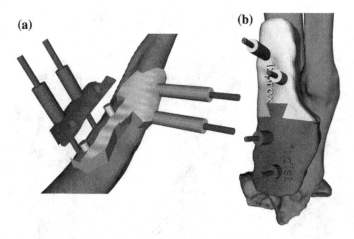

Fig. 11 Multi-part guides where each part can be removed separately in case of divergent K-wires. **a** Variant using cylinders that are plugged into corresponding holes in the opposite part. **b** Variant using a V-shaped connecting surface

Another approach is to directly utilize the implant (i.e., the osteosynthesis plate) for supporting the surgeon in the reduction task. The most straight-forward way is to manufacture a 1:1 replica of the reduced bone to prebend the plate according to the bone shape (see Fig. 12). Plate prebending is a common method in conventional orthopedics to define the transformation of the fragments in their reduced position. Although effective, the method has limited accuracy because it does not fully constrain all DoF.

Athwal et al. [23] introduced a more sophisticated and accurate technique based on a navigation system to realign the fragments using the screws of the fixation plate. We [32] and others [28, 29] have further developed this technique by applying patient-specific guides to avoid the use of a navigation system. The method requires the use of an implant based on angular-stable locking screws. Locking screws have the property that threads on the screw head lock into corresponding threads of the screw holes in the plate, resulting in an angular and axial

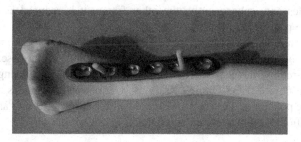

Fig. 12 Osteosynthesis plates can be prebent before surgery based on a real-size bone model. The model is generated by additive manufacturing the reduced bone after simulated osteotomy

(a)   **(b)**   **(c)**

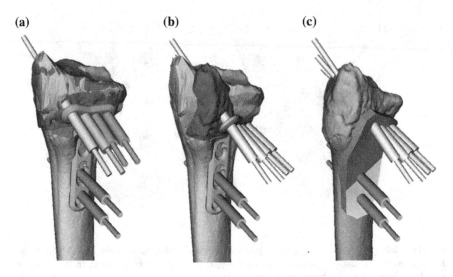

Fig. 13 Reduction via the screws of the implant. The cut bone fragments of the pathological radius are shown in *brown* and *yellow*, respectively. The contralateral mirrored bone is shown in *green*. **a** Implant (*cyan*) is positioned on the bone fragments after simulated osteotomy. Grey cylinders represent the angular-stable locking screws. **b** The screws are transformed back to uncorrected bone by applying the inverse reduction. **c** A guide with drilling sleeves is created based on the screws (Color figure online)

stable screw anchorage. Therefore, the plate maintains distance to the bone during fixation, in contrast to compressing plates where the fragment is pulled towards the plate. For this purpose, a locking system is particularly suited for integration into a computer-based surgical planning as the bone-plate-screw interface is uniquely defined. The preoperative planning method is similar to the one using reduction guides based on K-wires. After the plate model is positioned on the bone surface in postoperative configuration (Fig. 13a), the screw models are transformed back to their preoperative position by applying transformation T^{-1} (Fig. 13b), and used for creating a drilling guide which has drill sleeves for the screws (Fig. 13c). In the surgery, the guide is applied before the osteotomy to drill the screw holes, as shown in Fig. 14a. After cutting the bone, the plate is first fixed to one fragment using the predrilled holes and, subsequently, the other fragment is reduced as planned by fixation to the plate (Fig. 14b).

Using additive manufacturing, patient-specific instruments can be produced based on 3D models designed in the planning application. In our case, the guides are manufactured by Medacta SA (Castel San Pietro, Switzerland) using a selective laser sintering device (EOS GmbH, Krailling, Germany). The guides are made of medical-grade polyamide 12. They are cleaned with a surgical washer and sterilized using conventional steam pressure sterilization at our institution.

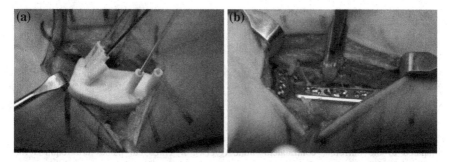

Fig. 14 Patient-specific guide for pre-drilling the screw holes. **a** The screw holes are pre-drilled before the osteotomy. **b** After osteotomy, the fragments are reduced as planned by fixation to the plate

4 Intraarticular Osteotomies

Corrective osteotomies of intra-articular malunions are among the most challenging orthopedic procedures with respect to both preoperative planning and surgery. The limited access to the joint surface makes an inside-out-osteotomy (i.e., cutting from the joint surface towards the shaft) difficult or, in certain settings, even impossible. An extra-articular, outside-in approach to the intra-articular malunion provides an alternative in the surgical treatment. However, such a technique requires extensive preoperative planning and, moreover, the surgeon must be guided intraoperatively because he has no direct view into the joint. We have developed a preoperative planning methodology combined with a closely-linked surgical technique [11], enabling the surgeon to perform an outside-in approach in a controlled way. We will describe the approach based on one of the most complex cases treated at our institution so far.

The 62 year old patient sustained a distal radius fracture that had been insufficiently treated at another institution by open reduction and internal fixation with palmar plating. As demonstrated in (Fig. 15a), the CT-reconstructed 3D model of the pathological radius showed steps and gaps up to 4 mm in the joint surface area, well visible, especially if compared to the opposite mirrored normal bone. The 3D analysis based on the former fracture lines and the contralateral bone identified 3 intraarticular fragments and an overall shortening of the radius. The coarse preoperative plan depicted in (Fig. 15b) intended to align the styloid fragment (denoted by the dark blue fragment) and the central palmar fragment (light blue fragment) to the lunate facet fragment (purple arrow), which was initially left fixated to the radius shaft. Before simulating the osteotomies, the exact cut planes had to be defined. To do so, consecutive line segments were specified by the surgeon along the former fracture lines of the joint surface. Thereafter, the corresponding 3D osteotomy planes were automatically generated by extrusion of the line segments. Next, the surgeon defined the extra-articular entry point of the cut planes respecting the best access through the soft tissue by rotating the osteotomy planes around the

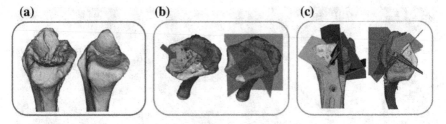

Fig. 15 Definition of the cut planes for an intra-articular osteotomy. **a** Pathological radius (*yellow*) compared to the mirrored, contralateral bone (*green*). **b** Three cut planes are required for mobilization of the fragments. The purple arrow denotes the lunate facet fragment. **c** Additional cut planes are calculated, necessary to remove bone parts that would overlap after reduction (Color figure online)

previously defined line segments. The final cut planes planned for fragment mobilization are shown in Fig. 15b. Thereafter, the fragments were interactively and consecutively aligned to the normal bone in their correct anatomical position to restore a congruent joint surface (Fig. 16). Based on the fragment positions in their original and reduced positions, the planning software CASPA also enables the automatic calculation of the offcut (Fig. 16a), which should be removed (i.e., closing wedge osteotomy). In total six cutting planes, i.e., three planes for mobilizing the fragments and three for removing the offcut, were required in this case as shown in Fig. 15c.

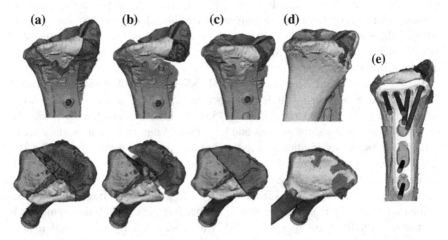

Fig. 16 Reduction of a complex intra- (**a–d**) and extra-articular (**e**) malunion. *Top row* Palmar view. *Bottom row* axial view. **a, b** Fragments in their pathological positions before and after removal of the offcut (*red*). **c** Fragments in their reduced positions. **d** Overlay with the mirrored contralateral bone (*green*), demonstrating joint congruency but a residual shaft deformity. **e** Result after intra- and extra-articular osteotomy compared to the contralateral bone (*green*, transparent). The *red cylinders* denote the directions of the angular stable locking screws of the osteosynthesis plate used for fixation (*grey*) (Color figure online)

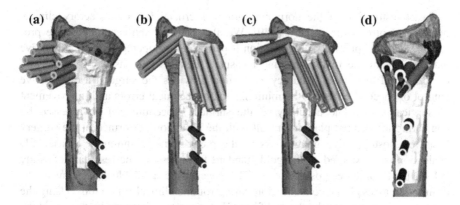

Fig. 17 Patient-specific guides designed for a combined intra- and extra-articular distal radius osteotomy. **a–c** Three drilling guides combining six cut planes were applied to guide the intra-articular osteotomies. **d** Reduction guide for the extra-articular osteotomy

After the joint area was reconstructed, the pathological bone still showed a malposition of the shaft (Fig. 16d), requiring a correction of the radius length and epiphysis orientation with an additional extra-articular osteotomy (Fig. 16e). Once the preoperative plan was completed, patient-specific guides were designed for the intra- and extra-articular osteotomies as depicted in Fig. 17. The key idea of our outside-in approach is to mobilize complex-shaped fragments in the surgery using a drill; i.e., the curved cut is perforated by consecutive drill holes (i.e., spaced by 5 mm) using a surgical drill instead of a saw. By doing so, the exact position, direction, and depth of the drill holes can be calculated preoperatively and integrated into a guide with corresponding drill sleeves (Fig. 17a–d). The heights of the sleeves were carefully matched to the length of the drill bit to avoid entering to far into the joint. In the surgery, the holes were drilled using three drilling guides, a K-wire was inserted into the holes, and a cannulated chisel was used to connect the holes to complete the osteotomy. An additional guide was used for supporting the extra-articular osteotomy (Fig. 17d) by predrilling the screw holes of the plates as previously described.

5 Results and Discussion

In this chapter, we have described a technique combining 3D preoperative planning of long bone deformity correction with additive-manufactured patient-specific instruments. So far, we have applied the approach to osteotomies of the radius, ulna, humerus, femur, tibia, and fibula. The method has enabled us to perform the osteotomies more accurately and in a more controlled fashion.

Exact restoration of the normal anatomy is crucial for a satisfactory clinical outcome of orthopedic surgeries [49]. We also consider the accuracy of the procedure, i.e., how precisely the reduction is performed compared to the preoperative plan, as one of the major outcome measures. Besides demonstrating efficacy of the presented technique, accuracy evaluation provides the surgeon a quantitative control of success, enabling the minimization of technical errors and improvement of surgical skills. The accuracy of the surgical procedure can be assessed by comparing the desired planning result with the reduction performed in the surgery based on postoperative images. As in the preoperative planning, CT-based 3D evaluation is considered as the gold standard for assessing the residual deformity [11, 50]. In our case, postoperative CT images were available in several cases, acquired to assess bone consolidation during routine clinical follow-ups. Using the postoperative bone model extracted from CT, the same registration method as in the preoperative planning can be applied to quantify the difference between planned and performed reduction in 3D. We have performed CT-based accuracy evaluation of 37 surgeries, comprising different types of osteotomies and anatomy. The residual rotation error is expressed by the 3D angle of the rotation in axis-angle representation. The residual translation error is given by the length of the 3D displacement vector.

On average, the accuracy in rotation and translation for the evaluated epi-/ diaphysial osteotomies (n = 27) was $4.3° \pm 3.6°$ and 1.9 ± 1.3 mm, respectively. For intra-articular osteotomies (n = 10) an average residual angulation and displacement $4.7° \pm 3.5°$ and 0.9 ± 0.5 mm, respectively, was assessed with the method described in [11]. Figure 18 demonstrates the comparison between the planning and the postoperative 3D model of the 4-part intra-articular distal radius osteotomy presented in Sect. 4. In this case, a rotational error of less than 2° and a translational error below 2 mm were achieved. One year after surgery, the patient had symmetric strength and range of motion of her wrist and was pain free. The measured errors are slightly higher compared to in-vitro experiments [27, 51], where an average

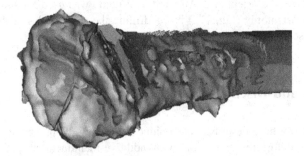

Fig. 18 Postoperative accuracy evaluation of the combined intra- and extra-articular osteotomy presented in Sect. 4. The preoperative plan (*brown* and *orange fragments*) are aligned with the postoperative bone model (*cyan*) obtained from postoperative CT scans. Note that the intra-articular fragments of the preoperative plan (*brown*) were virtually fused to make the registration more robust (Color figure online)

accuracy of 1°/1 mm was reported for distal radius osteotomies performed with a similar technique. Several other studies have evaluated the feasibility and accuracy in a clinical setting for the forearm bones [29–31], the humerus [52], and the lower extremities [34]. In these studies, an average deviation from the 3D preoperative plan between 1° and 5° was measured after surgery, but the evaluation was only performed on postoperative radiographs.

Computerized preoperative planning offers high accuracy and satisfactory results for surgical outcome; nevertheless, it still presents some challenges to be considered carefully. Preoperatively defining the reduction target (i.e., the goal model) correctly is a major contributor to a successful planning. So far, the contralateral bone has been proved to be the best reconstruction template available [12, 18] although considerable bilateral asymmetry may exist [39–41]. Even in case of perfect symmetry, soft tissues may have a considerable influence on the joint function. Consequently, a successful clinical outcome cannot always be ensured by merely relying on the contralateral bone anatomy, even if the reduction is performed precisely as planned. The integration of a pre- and postoperative motion simulation into the preoperative planning application may be the next step to better predict the functional outcome [53] for soft tissue injuries. Apart from this, the resolution of the image data used for planning and the subsequent segmentation process are additional technical factors, which may also influence the planning accuracy. Nevertheless, bone models extracted from CT scans with an axial resolution of 1 mm and using interactive segmentation methods, such as thresholding and region growing, have been shown to be sufficiently accurate for preoperative planning [54]. Intraoperatively, the correct placement of the guide(s) on the bone also has a major impact on the accuracy of the entire procedure, because the reduction is computed relative to the osteotomy site. Therefore, all means should be taken to ensure a precise identification of the intended guide position intra-operatively. One first step verification is to check the fit of the guide on real-size replica of the bone anatomy, manufactured with laser sintering devices. For a precise fit in the surgery, the bone should be debrided from periosteum as much as possible. Nevertheless, for finding a stable fit on regularly shaped surfaces such as the cylindrical mid-shaft region, manual measurements with respect to anatomical reference points may still be necessary, which is a limitation of the presented method.

We have had different experiences regarding the two different types of reduction guides described in Sect. 3. Using the technique based on pre-drilling the screw holes of the implant, we have observed some difficulties in achieving the correct reduction for osteotomies where extensive soft tissue tension was present (e.g., opening wedge distal radius or tibia osteotomies). Particularly in these cases, separate reduction guides appeared to be more suited and more accurate. An additional benefit of separate reduction guides is that dedicated parts of the guide body, such as wedges, can be used in addition to the K-wires to further improve the control of the reduction. However, guide generation is more time-consuming for multi-part reduction guides and the exposure of the bone must be larger. Due to the limited access to the bone, the separate reduction guide and additionally required K-wires may complicate surgical procedures such as sawing or implant fixation.

Table 1 Average time necessary for generating a computer-based preoperative plan for three different categories of complexity

	Category I	Category II	Category III
Surgical plan	Simple: opening-wedge or single-cut shaft osteotomies	Medium: closing wedge osteotomies, two-bone osteotomies (radius-ulna, tibia-fibula)	Complex: intra-articular osteotomies, double osteotomies
Guide design	Simple and standardized: one pre- and one postoperative guide	Medium: three or more guides	Complex: more than three, highly individualized guides
Average planning time (h)	2.1 ± 0.8 h (n = 16)	4.0 ± 0.9 h (n = 20)	9.1 ± 3.7 h (n =28)

In the future, patient-specific osteosynthesis plates with additional functionality, such as supporting the surgeon in the reduction, may be a promising alternative to combine the advantages of the present guiding techniques. Moreover, the possibility of using patient-specific osteosynthesis implants, integrated into the preoperative plan and fitted to the patient anatomy, would also improve the evident problem of the poor fitting of anatomical standard-plates. Recently, the application of a first patient-specific implant prototype has been demonstrated in-vitro [55]. However, its high manufacturing costs over 1000 USD limit its wide-spread clinical application. Selective laser melting, which has proven to be highly productive in other medical fields, may allow for a cost-effective fabrication in the future.

Long planning times and the necessary effort have been seen as disadvantages of computer-assisted preoperative planning approaches as the one presented herein. We have evaluated the time required for our preoperative planning including the guide design based on 64 cases. The cases were assigned to one of three different categories dependent on their complexity with respect to the surgical plan and/or the patient-specific instruments. Table 1 summarizes the average planning times for each category.

Compared to the conventional approach, an additional cost of €250 ($340 USD) per case arises due to guide manufacturing. Moreover, the radiation exposure for the patient may be higher due to the increased use of preoperative CT imaging. Nevertheless, the technique may reduce total fluoroscopy time in the surgery.

In conclusion, 3D planning has become an integral part of the preoperative assessment for long bone deformities at our institution. While simpler corrective osteotomies can be efficiently planned in a standardized way using dedicated planning tools, more complex cases still require a laborious manual effort, resulting in considerable personnel costs. Considering these cases, further research must be performed to reduce the preoperative planning time. Additive manufacturing has revolutionized computer-assisted surgery. Patient-specific surgical guides provide an efficient and accurate way of implementing a computer-based surgical plan in the surgery. As the preoperative plan and the surgical technique are closely linked, training of the surgeons in using surgical guides is essential. After gaining practical experience, the technique may reduce operation times and enables osteotomy corrections that were not possible before, e.g., complex intra-articular osteotomies [11]. As a next future step, the extension of the method to other anatomy, such as bones of the wrist or foot, will be studied.

References

1. Cambron J, King T (2006) The bone and joint decade: 2000 to 2010. J Manipulative Physiol Ther 29(2):91–92
2. Brooks PM (2006) The burden of musculoskeletal disease—a global perspective. Clin Rheumatol 25(6):778–781
3. Woolf AD, Pfleger B (2003) Burden of major musculoskeletal conditions. Bull World Health Organ 81(9):646–656
4. Wade RH, New AM, Tselentakis G, Kuiper JH, Roberts A, Richardson JB (1999) Malunion in the lower limb. A nomogram to predict the effects of osteotomy. J Bone Joint Surg Br 81 (2):312–316
5. Honkonen SE (1995) Degenerative arthritis after tibial plateau fractures. J Orthop Trauma 9(4):273–277
6. Nagy L, Jankauskas L, Dumont CE (2008) Correction of forearm malunion guided by the preoperative complaint. Clin Orthop Relat Res 466(6):1419–1428
7. Jayakumar P, Jupiter JB (2014) Reconstruction of malunited diaphyseal fractures of the forearm. Hand 9(3):265–273
8. Lustig S, Khiami F, Boyer P, Catonne Y, Deschamps G, Massin P (2010) Post-traumatic knee osteoarthritis treated by osteotomy only. Orthop Traumatol Surg Res 96(8):856–860
9. Espinosa N (2012) Total ankle replacement. Preface. Foot Ankle Clin 17(4):xiii–xiv
10. Parratte S, Boyer P, Piriou P, Argenson JN, Deschamps G, Massin P (2011) Total knee replacement following intra-articular malunion. Orthop Traumatol Surg Res 97(6 Suppl): S118–S123
11. Schweizer A, Fürnstahl P, Nagy L (2013) Three-dimensional correction of distal radius intra-articular malunions using patient-specific drill guides. J Hand Surg Am 38(12):2339–2347
12. Mast J, Teitge R, Gowda M (1990) Preoperative planning for the treatment of nonunions and the correction of malunions of the long bones. Orthop Clin North Am 21(4):693–714
13. Paley D (2002) Principles of deformity correction. Springer, Berlin. ISBN:354041665X
14. Marti RK, Heerwaarden RJ, Arbeitsgemeinschaft für Osteosynthesefragen (2008) Osteotomies for posttraumatic deformities: Thieme Stuttgart. ISBN:3131486716
15. Fernandez DL (1982) Correction of post-traumatic wrist deformity in adults by osteotomy, bone-grafting, and internal fixation. J Bone Joint Surg Am 64(8):1164–1178
16. Paley D (2011) Intra-articular osteotomies of the hip, knee, and ankle. Oper Tech Orthop 21(2):184–196
17. Ring D, Prommersberger KJ, Gonzalez del Pino J, Capomassi M, Slullitel M, Jupiter JB (2005) Corrective osteotomy for intra-articular malunion of the distal part of the radius. J Bone Joint Surg Am 87(7):1503–1509
18. Schweizer A, Fürnstahl P, Harders M, Szekely G, Nagy L (2010) Complex radius shaft malunion: osteotomy with computer-assisted planning. Hand 5(2):171–178
19. Bilic R, Zdravkovic V, Boljevic Z (1994) Osteotomy for deformity of the radius. Computer-assisted three-dimensional modelling. J Bone Joint Surg Br 76(1):150–154
20. Zdravkovic V, Bilic R (1990) Computer-assisted preoperative planning (CAPP) in orthopaedic surgery. Comput Methods Programs Biomed 32(2):141–146

21. Langlotz F, Bachler R, Berlemann U, Nolte LP, Ganz R (1998) Computer assistance for pelvic osteotomies. Clin Orthop Relat Res 354:92–102
22. Jaramaz B, Hafez MA, DiGioia AM (2006) Computer-assisted orthopaedic surgery. Proc IEEE 94(9):1689–1695
23. Athwal GS, Ellis RE, Small CF, Pichora DR (2003) Computer-assisted distal radius osteotomy. J Hand Surg Am 28(6):951–958
24. Wang G, Zheng G, Gruetzner PA, Mueller-Alsbach U, von Recum J, Staubli A et al (2005) A fluoroscopy-based surgical navigation system for high tibial osteotomy. Technol Health Care 13(6):469–483
25. Hufner T, Kendoff D, Citak M, Geerling J, Krettek C (2006) Precision in orthopaedic computer navigation. Orthopade 35(10):1043–1055
26. Sarment DP, Sukovic P, Clinthorne N (2003) Accuracy of implant placement with a stereolithographic surgical guide. Int J Oral Maxillofac Implants 18(4):571–577
27. Ma B, Kunz M, Gammon B, Ellis RE, Pichora DR (2014) A laboratory comparison of computer navigation and individualized guides for distal radius osteotomy. Int J Comput Assist Radiol Surg 9(4):713–724
28. Kunz M, Ma B, Rudan JF, Ellis RE, Pichora DR (2013) Image-guided distal radius osteotomy using patient-specific instrument guides. J Hand Surg Am. 38(8):1618–1624
29. Miyake J, Murase T, Moritomo H, Sugamoto K, Yoshikawa H (2011) Distal radius osteotomy with volar locking plates based on computer simulation. Clin Orthop Relat Res 469(6): 1766–1773
30. Miyake J, Murase T, Oka K, Moritomo H, Sugamoto K, Yoshikawa H (2012) Computer-assisted corrective osteotomy for malunited diaphyseal forearm fractures. J Bone Joint Surg Am 94(20):e150
31. Murase T, Oka K, Moritomo H, Goto A, Yoshikawa H, Sugamoto K (2008) Three-dimensional corrective osteotomy of malunited fractures of the upper extremity with use of a computer simulation system. J Bone Joint Surg Am 90(11):2375–2389
32. Schweizer A, Fürnstahl P, Nagy L (2014) Three-dimensional planing and correction of osteotomies in the forearm and the hand. Ther Umsch 71(7):391–396
33. Tricot M, Duy KT, Docquier PL (2012) 3D-corrective osteotomy using surgical guides for posttraumatic distal humeral deformity. Acta Orthop Belg 78(4):538–542
34. Victor J, Premanathan A (2013) Virtual 3D planning and patient specific surgical guides for osteotomies around the knee: a feasibility and proof-of-concept study. Bone Joint J 95-B (11 Suppl A):153–158
35. Koch PP, Muller D, Pisan M, Fucentese SF (2013) Radiographic accuracy in TKA with a CT-based patient-specific cutting block technique. Knee Surg Sports Traumatol Arthrosc 21(10):2200–2205
36. Lorensen WE, Cline HE (1987) Marching cubes: a high resolution 3D surface construction algorithm. In: ACM SIGGRAPH computer graphics, ACM. ISBN:0897912276
37. Chen Y, Medioni G (1992) Object modelling by registration of multiple range images. Image Vis Comput 10(3):145–155
38. Mount DM, Arya S (1998) ANN: library for approximate nearest neighbour searching
39. Dumont CE, Pfirrmann CW, Ziegler D, Nagy L (2006) Assessment of radial and ulnar torsion profiles with cross-sectional magnetic resonance imaging. A study of volunteers. J Bone Joint Surg Am 88(7):1582–1588
40. Matsumura N, Ogawa K, Kobayashi S, Oki S, Watanabe A, Ikegami H et al (2014) Morphologic features of humeral head and glenoid version in the normal glenohumeral joint. J Shoulder Elbow Surg 23(11):1724–1730
41. Vroemen JC, Dobbe JG, Jonges R, Strackee SD, Streekstra GJ (2012) Three-dimensional assessment of bilateral symmetry of the radius and ulna for planning corrective surgeries. J Hand Surg Am 37(5):982–988
42. Horn BK (1987) Closed-form solution of absolute orientation using unit quaternions. JOSA A 4(4):629–642

43. Meyer DC, Siebenrock KA, Schiele B, Gerber C (2005) A new methodology for the planning of single-cut corrective osteotomies of mal-aligned long bones. Clin Biomech (Bristol, Avon) 20(2):223–227
44. Miniaci A, Ballmer F, Ballmer P, Jakob R (1989) Proximal tibial osteotomy: a new fixation device. Clin Orthop Relat Res 246:250–259
45. FuJISAwA Y, Masuhara K, Shiomi S (1979) The effect of high tibial osteotomy on osteoarthritis of the knee. An arthroscopic study of 54 knee joints. Orthop Clin North Am 10(3):585–608
46. Schneider P, Eberly DH (2002) Geometric tools for computer graphics. Morgan Kaufmann, Burlington ISBN:0080478026
47. Moreland J, Bassett L, Hanker G (1987) Radiographic analysis of the axial alignment of the lower extremity. J Bone Joint Surg Am 69(5):745–749
48. Fabri A, Pion S (2009) CGAL: the computational geometry algorithms library. In: Proceedings of the 17th ACM SIGSPATIAL international conference on advances in geographic information systems, ACM. ISBN:1605586498
49. Schemitsch EH, Richards RR (1992) The effect of malunion on functional outcome after plate fixation of fractures of both bones of the forearm in adults. J Bone Joint Surg Am 74(7):1068–1078
50. Kendoff D, Lo D, Goleski P, Warkentine B, O'Loughlin PF, Pearle AD (2008) Open wedge tibial osteotomies influence on axial rotation and tibial slope. Knee Surg Sports Traumatol Arthrosc 16(10):904–910
51. Oka K, Murase T, Moritomo H, Goto A, Nakao R, Sugamoto K et al (2011) Accuracy of corrective osteotomy using a custom-designed device based on a novel computer simulation system. J Orthop Sci 16(1):85–92
52. Takeyasu Y, Oka K, Miyake J, Kataoka T, Moritomo H, Murase T (2013) Preoperative, computer simulation-based, three-dimensional corrective osteotomy for cubitus varus deformity with use of a custom-designed surgical device. J Bone Joint Surg Am 95(22):e173
53. Fürnstahl P, Schweizer A, Nagy L, Szekely G, Harders M (2009) A morphological approach to the simulation of forearm motion. In: Conference proceedings of IEEE engineering medicine biology society, pp 7168–71
54. Oka K, Murase T, Moritomo H, Goto A, Sugamoto K, Yoshikawa H (2009) Accuracy analysis of three-dimensional bone surface models of the forearm constructed from multidetector computed tomography data. Int J Med Robot 5(4):452–457
55. Omori S, Murase T, Kataoka T, Kawanishi Y, Oura K, Miyake J et al (2014) Three-dimensional corrective osteotomy using a patient-specific osteotomy guide and bone plate based on a computer simulation system: accuracy analysis in a cadaver study. Int J Med Robot 10(2):196–202

Preoperative Planning of Periacetabular Osteotomy (PAO)

Timo M. Ecker, Li Liu, Guoyan Zheng, Christoph E. Albers
and Klaus A. Siebenrock

Abstract Pelvic osteotomies improve containment of the femoral head in cases of developmental dysplasia of the hip or in femoroacetabular impingement due to acetabular retroversion. In the evolution of osteotomies, the Ganz Periacetabular Osteotomy (PAO) is among the complex reorientation osteotomies and allows for complete mobilization of the acetabulum without compromising the integrity of the pelvic ring. For the complex reorientation osteotomies, preoperative planning of the required acetabular correction is an important step, due to the need to comprehend the three-dimensional (3D) relationship between acetabulum and femur. Traditionally, planning was performed using conventional radiographs in different projections, reducing the 3D problem to a two-dimensional one. Known disturbance variables, mainly tilt and rotation of the pelvis make assessment by these means approximate at the most. The advent of modern enhanced computation skills and new imaging techniques gave room for more sophisticated means of preoperative planning. Apart from analysis of acetabular geometry on conventional x-rays by sophisticated software applications, more accurate assessment of coverage and congruency and thus amount of correction necessary can be performed on multi-planar CT images. With further evolution of computer-assisted orthopaedic surgery, especially the ability to generate 3D models from the CT data, examiners were enabled to simulate the in vivo situation in a virtual in vitro setting. Based on this ability, different techniques have been described. They basically all employ virtual definition of an acetabular fragment. Subsequently reorientation can be simulated using either 3D calculation of standard parameters of femoroacetabular morphology, or joint contact pressures, or a combination of both. Other techniques employ patient specific implants, templates or cutting guides to achieve the goal of safe

T.M. Ecker (✉) · C.E. Albers · K.A. Siebenrock
Department of Orthopaedic Surgery, Inselspital, University of Bern, Freiburgstrasse,
3010 Bern, Switzerland
e-mail: TimoMichael.Ecker@insel.ch

L. Liu · G. Zheng
Institute for Surgical Technology and Biomechanics, University of Bern,
Stauffacherstrasse 78, 3014 Bern, Switzerland

© Springer International Publishing Switzerland 2016
G. Zheng and S. Li (eds.), *Computational Radiology
for Orthopaedic Interventions*, Lecture Notes in Computational
Vision and Biomechanics 23, DOI 10.1007/978-3-319-23482-3_8

periacetabular osteotomies. This chapter will give an overview of the available techniques for planning of periacetabular osteotomy.

1 Evolution and Types of Pelvic Osteotomies

Pelvic osteotomies can basically be divided into two subgroups [11]. The first group consists of the osteotomies merely thought for acetabular augmentation. Sometimes these osteotomies are also referred to as salvage procedures and are used once joint congruity is largely lost and an irreducible (sub-)luxation of the joint and mismatch between the femoral head size and size of the acetabulum predominates. The second group of osteotomies is defined as acetabular reorientation osteotomies. These techniques are usually applied to a hip joint, which is still largely congruent and in which a concentric correction can be a goal of surgery. Reorientation osteotomies can be either simple or complex. The simple osteotomies have in common that most of these techniques utilize the not yet ossified portions of the acetabulum for correction. These osteotomies may thus become applicable in very young children (<2 years of age). On the contrary, the other reorientation osteotomies can be defined as complex. Whereas the above-named augmentation osteotomies and also the simple reorientation osteotomies only change certain parts of acetabular morphology, the complex reorientation osteotomies generally aim for redirection of the position of the whole acetabulum within the pelvis. The procedures are all technically demanding and hence associated with an increasing need for accurate preoperative planning, as shall be illustrated later in this chapter.

2 Augmentation Osteotomies

2.1 Chiari Osteotomy

The Chiari osteotomy [6] is an acetabular augmentation osteotomy which can be the treatment of choice in hip joints were congruency is already lost and acetabular reorientation osteotomies cannot be applied. It is a salvage procedure that utilizes a supraacetabular ilium osteotomy and breakage of the supraacetabular bone stock to achieve a medialization of the acetabulum. The containment of the head is achieved by the supracetabular bone stock and the interposition of capsular and periarticular soft tissue.

2.2 Shelf Osteotomy

The Shelf osteotomy [28] is an augmentation osteotomy using local reinforcement of the acetabular roof, rather than osteotomizing the periacetabular bone. Basically, a supraacetabular bone slot is created and subsequently impacted with structural bone graft. Thus the superolateral containment of the femoral head can be improved.

3 Simple Reorientation Osteotomies

3.1 Pemberton Osteotomy [18]

This osteotomy was first described in the 1960s. It is an osteotomy, which uses the non-ossified triradiate cartilage in children between 1–14 years of age for correction. The osteotomy starts the level of the anterior inferior iliac spine and towards the posterior aspect of the non-ossified triradiate cartilage. Hence, it is only applicable until the closure of the triradiate cartilage. Congruent size of the femoral head and the acetabulum is furthermore mandatory. Limitations for correction are dependent on the flexibility of the triradiate cartilage and commonly this osteotomy only permits for correction of containment in the anterior and lateral aspects of the hip joint.

3.2 Salter Osteotomy [25]

The Salter osteotomy was also first described in the 1960s. It consists of a single osteotomy along the innominate bone and is thus also referred to as the innominate osteotomy. Its correction possibilities are dependent on the flexibility of the pelvis encountered in young children, which puts the best age for surgery between 2 and 9 years of age. Once the osteotomy has been performed from the sciatic notch to the anterior inferior iliac spine, the correction is achieved by rotating the fragment over the femoral head. This shows that two further prerequisites for application of this procedure is a flexible symphysis and a hip joint that must be congruent without any dislocation associated with the underlying dysplasia. Correction possibilities are also limited to anterolateral improvement of containment.

3.3 Dega Osteotomy [7]

The Dega osteotomy is a similar osteotomy to the Pemberton osteotomy, using the non-ossified triradiate cartilage to correct deformity. In contrast to the Pemberton osteotomy, the bone cut is initiated at supraacetabular position in form of a transiliac osteotomy. The location of the osteotomy and the technical aspect of bone block interposition basically also allows for correction of posterior containment. As for the other simple reorientation osteotomies, the triradiate cartilage may not yet be closed and the hip joint must be largely congruent in order to improve containment.

4   Complex Reorientation Osteotomies

4.1   Sutherland Osteotomy

The Sutherland osteotomy [31] is a complex double osteotomy of the pelvis, where the innominate bone is osteotomized at the pubis and the ilium. A prerequisite for application is a congruent articulation. It is usually performed through a double incision technique, allowing for a Salter-like supraacetabular osteotomy and a second osteotomy of the pubic bone between the pubic tubercle and the symphysis with resection of a bone block. Consequently, the correction possibility due to the double osteotomy is increased in contrast to other simple reorientation osteotomies. The vicinity of important neurovascular structures, especially during the pubic osteotomy, makes this technique technically demanding.

4.2   Steele and Toennis Ostetomies

These osteotomies are technically related and both require three osteotomies of the innominate bone at the pubis, the ischium and the ilium. In contrast to the techniques described before, the triple osteotomies allow for a three-dimensional (3D) reorientation of the acetabulum. Both techniques have initially been performed using two incision techniques. The Steel osteotomy [30] requires an approach located inguinally over the pubic ramus through which after dissection of the pectineus muscle the pubic ramus and subsequently after dissection of the adductor magnus muscle, the ischium osteotomy can be absolved. The osteotomy of the ilium is then achieved in analogy to the Salter technique. The Toennis osteotomy [35] has initially been described as a two incision technique requiring a prone position with a direct posterior access for the ischium osteotomy and a supine position for the ilium and pubis cuts. As already stated, these triple osteotomies allow for larger degrees of freedom and hence bigger correction possibilities. The adverse effect of these invasive procedures is disruption of pelvic column continuity, associated risk for pseudarthrosis, avascular necrosis of the acetabulum as well as injuries to major anatomic structures in the vicinity of the osteotomies (sciatic nerve, inguinal neurovascular bundle, etc.).

4.3   Bernese Periacetabular (Ganz) Osteotomy (PAO)—Introduction and Surgical Technique

All the previously described pelvic osteotomies, although still used today under the correct indication, had several limitations. In summary, none of these osteotomies were truly "periacetabular" osteotomies. Moreover, they partly required multiple incisions, correction possibilities were limited due to soft-tissue restraints or cartilage properties and some techniques were not applicable to the mature skeleton with sequelae of developmental dysplasia of the hip. Finally, some surgical approaches and the localization of the bone cuts potentially threatened acetabular bloodflow posing a risk for avascular necrosis. In contrast to all the above-named pelvic osteotomies, the Bernese periacetabular osteotomy, also called Ganz Osteotomy, has several unique features. First described in 1983 by Ganz et al. [10], this osteotomy is performed through a single incision through a modified Smith-Petersen approach and it provides large 3D correction possibilities of the entire acetabulum. Furthermore, by preserving the posterior column of the acetabulum, it does not interfere with acetabular blood supply, does not disrupt pelvic continuity, nor change the true shape of the pelvis—an additional advantage in women of childbearing age. After establishment of the surgical approach through a modified Smith-Petersen approach, the first osteotomy to be performed is the osteotomy of the ischium. It is a blind osteotomy at the infracotyloid groove and it is an incomplete cut through the ischium leaving the posterior column intact. The second osteotomy is the pubis osteotomy medially to the pubic eminence, where the inguinal neurovascular bundle has to be protected. The final osteotomy is performed in two steps and is finalized with a controlled breakage of the acetabulum. The first limb of this osteotomy begins inferior to the anterior superior iliac spine and is conducted towards the posterior column, ending approximately 2 cm before the pelvic brim. The second limb of this osteotomy angles downward about 110° parallely to the posterior column and meets the primary ischium osteotomy. Insertion of a lamina spreader into this osteotomy aids is breaking the remaining bony bridges and completely mobilizes the acetabulum. After the acetabulum has been mobilized, the correction possibilities are given in all degrees of freedom (Fig. 1). Rotational movements can be performed in all three dimensions and the acetabular fragment can furthermore be medialized and lateralized. The only boundaries are given by the bony constraints of the pelvis towards the extremes of movement of the acetabular fragment. This is a large advantage compared to other complex reorientation osteotomies, where ligamentous restraints such as the sacro-spinous ligaments (Tonnis osteotomy), or remaining bone bridges (Sutherland osteotomy) limited the amount of achievable acetabular reorientation. The initial indication for PAO was for patients suffering from developmental dysplasia of the hip joint. In these patients, acetabular reorientation is generally achieved by flexion and external rotation of the acetabular fragment in order to improve containment of the femoral head. Later, the indication for PAO was extended to patients with femoroacetabular impingement due to a generally retroverted acetabulum. In these patients the excessive containment of the femoral head is corrected by extension and internal rotation of the acetabular fragment.

The amount of possible correction and the possibility to reorient the acetabulum in all degrees of freedom show why it is so important to plan this operation appropriately. It also shows the challenge the surgeon faces in trying to comprehend such a procedure in a 3D understanding. With the advent of more complex pelvic osteotomies thus, the necessity for appropriate planning became undeniable.

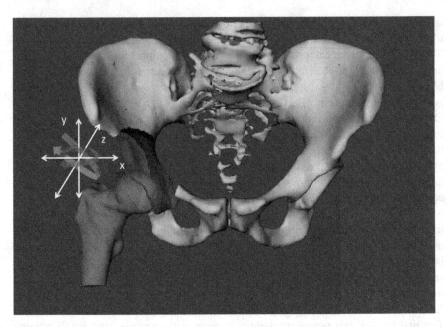

Fig. 1 A 3D model of the pelvis with the osteotomized acetabulum is depicted. The Ganz PAO allows for reorientation of the acetabulum in all degrees of freedom, since there are no ligamentous restraints limiting reorientation. The large correction possibilities and the complexity of such a procedure explain the need for proper preoperative planning

5 Basic Goals of Acetabular Reorientation

The goal of acetabular reorientation is to restore or to approximate normal acetabular geometry. Several parameters describing the configuration and geometry of the acetabulum have been defined and are recognized as standard parameters. Their importance in using these parameters for quantification of acetabular under- or over-coverage has been proven (Fig. 2) [32]. These parameters are generally used to estimate the amount and type of correction to aim for, when planning a PAO. The geometric complexity of this task becomes apparent when looking at all the factors to consider. The common parameter should be explained here briefly and are depicted in Fig. 2. The lateral and medial center-edge angles (LCE and MCE angle) as well as the acetabular arch (AA) are measurements that describe angles between a parallel vertical line to the longitudinal pelvic axis and another line starting at the center of the femoral head extending to either the lateral (LCE) or medial (MCE) edge of the acetabular sourcil, respectively the sum of both angles (AA = LCE + MCE). The extrusion index quantifies the amount of femoral head, which is covered, respectively uncovered by the acetabulum as measured in two dimensions in relation to the total horizontal femoral head diameter. The acetabular

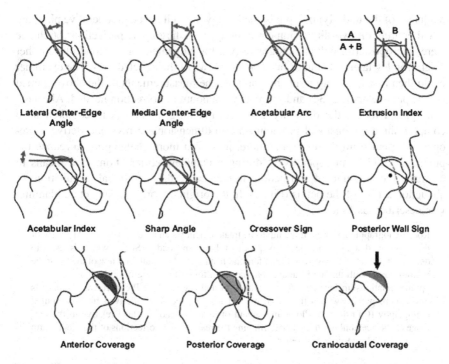

Fig. 2 The radiographic parameters quantifying acetabular geometry in terms of over- and under-coverage are shown. All of these parameters are assessable on plain radiographs. However, multiplanar imaging and modern software applications furthermore enable judgement in three-dimensions and aid in comprehending the complexity of the underlying disease. Moreover preoperative planning is largely enhanced. The goal of acetabular reorientation should be restoration or approximation of the normal values for these parameters. Reprinted with permission from Clin Orthop Relat Res (2015) 473:1234–1246

index (AI) and Sharp angle (SA) both measure the angle towards the edge of the superior acetabular rim, starting from either the most medial point of the sclerotic zone in the roof (AI) or a line parallel to the horizontal pelvic axis starting from the acetabular teardrop (SA). Finally, cross over sign, anterior and posterior wall sign and the anterior, posterior and craniocaudal coverage are all measurements depending on the relationship of the anterior and posterior acetabular rim towards each other. These parameters change significantly with general or local changes of version and inclination of the acetabulum.

6   Traditional Planning of PAO

When first described in 1983, the planning possibilities for pelvic osteotomies were limited. Multiplanar imaging was just introduced and the almost unlimited access to these imaging modalities as available today was not the case in these times. The diagnosis of the underlying hip joint pathology, as well as the preoperative planning relied entirely on two-dimensional radiographs. Radiographic projections of choice were anteroposterior pelvis radiographs and for judgement in a second plane either axial, frog leg lateral, Dunn or false profile views of the affected hips. The examiner could then assess the hip joints by applying the measurements of the above-named classic parameters [32, 34] and estimate the amount of correction needed. Although still applied today, the difficulty with this planning method is to comprehend a complex three-dimensional problem on two-dimensional x-rays and derive a pre-operative plan from these. Furthermore, it is even more challenging to realize the preoperative plan intraoperatively during acetabular reorientation. One means to estimate the amount of correction needed was the additional acquisition of abduction x-rays of the pelvis (Fig. 3). In the original work by Ganz et al. planning was described as follows:

> Roentgenographic analysis of the hip to be treated includes an anteroposterior (AP) view of the pelvis and a false profile (faux profil of Lequesne and deSeze), which is used to determine the anterior coverage of the femoral head. An additional AP view of the hip to be osteotomized, with the leg in abduction, has been found to be useful in demonstrating the optimal relationship between the femoral head and acetabulum. Preoperative planning is much more precise with a three-dimensional imaging technique derived from computed tomography (CT) data. It allows movement of femoral head cartilage relative to the cartilage of the acetabulum, thereby determining the real angular corrections of the acetabulum and, if needed, the proximal femur.

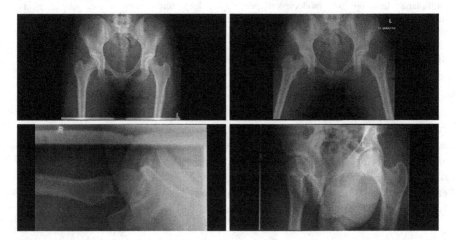

Fig. 3 A typical set of planning x-rays before periacetabular osteotomy is shown. The *left upper* image is a regular ap pelvis x-ray, the *lower left* and *right* images are axial, respectively false profile views of the hip joint. The *upper right* image is an abduction x-ray with both hip joints in 30° of abduction. In addition to the standard radiographs, the abduction view allows for estimation an increase in femoral head containment that might be achieved with the reorientation procedure

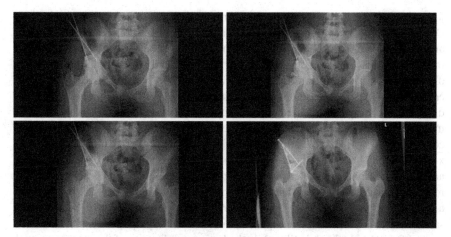

Fig. 4 Intra- and post-operative images of the same patient introduced for the planning x-rays are shown. In consecutive order from *upper left* to *lower right*, these images show how the acetabular reorientation is performed in a stepwise pattern. Once the periacetabular osteotomies have been performed, the acetabular fragment can be moved freely. According to the underlying pathology, the surgeon performs the correction. The fragment is then temporarily fixated using threaded Kirschner wires and intraoperative ap radiographs of the pelvis are obtained to control the orientation judged by the above-named parameters. This step can be repeated until the desired correction is achieved and the fragment can be definitely fixed by using 3.5 mm cortical screws (*lower right image*)

7 Intraoperative Verification of Reorientation

Intraoperatively, once the periacetabular osteotomies have been performed, the correct reorientation of the acetabulum is controlled with serial anteroposterior pelvic x-rays in analogy to preoperative planning. The above-mentioned parameters are re-assessed during this crucial step of the operation, while the acetabular fragment is temporarily secured with threaded Kirschner wires (Fig. 4). Once the desired position of the acetabulum has been achieved as verified by the intraoperative x-ray, the definitive fixation can then be performed with 3.5 mm cortical screws.

8 Assessment of Conventional X-rays Using Software Applications

With the advent of digital radiography and enhanced options of image processing, the methods for assessment of conventional x-rays and hence preoperative became more sophisticated. Analogous plain films have to be evaluated using rulers and markers, in order to measure the recognized radiographic parameters around the hip joint. Digital images—regardless the format—can be assessed using a computer. One of the things in common is the potential source of measurement error due to tilt and rotation of the pelvis during acquisition of the x-ray [27]. While correction on plain analogous films is not possible, one sophisticated application able to do on digital images has been introduced and validated. Hip^{2}Norm is a software application that allows for measurement of the above-named radiographic parameters on digital pelvic x-rays, normalizing these x-rays for tilt and rotation. Validated against measurements based on CT scans, the method proved to be an accurate, reliable, consistent and reproducible application for assessment of these parameters on plain x-rays [33]. In more detail, the software allows the correction of the projected acetabular rim and associated important parameters for individual pelvic tilt and rotation. It uses two linear distances, which are derived from an ap pelvis x-ray and a subsequent one time lateral pelvis x-ray. The vertical distance between the sacrococcygeal joint and the superior border of the symphysis is then used to control for individual pelvic tilt, while the horizontal distance of the sacrococcygeal joint and the symphysis is used to control for rotation. The examiner can then use an interface to define certain anatomic landmarks on the ap pelvis x-ray (Fig. 5). The computer then calculates the above-named classic geometrical parameters of the hip joint and provides a result sheet with the raw data (Fig. 6) and the data corrected for individual tilt and rotation of the patient's pelvis (Fig. 7). The potential of this software shows, how it can be used for more sophisticated preoperative planning of a PAO procedure. The examiner can assess the hip joint's configuration more accurate and base the correction during the intervention on this assessment. Furthermore, the software can be used to assess the postoperative x-rays in order to quantify the achieved amount of correction.

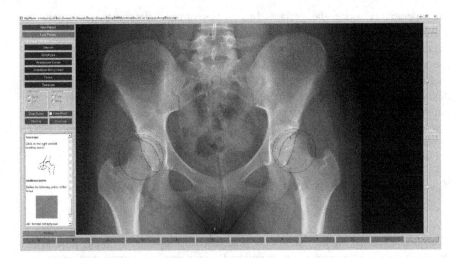

Fig. 5 The user interface of the software Hip^{2}Norm is shown. Using an ap pelvis x-ray and a one time lateral x-ray of the patients' pelvises, the surgeon can define a set of predefined anatomical landmarks. These will be used to calculate acetabular geometry in terms of the classic parameters

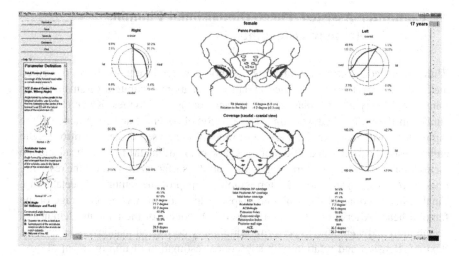

Fig. 6 The analysis report of the Hip^{2}Norm software is shown. After input of the anatomical landmarks on the ap pelvis x-ray, the computer calculates classic parameters describing hip joint morphology. This analysis shows the data, which is not yet corrected for the individual tilt and rotation of the patient's pelvis

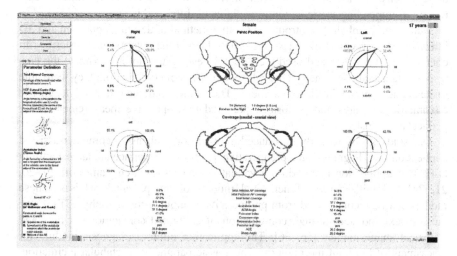

Fig. 7 The analysis report of Hip^{2}Norm of the same patient is shown, in this case now corrected for individual tilt and rotation. In order to achieve this correction, the software needs a one-time strict lateral x-ray of the patient's pelvis. The algorithm then uses the vertical distance between the sacrococcygeal joint and the superior border of the symphysis to control for individual pelvic tilt, while the horizontal distance of the sacrococcygeal joint and the symphysis is used to control for rotation. Note the differences in the values between uncorrected (Fig. 6) and corrected analysis

9   Computer-Assisted Three-Dimensional Planning
of Periacetabular Osteotomy

With enhanced multiplanar imaging methods, several authors have described methods of using the derived true 3D datasets for more accurate assessment of hip joint morphology and for planning of periacetabular osteotomies. Klaue et al. [12] already in 1988 published work on CT evaluation of coverage and congruency of the hip prior to osteotomy. In the absence of modern image segmentation software, the authors used the multiplanar slices of the CT scan to outline the femoral head and the facies lunata, as well as the cartilaginous acetabular surface. Afterwards, the resulting lines were backprojected to a 2D topographical map of the femur and acetabulum. Finally, intersections of those plotted maps could be defined as the coverage of the femoral head. Afterwards, the authors used 3D models of the patients' femur and acetabulum and were able to perform rudimentary correction maneuvers in version, flexion/extension and rotation with subsequent estimation of improvement of coverage. In another study, Millis and Murphy [16] used reconstructed CT scans for quantifying femoral head containment through measuring classic parameter such as lateral, anterior, and posterior center edge angles in true three dimensions. In order to judge the relationship of the acetabular rim and the femur, the acetabulum itself was modeled as a hemisphere using latitude angles. By assessing the hip joint in three-dimensions, the authors derived important information for the amount correction of impending reorientation procedures of different kinds.

Dutoit and Zambelli [8] introduced another method of deriving approximated 3D information from conventional x-rays. They defined the anterior and posterior rim contours on the x-rays and used the femoral head sphere to approximate the corresponding acetabular sphere. Through several steps of computation, they were able to calculate femoral head containment. They compared their method to classic coxometric measures and performed pre- and post-operative measurements after periacetabular osteotomies.

Our study group recently introduced another comprehensive computer-assisted application for diagnosis and preoperative planning of periacetabular osteotomy [14]. This application is based on a validated medical research framework [9, 20, 21, 24] and utilizes three-dimensional models of the patients' pelvises. These models can be reconstructed from preoperative computed-tomography (CT) scans, but also extrapolated through combination of calibrated conventional x-rays and statistical shape models [26, 36]. The application offers a diagnosis module for assessment of hip joint morphology. The classic parameters acetabular inclination, anteversion, LCE angle, Extrusion Index and Femoral Head Coverage are measured on the 3D models. This is possible by employment of an algorithm for automated detection of the acetabular rim [20] and by definition of the center of rotation of the hip joint on the 3D models, assuming an equidistant cartilage thickness. Definition of the acetabular rim allows for calculation of the acetabular opening plane. With this parameter established and after definition of the hip joint center of rotation, angular measurements can be performed in relation to the anterior pelvic plane.

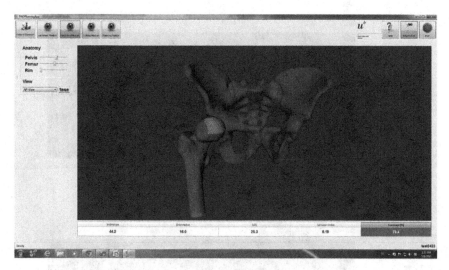

Fig. 8 A screenshot of the 3D preoperative planning software for PAO is depicted. The surgeon can visualize a model of the patient's pelvis and can assess the hip joint morphology by using the graphic information and by judging from numeric values of classic hip joint parameters, such as acetabular anteversion, inclincation, lateral center edge angle, Extrusion Index and Femoral Head Coverage. The image chosen shows a non-dysplastic pelvis exhibiting normal values

Furthermore, 3D coverage of the femoral head can be expressed in percent by differing the vertices of the femoral head model located inside the acetabulum from those outside, defined by the opening plane (Fig. 8). After the diagnosis module, the

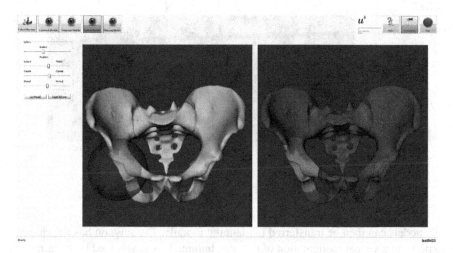

Fig. 9 Superimposing a sphere over the hip joint, which can be controlled for position and radius, creates the fragment. The acetabulum is then excised from the pelvis according to the predefined sphere

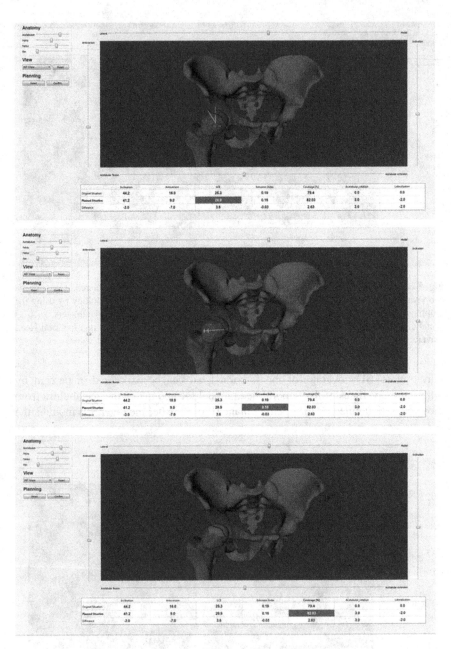

3D models can then be transferred to planning module. The surgeon has the chance to perform a virtual reorientation of the acetabulum. This is achieved by creating an acetabular fragment similar to the fragment in the impending surgery. Since the actual osteotomies do not need to be planned as an exact trajectory, superimposing

◄ **Fig. 10** A virtual reorientation of the acetabulum can be performed after it has been excised from the pelvis. The examiner can perform the typical movements in terms of flexion/extension, abduction/adduction and rotation of the fragment by using the levers depicted next to the 3D model. The changes made to the original situation become visible by real-time change of the graphic model. Furthermore, the numeric values of the above-mentioned values change and are depicted as pre- and post-planning values as well as the difference between both. This figure also shows how single values such as LCE angle, Extrusion Index and femoral head coverage can be displayed on the graphic model

a sphere over the acetabulum and excising the acetabular fragment according to this sphere simplify this process (Fig. 9). The examiner can now perform a virtual reorientation by following the typical intraoperative maneuvers including flexion/extension, abduction/adduction and rotation of the acetabular fragment. The change of parameters is displayed by the application in numeric values and real-time change of the 3D model (Fig. 10). The ability to use three-dimensional models to perform diagnosis and preoperative planning is an asset to the above-mentioned methods, because it allows for examination of a 3D problem on a 3D dataset and hence approximates reality more closely. In addition, preoperative planning performed with this application in future purposes could be transferred to a surgical navigation system and used for computer-assisted realization of the preoperative plan and control of the actual intraoperative reorientation procedure— eliminating the time consuming and rather error-prone acquisition of serial ap pelvis x-rays. Another stronghold of this application is the concomitant possibility to assess range of motion of the hip joint using a validated collision detection algorithm [9, 22]. Hence, co-existent cam impingement, which might be aggravated by acetabular reorientation, can be detected by pre- and post-operative range of motion detection after the virtual intervention.

10 Examination of Joint Contact Forces for Planning of Periacetabular Osteotomy

Apart from measuring the classic morphologic parameters of the hip joint, several authors have examined joint contact forces between femur and acetabulum, in order to detect difference in load bearing between normal and pathologic hip joints. Hipp et al. already in 1999 described a method for planning periacetabular osteotomies using joint contact pressures. They investigated a whole of 82 CT scans consisting of 70 dysplastic and 12 normal hip joints and were able to show relevant differences in contact areas and—pressure, when comparing these hip joints. After computerized virtual reorientation of the acetabulum, pressures were reassessed for different positions of the acetabulum and during different activities, such as gait, rotation, etc.

Mechlenburg et al. [15] used a stereologic method based on 3D CT scanning to measure the area of load bearing in 6 dysplastic hip joints before and after periacetabular osteotomy. These dysplastic hip joints were compared to a control group of 6 normal hip joints. While preoperatively the area of load bearing was significantly smaller in the dysplastic hips, the acetabular reorientation procedures significantly changed the load bearing area, approximating it to the load bearing area in normal hip joints.

Armiger, Lepistö and co-authors [2, 3, 13] published a series of papers about biomechanical guiding systems during periacetabular osteotomies. Ultimately in 2009 they presented a manuscript reporting on three-dimensional mechanical evaluation of joint contact pressure in 12 PAO patients with a 10-year followup. They measured radiologic angles and joint contact pressures in these patients pre- and post-operatively. The authors were able to show that 10 years postoperatively, peak contact pressures were reduced 1.7-fold and that lateral coverage increased in all patients.

Our study group has also implemented an algorithm for biomechanical optimization of planning of periacetabular osteotomy into the 3D planning application shown above using a constant thickness cartilage, which was validated for feasibility against a set of patient specific cartilage models. Three-dimensional models of the patients' anatomy are reconstructed using image segmentation software. Subsequently, classic parameters describing hip joint morphology are calculated by the application as introduced above. In order to employ improved assessment through measurement of contact forces, the three-dimensional polygonal models of the acetabulum and femur are imported into a special software application, which allows for generation of a surface mesh onto the joint couples, simulating the articular cartilage. This surface mesh can be chosen as a constant cartilage thickness or, as far as available as a patient specific cartilage thickness. As described before [37], after application of a predefined Young modulus and Poisson ratio to the cartilage mesh and the underlying cortical bone, another software application is used to calculate load distribution between the femoral and acetabular cartilage, neglecting friction coefficients and applying accepted boundary and loading conditions [4, 5, 19] (Fig. 11). This method can ultimately be used to assess any hip joint for force distribution along the bearing surfaces and especially be useful during planning of periacetabular osteotomy.

11   Using Patient Specific Implants and Mechanical Devices for Planning of Periacetabular Osteotomy

Several authors reported on the use of patient specific implants, templates, or mechanical devices for planning and conduction of periacetabular osteotomy. All of these studies rather focused on preoperative planning and intraoperative guidance of the osteotomies. They commonly do not enable for exact planning of the actual reorientation procedure.

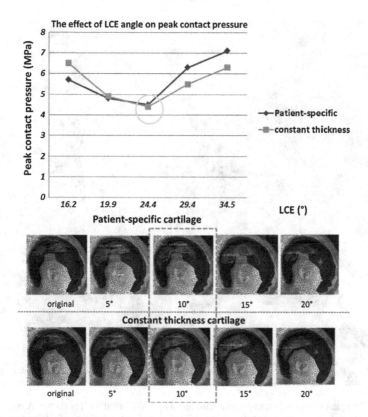

Fig. 11 The optimization of preoperative planning by employing biomechanical evaluation through calculation of joint contact pressures is depicted. The graph and the images show the feasibility of application of a constant cartilage thickness. Comparing the pressure distribution of deliberately chosen constant cartilage thickness and patient specific cartilage thickness showed similar results, when assessed under varying LCE angles. The graph correlates pressure to LCE angle and the images show the change in distribution for both groups with LCE angle change in 5° increments

Radermacher and Staudte [23, 29] report on the use of image-based individual templates for performance of triple osteotomy of the pelvis. They used individual templates, which were reconstructed and customized on the basis of 3D models of corresponding CT scans of the patients. Preoperatively, these templates were milled into or out of the 3D reconstruction of patient bone models. For triple osteotomy, the surgeon used a 3D model from CT data of the patient to define an acetabular sphere for safe zones of the periacetabular osteotomies and afterwards specified all three osteotomies in relation to this sphere. The acetabular fragment was created virtually and reoriented by the lateral and anterior center edge angles computed on the 3D model. With proper reorientation and anticipation of likely bony collision, the surgeon could then define for which of the osteotomies, a cutting guide should

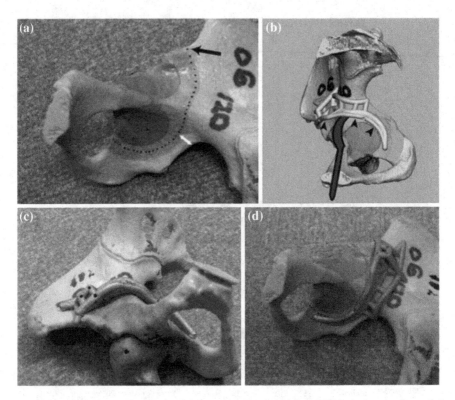

Fig. 12 The patient specific cutting guide introduced by Otsuki et al. [17] is shown. A color coded map was painted on a 3D model of the patients' pelvis in order to define the safe zones for a curved osteotomy (**a**). The cutting line was implemented in a virtual 3D model and a cutting guide was planned according to the zone of safe osteotomy (**b**). The cutting guide was then 3D-printed and intraoperatively fixed to the pelvis in order to guide the osteotome during the curved osteotomy (**c, d**). Reprinted with permission from Int Orthop (2013) Jun 37(6):1033–8

be milled using a 3D desktop milling device. The resulting polycarbonate templates could then be sterilized. The authors usually only used a template for the iliac osteotomy. The template was intraoperatively fixed to the bone by pins. Subsequently, the osteotomy was performed using a Jigsaw and guided by the template. The actual reorientation of the osteotomy was then again performed conventionally and measured by goniometer or x-ray assessment.

In a more recent study Otsuki et al. [17] introduced a novel custom cutting guide for curved periacetabular osteotomy. They used full-scale three-dimensional plaster models of the patients' anatomy, which were derived from preoperative CT scans. These models were color-coded spherically around the acetabulum representing three pre-defined radii originating from the femoral head center. It needs to be stated that only congruent hip joint with minimal deformity of the femoral head were chosen. Using these models, a cutting line was drawn manually into the color-coded spheres. It was chosen in order to avoid intraarticular penetration and to preserve a minimum 1.5 cm of posterior column. Subsequently, according to the manually reconfirmed cutting line, they used the 3D models to create a custom cutting guide employing a selective laser-melting machine using commercially available pure titanium powder. The cutting guide was equipped with holes for wire fixation to the pelvis and also a guiding fin which surface is parallel to the cutting surface in order to guide an osteotome along its curvature (Fig. 12). In surgery, after access, the cutting guide was fixed to the pelvis, the correct fit confirmed under fluoroscopy and the curved osteotomy performed under guidance of the cutting template.

12 Summary

Pelvic osteotomies have been established in order to correct deformities of the acetabulum encountered in diseases like developmental dysplasia of the hip or femoroacetabular impingement due to generalized acetabular retroversion. This group of surgical procedures has undergone an evolution, ranging from rather simple osteotomies, which are thought to reinforce and augment the acetabulum to complex three-dimensional reorientation osteotomies, allowing for large corrections of acetabular position. With increasing complexity of the procedure, preoperative planning has become more important—on the one hand to assure for safe con-duction of the osteotomies and more important for exact prediction and planning of acetabular reorientation. The quality of this reorientation during surgery may be the most important outcome modifier in the long-term outcome of these procedures [1]. While traditional planning is performed using conventional x-rays, failing to comprehend the 3D complexity of the underlying problem, more modern approa-ches to preoperative planning include employment of intelligent software applica-tions to assess geometric parameters of the hip joint on conventional x-rays, as well as utilization of enhanced multiplanar imaging and 3D model reconstructions in order to virtually simulate the impending reorientation procedure and control the surgery using navigation systems. Other approaches assess contact pressures between femur and acetabulum in order to guide towards successful reorientation. Finally, the science of patient specific implants/guides is also applied to the field, by creation of cutting templates, which are supposed to aid during the partly complex osteotomies.

In conclusion, pelvic osteotomies comprise a number of technically demanding procedures. In case of the reorientation osteotomies, preoperative planning is cru-cial for beneficial postoperative outcome. While several modern and sophisticated methods have been introduced, the subject of future research will be refinement of these methods, utilizing advanced methods of computation, imaging and 3D templating.

References

1. Albers CE, Steppacher SD, Ganz R, Tannast M, Siebenrock KA (2013) Impingement adversely affects 10-year survivorship after periacetabular osteotomy for DDH. Clin Orthop Relat Res 471:1602–1614
2. Armand M, Lepisto J, Tallroth K, Elias J, Chao E (2005) Outcome of periacetabular osteotomy: joint contact pressure calculation using standing AP radiographs, 12 patients followed for average 2 years. Acta Orthop 76:303–313
3. Armiger RS, Armand M, Tallroth K, Lepisto J, Mears SC (2009) Three-dimensional mechanical evaluation of joint contact pressure in 12 periacetabular osteotomy patients with 10-year follow-up. Acta Orthop 80:155–161
4. Bergmann G, Deuretzbacher G, Heller M, Graichen F, Rohlmann A, Strauss J, Duda GN (2001) Hip contact forces and gait patterns from routine activities. J Biomech 34:859–871
5. Caligaris M, Ateshian GA (2008) Effects of sustained interstitial fluid pressurization under migrating contact area, and boundary lubrication by synovial fluid, on cartilage friction. Osteoarthritis Cartilage 16:1220–1227
6. Chiari K (1953) Results of the earliest treatment of congenital hip joint luxation. Arch Orthop Unfallchir 45:644–653
7. Dega W (1966) Anatomical and functional restitution in congenital hip dislocation by one-stage surgical procedure. Arch Orthop Unfallchir 60:16–29
8. Dutoit M, Zambelli PY (1999) Simplified 3D-evaluation of periacetabular osteotomy. Acta Orthop Belg 65:288–294
9. Ecker TM, Puls M, Steppacher SD, Bastian JD, Keel MJ, Siebenrock KA, Tannast M (2012) Computer-assisted femoral head-neck osteochondroplasty using a surgical milling device an in vitro accuracy study. J Arthroplasty 27:310–316
10. Ganz R, Klaue K, Vinh TS, Mast JW (1988) A new periacetabular osteotomy for the treatment of hip dysplasias. Technique and preliminary results. Clin Orthop Relat Res 232:26–36
11. Jager M, Westhoff B, Zilkens C, Weimann-Stahlschmidt K, Krauspe R (2008) Indications and results of corrective pelvic osteotomies in developmental dysplasia of the hip. Orthopade 37:556–570, 572–574, 576
12. Klaue K, Wallin A, Ganz R (1988) CT evaluation of coverage and congruency of the hip prior to osteotomy. Clin Orthop Relat Res 232:15–25
13. Lepisto J, Armand M, Armiger RS (2008) Periacetabular osteotomy in adult hip dysplasia— developing a computer aided real-time biomechanical guiding system (BGS). Suom Ortoped Traumatol 31:186–190
14. Liu L, Ecker T, Schumann S, Siebenrock K, Nolte L, Zheng G (2014) Computer assisted planning and navigation of periacetabular osteotomy with range of motion optimization. Med Image Comput Comput Assist Interv 17:643–650
15. Mechlenburg I, Nyengaard JR, Romer L, Soballe K (2004) Changes in load-bearing area after Ganz periacetabular osteotomy evaluated by multislice CT scanning and stereology. Acta Orthop Scand 75:147–153
16. Millis MB, Murphy SB (1992) Use of computed tomographic reconstruction in planning osteotomies of the hip. Clin Orthop Relat Res 154–159
17. Otsuki B, Takemoto M, Kawanabe K, Awa Y, Akiyama H, Fujibayashi S, Nakamura T, Matsuda S (2013) Developing a novel custom cutting guide for curved peri-acetabular osteotomy. Int Orthop 37:1033–1038
18. Pemberton PA (1965) Pericapsular osteotomy of the ilium for treatment of congenital subluxation and dislocation of the hip. J Bone Joint Surg Am 47:65–86
19. Phillips AT, Pankaj P, Howie CR, Usmani AS, Simpson AH (2007) Finite element modelling of the pelvis: inclusion of muscular and ligamentous boundary conditions. Med Eng Phys 29:739–748

20. Puls M, Ecker TM, Steppacher SD, Tannast M, Siebenrock KA, Kowal JH (2011a) Automated detection of the acetabular rim using three-dimensional models of the pelvis. Comput Biol Med 41:285–291
21. Puls M, Ecker TM, Steppacher SD, Tannast M, Siebenrock KA, Kowal JH (2011) Automated detection of the osseous acetabular rim using three-dimensional models of the pelvis. Comput Biol Med 41:285–291
22. Puls M, Ecker TM, Tannast M, Steppacher SD, Siebenrock KA, Kowal JH (2010) The equidistant method—a novel hip joint simulation algorithm for detection of femoroacetabular impingement. Comput Aid Surg 15:75–82
23. Radermacher K, Portheine F, Anton M, Zimolong A, Kaspers G, Rau G, Staudte HW (1998) Computer assisted orthopaedic surgery with image based individual templates. Clin Orthop Relat Res 28–38
24. Rudolph T, Puls M (2008) MARVIN: a medical research application framework based on open source software. Comput Methods Programs Biomed 91:165–174
25. Salter RB (1966) Role of innominate osteotomy in the treatment of congenital dislocation and subluxation of the hip in the older child. J Bone Joint Surg Am 48:1413–1439
26. Schumann S, Tannast M, Nolte LP, Zheng G (2010) Validation of statistical shape model based reconstruction of the proximal femur—A morphology study. Med Eng Phys 32:638–644
27. Siebenrock KA, Kalbermatten DF, Ganz R (2003) Effect of pelvic tilt on acetabular retroversion: a study of pelves from cadavers. Clin Orthop Relat Res 241–248
28. Staheli LT (1981) Slotted acetabular augmentation. J Pediatr Orthop 1:321–327
29. Staudte HW, Schkommodau W, Honscha M, Portheine F, Radermacher K (2004) Pelvic osteotomy with template navigation. In: Stiehl JB, Konermann WH, Haaker RG (eds) Navigation and robotics in total joint and spine surgery. Springer, Berlin
30. Steel HH (1973) Triple osteotomy of the innominate bone. J Bone Joint Surg Am 55:343–350
31. Sutherland DH, Greenfield R (1977) Double innominate osteotomy. J Bone Joint Surg Am 59:1082–1091
32. Tannast M, Hanke MS, Zheng G, Steppacher SD, Siebenrock KA (2015) What are the radiographic reference values for acetabular under- and over-coverage? Clin Orthop Relat Res 473:1234–1246
33. Tannast M, Mistry S, Steppacher SD, Reichenbach S, Langlotz F, Siebenrock KA, Zheng G (2008) Radiographic analysis of femoroacetabular impingement with Hip^2Norm-reliable and validated. J Orthop Res 26:1199–1205
34. Tannast M, Siebenrock KA, Anderson SE (2008) Femoroacetabular impingement: radiographic diagnosiswhat the radiologist should know. Radiologia 50:271–284
35. Tonnis D, Behrens K, Tscharani F (1981) A new technique of triple osteotomy for turning dysplastic acetabula in adolescents and adults (author's transl). Z Orthop Ihre Grenzgeb 119:253–265
36. Xie W, Franke J, Chen C, Grutzner PA, Schumann S, Nolte LP, Zheng G (2014) Statistical model-based segmentation of the proximal femur in digital antero-posterior (AP) pelvic radiographs. Int J Comput Assist Radiol Surg 9:165–176
37. Zou Z, Chavez-Arreola A, Mandal P, Board TN, Alonso-Rasgado T (2013) Optimization of the position of the acetabulum in a ganz periacetabular osteotomy by finite element analysis. J Orthop Res 31:472–479

Computer Assisted Diagnosis and Treatment Planning of Femoroacetabular Impingement (FAI)

Christoph E. Albers, Markus S. Hanke, Timo M. Ecker,
Pascal C. Haefeli, Klaus A. Siebenrock, Simon D. Steppacher,
Corinne A. Zurmühle, Joseph M. Schwab and Moritz Tannast

Abstract Femoroacetabular impingement (FAI) is a dynamic conflict of the hip defined by a pathological, early abutment of the proximal femur onto the acetabulum or pelvis. In the past two decades, FAI has received increasing focus in both research and clinical practice as a cause of hip pain and prearthrotic deformity. Anatomical abnormalities such as an aspherical femoral head (cam-type FAI), a focal or general overgrowth of the acetabulum (pincer-type FAI), a high riding greater or lesser trochanter (extra-articular FAI), or abnormal torsion of the femur have been identified as underlying pathomorphologies. Open and arthroscopic treatment options are available to correct the deformity and to allow impingement-free range of motion. In routine practice, diagnosis and treatment planning of FAI is based on clinical examination and conventional imaging modalities such as standard radiography, magnetic resonance arthrography (MRA), and computed tomography (CT). Modern software tools allow three-dimensional analysis of the hip joint by extracting pelvic landmarks from two-dimensional antero-posterior pelvic radiographs. An object-oriented cross-platform program (Hip^{2}Norm) has been developed and validated to standardize pelvic rotation and tilt on conventional AP pelvis radiographs. It has been shown that Hip^{2}Norm is an accurate, consistent, reliable and reproducible tool for the correction of selected hip parameters on conventional radiographs. In contrast to conventional imaging modalities, which provide only static visualization, novel computer assisted tools have been developed to allow the dynamic analysis of FAI pathomechanics. In this context, a validated, CT-based software package (HipMotion) has been introduced. HipMotion is based on polygonal three-dimensional models of the patient's pelvis and femur. The software includes simulation methods for range of motion, collision detection and accurate mapping of impingement areas.

C.E. Albers (✉)
Department of Orthopaedic Surgery, Fiona Stanley Hospital, 11Robin Warren Drive,
Murdoch WA 6150, Melville, Australia
e-mail: christoph.albers@insel.ch

M.S. Hanke · T.M. Ecker · P.C. Haefeli · K.A. Siebenrock · S.D. Steppacher
C.A. Zurmühle · J.M. Schwab · M. Tannast
Department of Orthopaedic Surgery, Inselspital, University of Bern, Freiburgstrasse, 3010,
Bern, Switzerland

© Springer International Publishing Switzerland 2016
G. Zheng and S. Li (eds.), *Computational Radiology
for Orthopaedic Interventions*, Lecture Notes in Computational
Vision and Biomechanics 23, DOI 10.1007/978-3-319-23482-3_9

A preoperative treatment plan can be created by performing a virtual resection of any mapped impingement zones both on the femoral head-neck junction, as well as the acetabular rim using the same three-dimensional models. The following book chapter provides a summarized description of current computer-assisted tools for the diagnosis and treatment planning of FAI highlighting the possibility for both static and dynamic evaluation, reliability and reproducibility, and its applicability to routine clinical use.

1   Introduction

Femoroacetabular impingement (FAI) is a dynamic conflict of the hip caused by an early abutment of the proximal femur onto the acetabulum. FAI typically leads to hip pain and osteoarthritis of the hip if left untreated [1]. In the past two decades, FAI has received increasing focus in both research and clinical practice. FAI is the result of a morphological abnormality of the femur, the acetabulum, or both. Two types of intraarticular impingement are described. Cam type FAI is predominantly the result of an aspherical contour at the antero-superior femoral head-neck junction that—when entering the acetabulum—applies compression and shearing forces at the chondro-labral junction (Fig. 1). This can lead to chondro-labral separation, degeneration of the labrum and detachment of the cartilage from the subchondral bone [2]. Pincer type FAI, in contrast, is characterized by a focal or general overgrowth of the acetabular rim leading to compression of the labrum between the proximal femur and the acetabular rim (Fig. 1). Pincer type impingement is typically seen in hips with acetabular protrusio (general acetabular overgrowth) or retroversion (focal acetabular overgrowth, [3]). Both cam and pincer type FAI frequently occur concomitantly, which is referred to as "mixed type" FAI. Extraarticular impingement is defined by an extracapsular contact of the proximal femur and the pelvis. Areas of contact are located between the lesser trochanter and the ischium (ischiofemoral impingement), the greater trochanter and the supraacetabular region of the ilium (iliofemoral impingement), or a prominent anterior inferior iliac spine and the anterior femoral neck (subspinous impingement). While FAI describes the pathomechanism, the abnormal morphology is caused by various underlying conditions such as Legg-Calvé-Perthes disease [4, 5], slipped capital femoral epiphysis [6, 7], or post-traumatic deformities [8]. In most cases, however, the etiology of FAI is developmental, or idiopathic.

The diagnosis of FAI is based on clinical examination and imaging modalities such as conventional radiography, MR-arthrography, and computed tomography. Conventional radiography has the advantage of almost unlimited availability, cost effectiveness, and low radiation. Many conventional radiographic parameters have been established that allow the assessment and characterization of morphological parameters of the hip, both normal and abnormal, in patients with FAI (Table 1) [9]. However, conventional radiography allows only static evaluation of the underlying

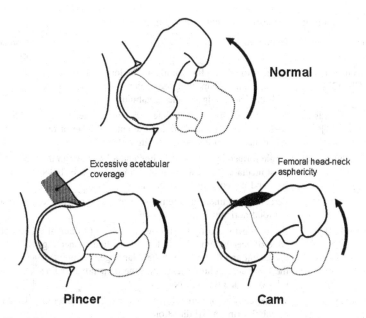

Fig. 1 The different pathomorphologies and pathobiomechanics of intraarticular FAI are shown. *Top* Normal hip with sufficient clearance for range of motion. *Bottom left* Pincer impingement: early contact of the femoral head and the acetabular rim caused by the excessive acetabular coverage or abnormal orientation of the acetabulum. *Bottom right* Cam impingement: the femoral head-neck asphericity is jammed into the acetabulum leading to labral as well as chondral damage of the acetabulum and femoral head. The majority of FAI hips exhibit features ofa combined cam-pincer pathomorphology

pathology. In addition, it is a two-dimensional representation of a three-dimensional structure, and relies on correct positioning during image acquisition. To counteract malpositioning and subsequent malorientation of the pelvis on the x-ray, software packages have been developed which correct pelvic tilt and rotation after the radiograph has been performed [10, 11]. MR-arthrograms are used for the three-dimensional assessment of bone, as well as periarticular soft tissue such as cartilage, the acetabular labrum, the hip capsule, and the surrounding muscular envelope [12–14]. Similar to conventional radiography, MR-arthrography provides only static impressions of the articulation, while it fails in the assessment of the underlying dynamic problem in FAI. Novel computer-assisted image-guiding software tools have therefore been developed to face this problem [15, 16]. CT is the method of choice for three-dimensional bony reconstruction and dynamic assessment of the hip joint due to its superior contrast properties compared to MRI [17]. The computer-assisted assessment of the dynamic conflict involved in FAI allows simulated range of motion, collision detection and accurate mapping of impingement areas. In addition, treatment planning can be achieved by performing a virtual resection of the previously detected impingement zones at the femoral head-neck junction. This book chapter gives an overview on selected

Table 1 Definition of the investigated radiographic hip parameters

Parameter	Definition	Normal values [46]
Lateral center-edge angle (LCE) [27]	Angle formed by a line parallel to the longitudinal pelvic axis and a line connecting the center of the femoral head with the lateral edge of the acetabulum	23°–33°
Anterior center-edge angle (ACE) [24]	Angle formed by a vertical line and a line connecting the center of the femoral head and the anterior edge of the acetabular rim at 25° to the sagittal plane	>25°
Acetabular index [28]	Angle formed by a horizontal line and a line through the most medial point of the sclerotic zone of the acetabular roof and the edge of the acetabulum	3°–13°
Extrusion index [14]	Percentage of uncovered femoral head in comparison to the total horizontal head diameter	70–100 %
ACM angle [23]	Angle constructed by the following points: (A) lateral edge of acetabulum, (M) midpoint of a line connecting the lateral and the inferior acetabular edge, (C) point of the bony acetabulum intersecting the perpendicular line relative to line through point M	40°–50°
Anterior coverage [11]	The percentage of femoral head covered by the anterior acetabular rim in AP direction	15–26 %
Posterior coverage [11]	The percentage of femoral head covered by the posterior acetabular rim in AP direction	36–47 %
Craniocaudal coverage [11]	The percentage of femoral head covered by the acetabular rim in craniocaudal direction	70–83 %
Crossover sign [3]	Positive if the projected anterior wall crosses the posterior wall	Negative
Retroversion index [11]	Ratio of length of retroverted acetabular opening to the entire length of the acetabular opening	
Posterior wall sign [3]	Positive if the posterior acetabular rim is projected medial of the center of the hip	Negative

computer-assisted two- and three-dimensional tools and their role for the diagnosis and treatment plan in patients with FAI. The possibility for both static and dynamic evaluation, method reliability and reproducibility, and the applicability in the routine clinical setting are outlined.

2   Two-Dimensional Computer-Assisted Tools

Evaluation of acetabular pathomorphology for pincer-type impingement (i.e. acetabular overcoverage either from retroversion or protrusion) is routinely performed using conventional two-dimensional radiographs such as an AP pelvis view. However, many radiographic parameters used to assess acetabular orientation and depth are highly dependent on the positioning of the patient during image acquisition [9, 18, 19]. Although AP pelvic radiographs are usually obtained according to a standardized protocol [14], variations of pelvic orientation are a natural consequence of the patient's posture, body habitus, and level of discomfort with certain positions. These variations are difficult to accurately assess and correct while the image is being acquired without adding additional radiation burden to the patient. If pelvic tilt and rotation are not accounted for during analysis, this can subsequently lead to misinterpretation of radiographic parameters of the hip and, ultimately, incorrect treatment decisions. Variations of pelvic tilt and rotation have been identified as the most important factors influencing the interpretation of radiographic hip parameters [18, 19]. In this context, a recent study evaluated the effect of pelvic rotation and tilt on 11 common radiographic hip parameters (Table 2) [9]. The study investigated potential deviations in an experimental setting involving 20 cadaver pelves. The pelves were placed in a neutral position (neutral rotation and 60° pelvic inclination [20]) prior to acquisition of a standardized AP radiograph. The radiograph was then virtually rotated and tilted in predefined increments. The authors found that five of the eleven parameters (anterior acetabular coverage, posterior acetabular coverage, cross-over sign, retroversion index, and posterior wall sign) changed with increasing pelvic malorientation while the other six evaluated parameters (lateral center-edge angle, acetabular index, extrusion index, ACM angle, Sharp angle, and craniocaudal coverage) exhibited no significant changes indicating to be inert to pelvic malorientation (Table 2).

The need to correct for pelvic tilt and rotation lead to the development of a software package called Hip^{2}Norm [10]. Similar to computed tomography, Hip^{2}Norm allows to accurately calculate femoral head coverage from a conventional antero-superior radiograph of the pelvis. The program extrapolates three-dimensional information about the hip joint morphology from two-dimensional AP pelvis radiographs. This is achieved by interactively digitizing the following landmarks from two-dimensional radiographic images: the inferior margins of the tear drops as the horizontal reference, the contour of the projected anterior and posterior acetabular rim, the middle of the sacrococcygeal joint and the upper border of the symphysis as the vertical reference (Fig. 2). Additionally, the femoral head as well as acetabular center and radius are obtained from a fitting

Table 2 Table outlining the parameters that are inert to pelvic rotation and tilt (restricted to a range of ±24° tilt and ±12° rotation, maximum deviation) [9]

Radiographic hip parameters that are inert to pelvic rotation and tilt	Radiographic hip parameters that change relevantly with pelvic rotation and tilt
Lateral center-edge angle [27]	Anterior acetabular coverage [11]
Acetabular index [28]	Posterior acetabular coverage [11]
Extrusion index [14]	Crossover sign [3]
ACM angle [23]	Retroversion index [18]
Sharp angle [47]	Posterior wall sign [3]
Craniocaudal coverage [46]	

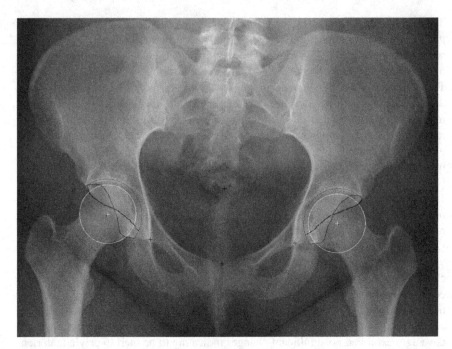

Fig. 2 The graphical user interface of Hip^{2}Norm is shown. For the calculation of radiographic hip parameters (Table 1) and the correction of deviations of pelvic tilt and rotation, the following landmarks are manually plotted [1]. The mid-point of the sacrococcygeal joint (*upper blue cross*), the upper border of the symphysis (*lower blue cross*), the inferior margins of the teardrops (*red crosses*) and the contours of the projected anterior (*blue line*) and posterior (*red line*) acetabular rim. Additionally, the femoral head center (*pink cross*) and the femoral head circumference (*pink circle*) and the acetabular center (*green cross*) and acetabular circumference (*green circle*) are obtained by fitting a circle to three points specified by the user (Color figure online)

circle that is created from three points drawn by the user. The projected anterior and posterior acetabular rims are also manually defined by the user (Fig. 2). The technique is based on a cone-beam projection model presuming the source to film distance, that is extracted directly from the technical documentation of the X-ray machine (Fig. 3), and the object to film distance (representing the surrounding soft tissue envelope) that has previously been shown to be 30 mm on average [19]. The software also requires sphericity of both the acetabulum and the femoral head. The reconstructed coordinate system from the defined landmarks allows for calibrating the acetabular rim back to a neutral, pelvic orientation consisting of pelvic tilt (around the transverse axis), pelvic rotation (around the longitudinal axis), and pelvic obliquity (around the sagittal axis). The latter is corrected by the inter-teardrop line. Pelvic rotation is restored accurately by adjusting the midpoint of the sacroiliac joint and the upper border of the symphysis to meet on a vertical

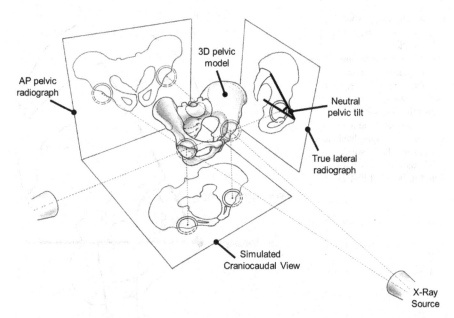

Fig. 3 Hip^{2}Norm calculates the three-dimensional configuration of the hip joint from a conventional AP pelvic radiograph. The software uses a cone projection model allowing to determine cranio-caudal, anterior and posterior femoral coverage. In addition, alterations of pelvic tilt, rotation and obliquity can be corrected with the help of vertical/horizontal distances between the symphysis and the sacrococcygeal joint (pelvic tilt and rotation) and the inter-teardrop line (pelvic obliquity). In order to assess pelvic tilt, an additional one-time strong lateral radiograph is necessary to normalize the absolute horizontal distance to the pelvic inclination angle. Alternatively, the individual pelvic tilt can be estimated with mean values for this vertical distance in a normal population (32 mm male; 47 mm female) [11, 18]

line [10]. Pelvic tilt is corrected with a one-time strong lateral pelvic radiograph and the pelvic inclination angle. The pelvic inclination angle is defined by the inter-section of a line connecting the anterior boarder of the sacral promontory with the upper border of the symphysis and a horizontal line (Fig. 4). Normal pelvic inclination has been found to be 60° [20–22]. Pelvic inclination strongly correlates with the change of the distance between the upper border of the symphysis and the midpoint of the sacrococcygeal joint. Thus, one time calibration of the distance with a lateral pelvic radiograph is recommended. However, as a strong lateral pelvic view is rarely obtained in the routine clinical setup, the individual pelvic tilt can be estimated with the vertical distance between the sacrococcygeal joint and the upper border of the symphysis. The mean values for this vertical distance in a normal population have been shown to be 32 mm in men and 47 mm in women [18]. After virtual neutralization of pelvic tilt and rotation, Hip^{2}Norm calculates the corrected coverage of the femoral head (total anterior coverage, total posterior coverage, total

Fig. 4 The image shows a
schematic of strong lateral
pelvic radiograph
(superimposition of both
femoral heads). Pelvic
inclination is measured as the
angle between a line
connecting the sacral
promontory and the anterior
boarder of the symphysis and
a horizontal line. Normal
pelvic inclination has been
found to be 60° [20–22]

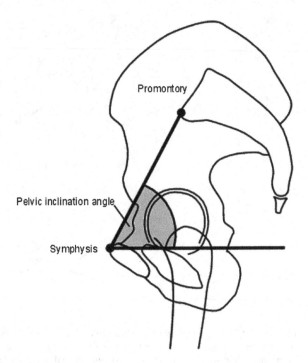

cranio-caudal coverage), as well as acetabular version (cross-over sign, retroversion index, posterior wall sign) and further common hip parameters (lateral center edge angle, acetabular index, ACM-angle, extrusion index, anterior center edge angle; Table 1) in this neutralized orientation (Fig. 5) [3, 18, 23–27]. Hip²Norm has been validated and showed high consistency, accuracy, reliability and reproducibility [11]. The mean accuracy to correct for pelvic malpositioning ranged from 0.1° to 0.7° for the angular measurements and from 0.4 to 2.0 % for the relative units/acetabular coverage when compared to the gold standard (defined by computed tomography and conventional radiographs of cadaver pelves in a neutral orientation [11]). The least accurate measurements were found for the lateral center edge angle (ICC 0.53 [0.34–0.77]) [27], the acetabular index (intraclass correlation coefficient [ICC] 0.49 [0.29–0.67]) [28], and anterior center edge angle (ICC 0.54 [0.40–0.81]) [24]. Consistency of the software was assessed by re-transforming radiographic hip parameters obtained from radiographs with the pelvis in predefined malrotation and -tilt back to the neutral orientation (correction algorithm). All parameters showed good to excellent consistency (ICC > 0.6) except the anterior center edge angle (ICC 0.61 [0.29–0.91]). A good to very good reproducibility and reliability (ICC > 0.6) was found for all parameters except for the reliability of the retroversion index (ICC 0.56 [0.46–0.65]) [11].

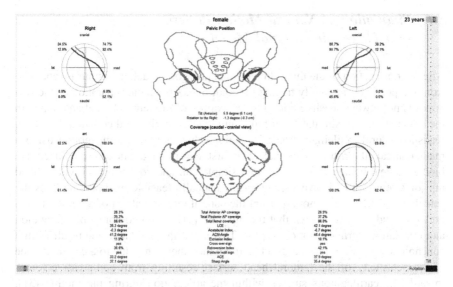

Fig. 5 The result sheet of the Hip²Norm analysis is shown. After definition of the osseous landmarks (Fig. 3), the software computes common radiographic hip parameters (Table 1). After deviations of pelvic tilt and rotation have been normalized, the software corrects all parameters that show variations depending on pelvic orientation (Table 2) [9]. High software accuracy, consistency and reliability has previously been proven in a validation study [11]

3   Three-Dimensional Computer-Assisted Tools

While conventional imaging modalities provide only static impressions of the dynamic problem in FAI, the data sets of tomographic imaging methods using image guided applications bear high potential for the dynamic assessment of FAI allowing to perform preoperative simulation of range of motion, collision detection and accurate visualization of impingement areas. In this respect, the software HipMotion has been developed to perform a CT based, three-dimensional kinematics analysis of the hip joint. The application uses three-dimensional models of the patient's anatomy based on polygonal meshes that are created from DICOM (digital imaging and communications in medicine) volume data. CT data originate from a native computed tomography scan without contrast with a minimum slice thickness of 2 mm. The CT scan of the pelvis has to include the proximal femur including the greater and lesser trochanter and distal femur exposing the femoral condyles. Although there is a potential hazard of radiation exposure, the advantages of CT data for three-dimensional model segmentation outweigh those of magnetic resonance imaging. This is mainly due to the sharp contrast between osseous structures and soft tissue. The attempts to use alternative tomographic modalities without radiation such as MRI have been unsatisfactory until now because automated segmentation of three-dimensional models remains a challenge as a result of the morphological complexity and large signal-to-noise ratio of MRI.

3.1   Simulation of Natural Motion Pattern
in a Three-Dimensional Model

After reconstruction of the three-dimensional polygon model, the femur is separated
from the pelvis by manually tracing the joint line. Automatic separation of the two
bones is not always feasible due to the proximity of the femoral head and pelvis in
the articulation. For simulation of hip joint range of motion and collision detection,
realistic, natural implementation of the biomechanics of hip joint motion using a
mathematical algorithm is crucial. The first described applications reduced the
biomechanical behavior of the hip to a simple ball and socket model with a fixed
hip joint center [29]. However, this mathematically feasible method simplifies the
calculation while it does not represent the natural motion pattern of a hip joint. In a
previous study, it was shown that the hip joint performs a combination of rotation
and translation during hip motion [30]. This is, on the one hand, the result from a
conchoid shape of the femoral head, and, on the other hand, due to a certain degree
of viscoelasticity of the femoral head surface resulting from compressive forces
towards the cartilaginous surface within the articulation during hip motion [31].
A modification to the simple "fixed center" algorithm is the constrained method that
evaluates impingement areas within a maximum perimeter of 5 mm from the
acetabular rim [32]. However, with this method intraarticular impingement is
neglected. Another algorithm, the translated method, computes a displacement
vector according to the detected impingement and performs an additional translation to the rotational center perpendicular to the detected intraarticular impingement
zone. In order to achieve higher accuracy and reliability for the location and extent
of osseous impingement areas, the equidistant method has been introduced [15].
This algorithm preserves a constant joint space imitating the cartilage lining at any
functional hip position. This is achieved by superimposing the acetabular and
femoral sphere center based on the individual joint anatomy. The result is a
dynamic joint center during virtual hip motion allowing combined rotational and
translational motion patterns (Fig. 6). In an in vitro study, the equidistant method
has been proven to be superior to the conventional methods in terms of accuracy
and reliability representing the highest resemblance with the natural characteristics
of the hip joint (Fig. 7). Thus, HipMotion operates with the equidistant method for
simulation of hip motion, collision detection and treatment planning.

The detection of the acetabular rim plays a crucial role with respect to collision
detection in FAI. The crude geometry can be obtained from the two-dimensional
x-ray [33], and tomography provides additional information about its
three-dimensional orientation. However, the exact knowledge of the contours of the
acetabular rim is crucial to determine the dimensional orientation of the acetabulum
in relation to the anterior pelvic plane [34, 35]. This is important for the accurate
detection of impingement conflicts with the femur. In the early years of
three-dimensional hip modeling, the acetabular rim was defined manually
point-by-point which is both time consuming and error-prone. In addition, the
available methods were neither accurate enough [36], nor explicitly tested to detect

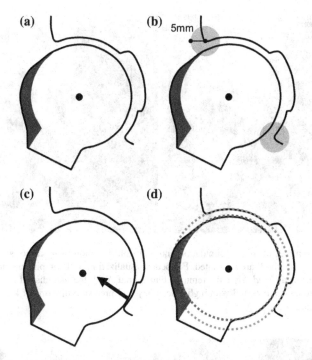

Fig. 6 Schematics of the femoral head and acetabular socket illustrating the determination of the hip joint center for purposes of hip motion simulation are shown. **a** The 'simple method' with a fixed and predefined center of rotation. **b** The 'constrained method' complying a 5 mm detection area. **c** The 'translated method' based on the simple method with the implementation of an additional vector (*arrow*) pointing in the perpendicular direction of the detected collision [4]. The 'equidistant method' computing an acetabular (*green dotted*) and femoral (*red dotted*) fitting sphere. These two spheres maintain in an equal distance to each other at any position during virtual motion simulation. This constant joint space intends to mimic the cartilaginous layers. The hip joint center is dynamic and is reassessed at any motion position (Color figure online)

the acetabular rim [37], nor properly validated [38]. HipMotion uses a recently introduced algorithm for automated rim detection on three-dimensional polygon models. The algorithm defines the acetabular opening plane with an acetabular inclination angle of 45° and an anteversion angle of 15° in a pelvic coordinate system. The acetabular rim is constructed in a stepwise fashion using vector calculation, distance mapping and back projection of acetabular vertices to the polygon model in relation to the joint center (Fig. 8). In a validation study this algorithm proved to be accurate and reliable [16].

After the three-dimensional model has been constructed, the diagnosis application is launched to conduct the animation. To calculate the angles, two reference coordinate systems based on the patients osseous anatomy are created on the individually reconstructed three-dimensional models (Fig. 9). The user manually defines landmarks for the pelvic reference system (anterior pelvic plane) by identifying the anterior superior iliac spines and the pubic tubercles [32, 34, 35].

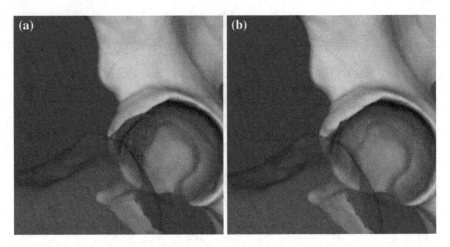

Fig. 7 Two examples of motion simulation and collision detection using **a** the 'simple method' and **b** 'equidistant method' are illustrated. For better visualization of the impingement zones at the acetabulum (*yellow* and *red dots*), the femur is blurred out. With the 'equidistant method' (**b**), the impingement zones are detected with higher accuracy and precision in terms of location and size [15] (Color figure online)

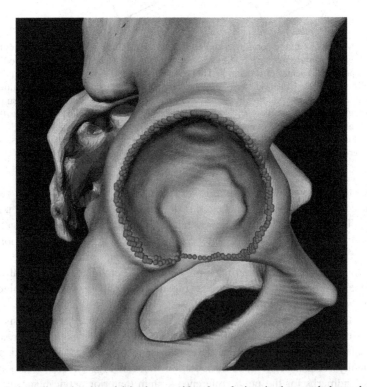

Fig. 8 A three-dimensional model is shown with a lateral view in the acetabular socket. The implementation of a previously described algorithm allows the HipMotion software to automatically detect the acetabular rim contours (*red dots*). In addition, the spatial orientation of the acetabular opening plane is determined [16] (Color figure online)

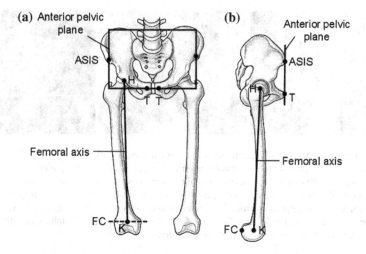

Fig. 9 The definition of the pelvic and femoral coordinate system based on anatomic landmarks is shown in the **a** antero-posterior and **b** the lateral plane. The anatomic reference points for the definition of anterior pelvic plane (pelvic coordinate system) are located at both pubic tubercles and anterior superior iliac spines. The femoral coordinate system is based the longitudinal axis between the femoral head center and he knee center (midpoint between a line connecting the medial and lateral epicondyles), and the transverse axis tangential to both posterior condyles [32]

The femoral coordinate system is constructed by defining the anatomical landmarks for the mechanical axis with the hip center and knee center as well as the posterior aspect of the femoral condyles as a reference for the coronal plane (Fig. 6; [39]). Neutral orientation of the leg in relation to the pelvis is based on these two coordinate systems and all motion patterns are performed using these two systems as direct references.

3.2 Range of Motion Simulation and Collision Detection

The software can calculate common acetabular reference values such as relative coverage of the femoral head [10, 11] or common angles (lateral center edge angle) [27], acetabular version [3, 39], inclination [40], or acetabular extrusion [41]. Assessment of virtual range of motion and collision detection is performed according to the previously described algorithm [42]. Collisions are visualized in real time on the three-dimensional model (Figs. 10, 11 and 12). The software simulates the entire spectrum of range of motion thereby visualizing any potential impingement conflict. The user can determine individual minimum and maximum values for range of motion parameters (flexion/extension; abduction/adduction; internal rotation/external rotation) as well as the degree of stepwise increments.

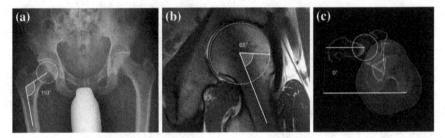

Fig. 10 The preoperative radiographs of a 38 years old male patient are shown. The patients history reveals groin pain on the right side since many years. Clinically, the patient has decreased internal rotation in flexion and increased external rotation in extension. The preoperative AP pelvic radiograph (**a**) shows a decreased CCD angle indicating a varus deformity. Next to hypertrophy of the labrum, the radial sequences around the femoral neck axis of the MR-arthrogram (**b**) show a reduction of antero-superior femoral head-neck offset with an increased alpha angle [48]. The rotation scanner CT (**c**) reveals a drop of femoral torsion (0°)

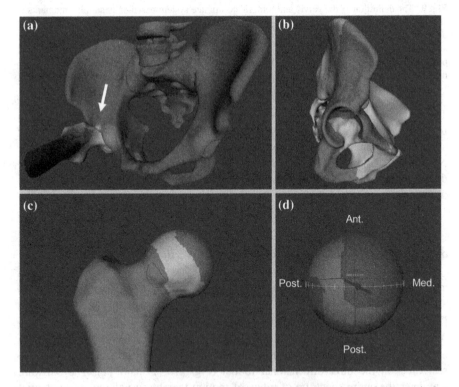

Fig. 11 Hip motion analysis of the same patient (Fig. 10) shows an early intraarticular impingement conflict (**a**, *arrow*) between the antero-superior acetabular rim (**b**) and the anterior femoral head neck junction (**c**). The motion sphere (**d**) graphically illustrates the impingement analysis with a color code map. The femur is represented by a cylinder. *Red areas* On the surface of the sphere visualize femoral positions causing impingement. *Green areas* Display impingement-free motion steps (Color figure online)

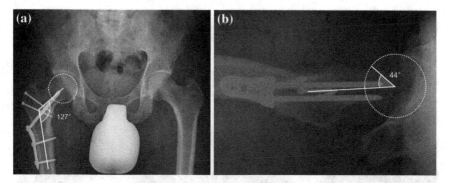

Fig. 12 The same patient (Fig. 10 or 11) was treated with a subtrochanteric, rotational-valgus osteotomy and concomitant surgical hip dislocation with head-neck offset correction resulting in **a** a restored CCD angle (127°) and **b** improved femoral head-neck offset (alpha angle 44°; [48]). One year post-operatively, the patient presented without symptoms, back to work and sports. Clinical examination showed physiological internal rotation in flexion (35°) and external rotation in extension (30°)

Using the user-supplied motion criteria, the software calculates all possible combinations of motion parameters within the defined ranges. Apart from the assessment of intraarticular impingement conflicts, the software also allows evaluation of periarticular osseous structures and detects extraarticular impingement conflicts. This is predominantly of interest in patients with complex deformities such as sequelae of Legg-Calvé-Perthes disease or Perthes-like deformities [5], or in hips with torsional deformities (Figs. 13, 14 and 15) [43] or a prominent inferior iliac

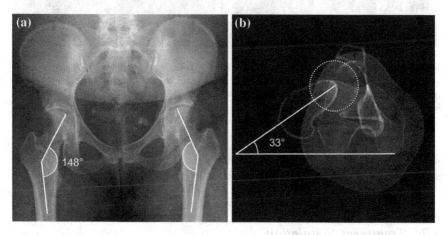

Fig. 13 The preoperative AP radiograph (**a**) and rotation scanner CT of a 36 years old female patient with hip pain, reduced external rotation in extension, increased internal rotation in flexion, and a positive anterior and posterior impingement test is shown. **a** On the AP radiograph an increased center-collum-diaphyseal (CCD) angle of 148° is present indicating a valgus deformity of the hip. **b** In addition, an abnormally high antetorsion of 33° is revealed on the rotation scanner CT

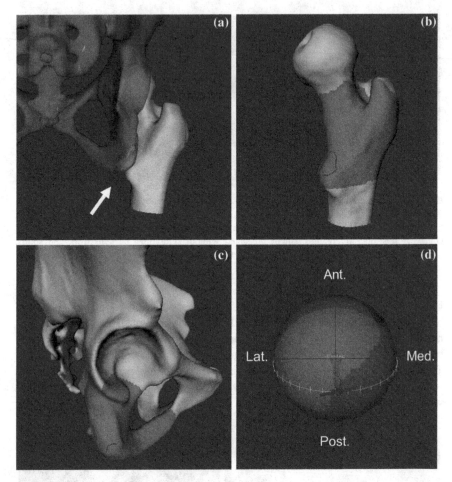

Fig. 14 Motion simulation in the same patient (Fig. 13) revealed an ischio-femoral impingement conflict (*arrow*) with a pathological abutment of the lesser trochanter (**b**) with the ischial tuberosity (**c**). The motion sphere (**d**) graphically visualizes the motion analysis distinguishing impingement-free motion steps (*green*) and positions resulting in intra-/extraarticular collisions between the acetabulum and the proximal femur (*red*) (Color figure online)

spine (Figs. 16, 17 and 18). During motion simulation and collision detection, the software does not assess interposing soft tissue or soft tissue restraints.

3.3   Treatment Simulation

Once any intraarticular and extraarticular impingement conflicts are identified and mapped, a surgical treatment plan for FAI hips can be created using a tool within

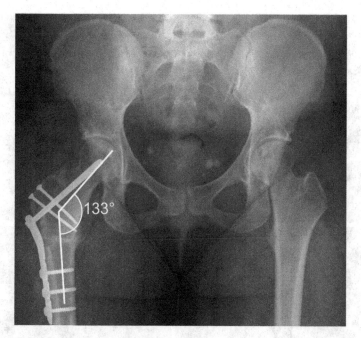

Fig. 15 The same patient (Figs. 13 and 14) was treated with a derotational-varus, subtrochanteric ostoeomy with concomitant surgical hip dislocation. At 12 months follow-up, the patient was pain free and normal range of motion was restored indicating correct postoperative femoral torsion. The osteotomies are completely consolidated

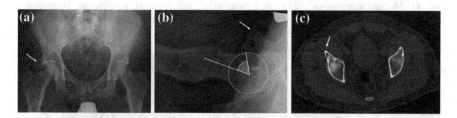

Fig. 16 The preoperative radiographic workup of a 47 years old male patient presenting with hip pain in flexion and internal rotation is shown. The patient had previously suffered from an avulsion fracture of the antero-inferior iliac spine on the right side that was treated conservatively with subsequent mal-union resulting in a protruding, prominent spine (*arrow*). **a** AP pelvic radiograph; **b** cross-table view right hip showing the prominent antero-inferior iliac spine (*arrow*) and an additional reduction of femoral head neck offset (increased alpha angle); **c** transverse section of a CT scan of the pelvis revealing the unilaterally prominent antero-inferior iliac spine (*arrow*)

HipMotion that can perform a virtual resection on the three-dimensional models. Based on the extension of the acetabular impingement zone obtained from motion simulation and collision detection, the acetabular rim is marked with a clockwise template and the osseous segment of the acetabular rim to be virtually resected

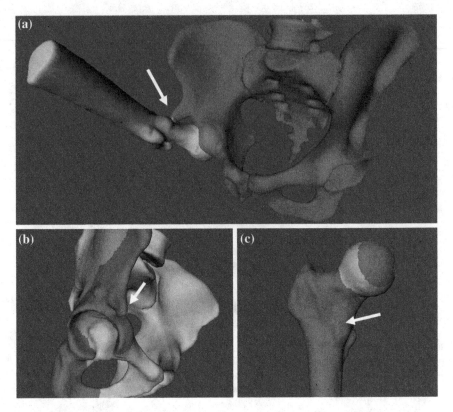

Fig. 17 Motion analysis in the same patient (Fig. 16) revealed sub-spinous impingment (**a**) between the prominent antero-inferior iliac spine (**b**) and the anterior femoral neck (**c**)

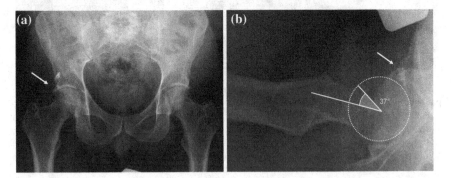

Fig. 18 In the same patient (Fig. 16 or 17) an open resection of the antero-inferior iliac spine was performed to allow for impingement-free range of motion. The origin of the rectus femoris muscle was detached for trimming down the bony protuberance and ultimately reattached with a suture anchor (**a**). In addition, open resection of the asphericity of the femoral head was performed resulting in a decreased alpha angle (**b**)

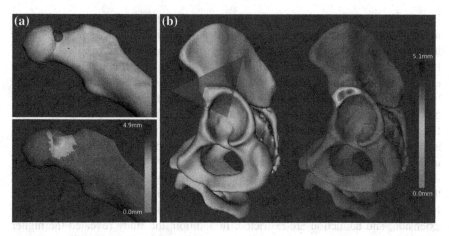

Fig. 19 Treatment simulation by virtual resection of **a** the asphericity at the femoral head neck junction and **b** the acetabular rim is illustrated. **a** On the femoral side, the virtual resection of the identified impingement area is simulated with manual mouse interactions over the bony area (*red ball* representing a burr). Visualization of the size and extend of resection is achieved by a color map indicating the resection depth. The implementation of a resection algorithm allows smooth phasing out at the resection edges. **b** On the acetabular side, the user determines the cutting distance by defining two points at the outer edges of the acetabular rim to be resected. The highlighted area (*red triangle*) can thereafter be modified to refine size and position of acetabular rim to be resected. The algorithm identifies all vertices within the selected acetabular area that are located outside of the hemisphere of the predefined acetabular opening plane contributing to acetabular overcoverage. The depth of the resection is illustrated with a color map (Color figure online)

quantified by measuring width in millimeters (Fig. 19) [29]. On the femoral side, femoral head sphericity is restored by performing a virtual head-neck osteochon-droplasty in a stepwise fashion (Fig. 19). After virtual correction of the impinge-ment, the volumetric data can be reevaluated for improved range of motion until the desired motion pattern is achieved. This helps the surgeon in decision making of the extent of surgical correction.

3.4 Research Applications

HipMotion is not only useful in the clinical setting, but has research applications as well. Several studies have been conducted to analyze the kinematics of the hip with different deformities leading to a better understanding of the biomechanics of the hip. Kubiak-Langer et al. investigated the effect of virtual osteochondroplasty of the femoral head neck junction and the acetabular rim trimming on internal rotation in hips with idiopathic cam and/or pincer type deformities [29]. The results were compared to those of normal/asymptomatic hips. The authors found decreased flexion, internal rotation and abduction for all FAI hips compared to normal hips.

By virtually correcting the deformity in hips with FAI, range of motion was restored to the degree of normal hips (Fig. 20) [29]. Another study evaluated range of motion in hips with complex deformities of the proximal femur and from sequelae of Legg-Calvé-Perthes disease or Perthes-like deformities [5]. It could be shown that range of motion in any dimension was diminished compared to both normal hips and hips with idiopathic FAI. In addition, collision detection revealed different locations and frequency of intra- as well as extraarticular impingement zones in hips with Legg-Calvé-Perthes disease. A recent study based on noninvasive three-dimensional navigation shed light on the necessity to consider torsional deformities of the femur in the setting of FAI [44]. In this context, it was shown that valgus hips with high antetorsion had distinctly different motion patterns when analyzed by means of three-dimensional simulation. Compared to normal hips and hips with idiopathic FAI, it was shown that internal rotation is increased while external rotation, extension, and adduction are restricted. In addition, the study revealed the higher incidence of posterior extraarticular impingement in valgus hips with high antetorsion [44]. Another recent study investigated the effect of acetabular reorientation in dysplastic hip on simulated range of motion [45]. In their study, the authors revealed decreased extension and internal rotation in flexion while external rotation was increased after surgery. In addition, a higher prevalence of subspine impingement was found compared to normal hips [45]. These new insights obtained from three-dimensional, noninvasive hip animation allow for a better understanding of the biomechanics of the hip and subsequent treatment planning.

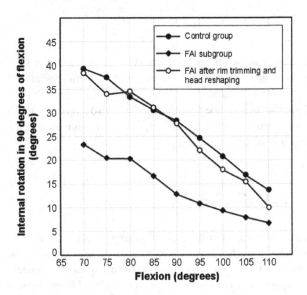

Fig. 20 The graph shows the effect of treatment planning by virtual resection of the head-neck asphericity and acetabular rim resection [29]. For all groups, internal rotation decreases with increasing flexion. However, internal rotation is significantly lower in hips with FAI compared to a control group. After virtual resection of the aspherical part of the femoral head neck junction and of the acetabular rim, internal rotation in hips with FAI is restored to the degree of the control group [29]

3.5 Future Perspectives

Noninvasive three-dimensional assessment of FAI has shown to improve the understanding of the dynamic conflict in FAI thereby achieving sophisticated preoperative treatment planning. However, developing intraoperative navigation tools to execute the preoperative plan is the subject of ongoing research. Intraoperative navigation is widely established for common surgical procedures such as total hip or knee arthroplasty. However, a comprehensive solution including a tool for the accurate diagnosis, preoperative treatment planning, virtual treatment performance and finally intraoperative navigation for patients with FAI has not been introduced yet. Another field of interest is the development of new automated segmentation algorithms as well as intraoperative non-invasive registration methods such as statistical shape models or fluoroscopically obtained images. These may open the possibility for less invasive navigational approaches in FAI surgery including open and arthroscopic procedures. The fact that current computer assisted tools rely on imaging datasets obtained from CT reveals several disadvantages. These include radiation exposition of the patient during image acquisition, limited availability and increased costs. Different attempts have been made to base segmentation of sectional imaging on alternative modalities such as MRI. However, automated image segmentation on the basis of MRI has not yet reached the standard of that attained with CT datasets. This is mainly due to the associated morphological complexity of MRI and the low contrast thresholds between bony and soft tissue structures on MRI. Modern MRI techniques with stronger magnetic fields and novel sequencing methods may open up a new field for sophisticated segmentation techniques based on MRI.

4 Limitations of Computer Assisted Diagnosis and Treatment Planning of FAI

Although two- and three-dimensional computer assisted tools facilitate the assessment of the pathomorphology as well as the associated pathobiomechanics in patients suffering FAI, there are several limitations to both methods. While Hip^{2}Norm allows to accurately determine three-dimensional femoral head coverage, the assessment of a potential a cam deformity on the femoral side is neglected. Thus, Hip^{2}Norm reveals pincer type FAI (i.e. acetabular retroversion) while the more commonly found cam-type deformity is not observed. In addition to antero-posterior pelvic radiographs, cross-table radiographs or radial sequences of MR-arthrography are needed to determine pathomorphologies on the femoral side contributing to the FAI. Another drawback is the lack of dynamic assessment of FAI using two-dimensional tools. In this context, extra-articular impingement conflicts are easily missed and acetabular pathomorphologies contributing to a pincer type deformity may be under- or overestimated. In contrast, three-dimensional techniques such as HipMotion provide the possibility of dynamic assessment. Intraarticular impingement conflicts, femoro-pelvine abutments, or torsional abnormalities of the femur can be easily detected. However, HipMotion relies on CT datasets that expose the patient to high amounts of radiation during acquisition. Another limitation using HipMotion is the neglect of soft-tissue restraints during motion simulation and collision detection. This may lead to an overestimation of FAI conflicts due to unphysiologically extensive range of motion. In contrast, the lack of detection of soft-tissue interposition during motion simulation and collision detection may lead to an underestimation of the underlying impingement conflict as interposed soft tissue between the osseous structures is not respected.

In conclusion, both two- and three-dimensional computer assisted tools for the diagnosis and treatment planning of FAI facilitate the preoperative evaluation. However, neither one is superior and both exhibit limitations. These tools should therefore only be used as an adjunct and do notreplace the assessment of conventional imaging modalities.

References

1. Ganz R, Parvizi J, Beck M, Leunig M, Notzli H, Siebenrock KA (2003) Femoroacetabular impingement: a cause for osteoarthritis of the hip. Clin Orthop Relat Res 417:112–120 Epub 2003/12/04
2. Beck M, Kalhor M, Leunig M, Ganz R (2005) Hip morphology influences the pattern of damage to the acetabular cartilage: femoroacetabular impingement as a cause of early osteoarthritis of the hip. J Bone Joint Surg Br 87(7):1012–1018 Epub 2005/06/24
3. Reynolds D, Lucas J, Klaue K (1999) Retroversion of the acetabulum. A cause of hip pain. J Bone Joint Surg Br 81(2):281–288 Epub 1999/04/16
4. Albers CE, Steppacher SD, Ganz R, Siebenrock KA, Tannast M (2012) Joint-preserving surgery improves pain, range of motion, and abductor strength after Legg-Calve-Perthes disease. Clin Orthop Relat Res 470(9):2450–2461 Epub 2012/04/25
5. Tannast M, Hanke M, Ecker TM, Murphy SB, Albers CE, Puls M (2012) LCPD: reduced range of motion resulting from extra- and intraarticular impingement. Clin Orthop Relat Res 470(9):2431–2440 (Epub 2012/04/12)
6. Albers CE, Steppacher SD, Haefeli PC, Werlen S, Hanke MS, Siebenrock KA, et al (2014) Twelve percent of hips with a primary cam deformity exhibit a slip-like morphology resembling sequelae of slipped capital femoral epiphysis. Clin Orthop Relat Res (Epub 2014/12/03)
7. Wensaas A, Gunderson RB, Svenningsen S, Terjesen T (2012) Femoroacetabular impingement after slipped upper femoral epiphysis: the radiological diagnosis and clinical outcome at long-term follow-up. J Bone Joint Surg Br 94(11):1487–1493 Epub 2012/10/31
8. Eijer H, Myers SR, Ganz R (2001) Anterior femoroacetabular impingement after femoral neck fractures. J Orthop Trauma 15(7):475–481 Epub 2001/10/17
9. Tannast M, Fritsch S, Zheng G, Siebenrock KA, Steppacher SD (2014) Which radiographic hip parameters do not have to be corrected for pelvic rotation and tilt? Clin Orthop Relat Res (Epub 2014/09/19)
10. Zheng G, Tannast M, Anderegg C, Siebenrock KA, Langlotz F (2007) Hip²Norm: an object-oriented cross-platform program for 3D analysis of hip joint morphology using 2D pelvic radiographs. Comput Methods Programs Biomed 87(1):36–45 Epub 2007/05/15

11. Tannast M, Mistry S, Steppacher SD, Reichenbach S, Langlotz F, Siebenrock KA et al (2008) Radiographic analysis of femoroacetabular impingement with Hip²Norm-reliable and validated. J Orthop Res (Official publication of the Orthopaedic Research Society) 26 (9):1199–1205 (Epub 2008/04/12)

12. Klenke FM, Hoffmann DB, Cross BJ, Siebenrock KA (2015) Validation of a standardized mapping system of the hip joint for radial MRA sequencing. Skeletal Radiol 44(3):339–343 Epub 2014/10/14

13. Leunig M, Werlen S, Ungersböck A, Ito K, Ganz R (1997) Evaluation of the acetabular labrum by MR arthrography. J Bone Joint Surg Br 79(2):230–234 Epub 1997/03/01

14. Tannast M, Siebenrock KA, Anderson SE (2007) Femoroacetabular impingement: radiographic diagnosis-what the radiologist should know. AJR Am J Roentgenol 188 (6):1540–1552 Epub 2007/05/23

15. Puls M, Ecker TM, Tannast M, Steppacher SD, Siebenrock KA, Kowal JH (2010) The equidistant method—a novel hip joint simulation algorithm for detection of femoroacetabular impingement. Comput Aided Surg off J Int Soc Comput Aided Surg 15(4–6):75–82 Epub 2010/11/12

16. Puls M, Ecker TM, Steppacher SD, Tannast M, Siebenrock KA, Kowal JH (2011) Automated detection of the osseous acetabular rim using three-dimensional models of the pelvis. Comput Biol Med 41(5):285–291 Epub 2011/04/06

17. Ecker TM, Tannast M, Murphy SB (2007) Computed tomography-based surgical navigation for hip arthroplasty. Clin Orthop Relat Res 465:100–105 Epub 2007/09/19

18. Siebenrock KA, Kalbermatten DF, Ganz R (2003) Effect of pelvic tilt on acetabular retroversion: a study of pelves from cadavers. Clin Orthop Relat Res 407:241–248 Epub 2003/02/05

19. Tannast M, Zheng G, Anderegg C, Burckhardt K, Langlotz F, Ganz R et al (2005) Tilt and rotation correction of acetabular version on pelvic radiographs. Clin Orthop Relat Res 438:182–190 Epub 2005/09/01

20. Williams PL (1989) The skeleton of the lower limb. In: Williams PL, Warkick R, Dyson M, Bannister LH (eds) Gray's anatomy. Churchill Livingstone, Edinburgh

21. Drenckhahn D, Eckstein F (2003) Untere Extremität. In: Drenckhahn D (ed). Urban & Fischer, München

22. Lierse W (1988) Praktische Anatomie. In: Lantz T, Wachsmuth W (ed). Springer, Berlin

23. Idelberger K, Frank A (1952) (A new method for determination of the angle of the pevic acetabulum in child and in adult). Z Orthop Ihre Grenzgeb 82(4):571–577 (Epub 1952/01/01). Uber eine neue Methode zur Bestimmung des Pfannendachwinkels beim Jugendlichen und Erwachsenen

24. de Lequesne MS (1961) False profile of the pelvis. A new radiographic incidence for the study of the hip. Its use in dysplasias and different coxopathies. Revue du rhumatisme et des maladies osteo-articulaires 28:643–652 (Epub 1961/12/01)

25. Murphy SB, Ganz R, Muller ME (1995) The prognosis in untreated dysplasia of the hip A study of radiographic factors that predict the outcome. J Bone Joint Surg Am 77(7):985–989 Epub 1995/07/01

26. Toennis D (1962) On changes in the acetabular vault angle of the hip joint in rotated and tilted positions of the pelvis in children. Z Orthop Ihre Grenzgeb 96:462–478 Epub 1962/12/01

27. Wiberg G (1939) The anatomy and roentgenographic appearance of a normal hip joint. Acta Chir Scand 83:7–38

28. Tönnis D, Heinecke A (1999) Acetabular and femoral anteversion: relationship with osteoarthritis of the hip. J Bone Joint Surg Am 81(12):1747–1770 Epub 1999/12/23

29. Kubiak-Langer M, Tannast M, Murphy SB, Siebenrock KA, Langlotz F (2007) Range of motion in anterior femoroacetabular impingement. Clin Orthop Relat Res 458:117–124 Epub 2007/01/09

30. Gilles B, Christophe FK, Magnenat-Thalmann N, Becker CD, Duc SR, Menetrey J et al (2009) MRI-based assessment of hip joint translations. J Biomech 42(9):1201–1205 Epub 2009/04/28

31. Menschik F (1997) The hip joint as a conchoid shape. J Biomech 30(9):971–973 Epub 1997/09/26
32. Tannast M, Kubiak-Langer M, Langlotz F, Puls M, Murphy SB, Siebenrock KA (2007) Noninvasive three-dimensional assessment of femoroacetabular impingement. J Orthop Res (Official publication of the Orthopaedic Research Society) 25(1):122–131 Epub 2006/10/21
33. Siebenrock KA, Kistler L, Schwab JM, Buchler L, Tannast M (2012) The acetabular wall index for assessing antero-posteriorantero-posterior femoral head coverage in symptomatic patients. Clin Orthop Relat Res 470(12):3355–3360 Epub 2012/07/17
34. Tannast M, Langlotz U, Siebenrock KA, Wiese M, Bernsmann K, Langlotz F (2005) Anatomic referencing of cup orientation in total hip arthroplasty. Clin Orthop Relat Res 436:144–150 Epub 2005/07/05
35. DiGioia AM, Jaramaz B, Blackwell M, Simon DA, Morgan F, Moody JE et al (1998) The Otto Aufranc Award. Image guided navigation system to measure intraoperatively acetabular implant alignment. Clin Orthop Relat Res 355:8–22 Epub 1999/01/26
36. Ehrhardt J, Handels H, Malina T, Strathmann B, Plotz W, Poppl SJ (2001) Atlas-based segmentation of bone structures to support the virtual planning of hip operations. Int J Med Informatics 64(2–3):439–447 Epub 2001/12/06
37. Subburaj K, Ravi B, Agarwal M (2009) Automated identification of anatomical landmarks on 3D bone models reconstructed from CT scan images. Comput Med Imaging Graph (Official journal of the Computerized Medical Imaging Society) 33(5):359–368 Epub 2009/04/07
38. Tan S, Yao J, Yao L, Summers RM, Ward MM (2008) Acetabular rim and surface segmentation for hip surgery planning and dysplasia evaluation. Proc SPIE
39. Murphy SB, Simon SR, Kijewski PK, Wilkinson RH, Griscom NT (1987) Femoral anteversion. J Bone Joint Surg Am 69(8):1169–1176 (Epub 1987/10/01)
40. Little NJ, Busch CA, Gallagher JA, Rorabeck CH, Bourne RB (2009) Acetabular polyethylene wear and acetabular inclination and femoral offset. Clin Orthop Relat Res 467(11):2895–2900 Epub 2009/05/05
41. Heyman CH, Herndon CH (1950) Legg-Perthes disease; a method for the measurement of the roentgenographic result. J Bone Joint Surg Am 32 A(4):767–778 (Epub 1950/10/01)
42. Hu Q, Langlotz U, Lawrence J, Langlotz F, Nolte LP (2001) A fast impingement detection algorithm for computer-aided orthopedic surgery. Computer Aided Surg (Official journal of the International Society for Computer Aided Surgery) 6(2):104–110 Epub 2001/09/25
43. Siebenrock KA, Schoeniger R, Ganz R (2003) Anterior femoro-acetabular impingement due to acetabular retroversion Treatment with periacetabular osteotomy. J Bone Joint Surg Am 85-A (A2):278–286 Epub 2003/02/07
44. Siebenrock KA, Steppacher SD, Haefeli PC, Schwab JM, Tannast M (2013) Valgus hip with high antetorsion causes pain through posterior extraarticular FAI. Clin Orthop Relat Res 471 (12):3774–3780 Epub 2013/03/07
45. Steppacher SD, Zurmuhle CA, Puls M, Siebenrock KA, Millis MB, Kim YJ et al (2014) Periacetabular osteotomy restores the typically excessive range of motion in dysplastic hips with a spherical head. Clin Orthop Relat Res (Epub 2014/12/10)
46. Tannast M, Hanke MS, Zheng G, Steppacher SD, Siebenrock KA (2014) What are the radiographic reference values for acetabular under- and overcoverage? Clin Orthop Relat Res (Epub 2014/11/12)
47. Sharp IK (1961) Acetabular dysplasia: the acetabular angle. J Bone Joint Surg Br 43:268–272
48. Notzli HP, Wyss TF, Stoecklin CH, Schmid MR, Treiber K, Hodler J (2002) The contour of the femoral head-neck junction as a predictor for the risk of anterior impingement. J Bone Joint Surg Br 84(4):556–560 Epub 2002/06/05

2D-3D Reconstruction-Based Planning of Total Hip Arthroplasty

Guoyan Zheng, Steffen Schumann, Steven Balestra, Benedikt Thelen and Lutz-P. Nolte

Abstract This chapter proposed a personalized X-ray reconstruction-based planning and post-operative treatment evaluation framework called iJoint for advancing modern Total Hip Arthroplasty (THA). Based on a mobile X-ray image calibration phantom and a unique 2D-3D reconstruction technique, iJoint can generate patient-specific models of hip joint by non-rigidly matching statistical shape models to the X-ray radiographs. Such a reconstruction enables a true 3D planning and treatment evaluation of hip arthroplasty from just 2D X-ray radiographs whose acquisition is part of the standard diagnostic and treatment loop. As part of the system, a 3D model-based planning environment provides surgeons with hip arthroplasty related parameters such as implant type, size, position, offset and leg length equalization. With this newly developed system, we are able to provide true 3D solutions for computer assisted planning of THA using only 2D X-ray radiographs, which is not only innovative but also cost-effective.

1 Introduction

Total Hip Arthroplasty (THA) has high social-economic impact. In developed countries including South Korea and Japan, more than 1.1 million THAs were operated in 2006 [1]. Among them, more than 650,000 hip joint replacements and hip revision surgeries were performed in the European Union with a population of 3,800,000. USA with a population of 2,910,000 contributed another 420,000 hip joint arthroplasties.

Meticulous pre-operative planning and templating was advocated as an integral part of THA already by its pioneers, Sir Charnley [2] and Müller [3], and is still an indispensable part of the surgical procedure. When pre-operatively planning and templating a THA the surgeon searches for an optimal fit of the hip implant

G. Zheng (✉) · S. Schumann · S. Balestra · B. Thelen · L.-P. Nolte
Institute for Surgical Technology and Biomechanics, Bern University, Bern, Switzerland
e-mail: guoyan.zheng@ieee.org

© Springer International Publishing Switzerland 2016
G. Zheng and S. Li (eds.), *Computational Radiology*
for Orthopaedic Interventions, Lecture Notes in Computational
Vision and Biomechanics 23, DOI 10.1007/978-3-319-23482-3_10

components and for the best technique to reconstruct leg length and the position of the center of rotation, both of which are dependent on the implant size and positioning. Successful pre-operative planning can prevent the use of undersized and oversized hip implants while inadequate pre-operative planning and inaccurate templating may lead to various complications including femoral fractures, limb length inequality, insufficient offset, instability and failure to achieve ingrowth. Pre-operative planning also provides the surgeon with a tool to ascertain that the correct prosthetic component sizes are available and can be of assistance in logistic and stock management for the operating theatres.

In the past, both digital and analogue radiography-based pre-operative planning systems have been introduced into the market. However, the reported accuracy for these systems varies. Gonzalez Della Valle et al. [4] showed that the template size corresponded to the actual component used in approximately 78 and 83 % of cases for cemented femoral prostheses and combined cemented and cementless acetabular components, respectively. Eggli et al. [5] reported similar results, where more than 90 % of the cases in the series used cement fixation. For cementless prostheses, Carter et al. [6] reported that the exact size of the femoral components was predicted in approximately 50 % of 74 cases, and Unnanuntana et al. [7] reported that the size of the prosthesis was exactly predicted in 42 % for acetabular components and 68.8 % for femoral components. The error in determining accurate radiography magnification and the projection characteristics of the 2D radiography contributed significantly to the prediction errors of these systems. In addition, the insufficient definition of the intramedullary anatomy on plain radiograph also reduce the accuracy of proper implant selection. There is a trend to do pre-operative planning of total hip arthroplasty in 3D Computed Tomography (CT) data.

In comparison with plain radiograph, CT-based 3D planning of THA offers several advantages [8–13] such as avoiding errors resulting from magnification and incorrect patient positioning, providing true 3D depiction of the underlying anatomy and offering accurate information on bone quality. Pre-operative planning for THA using 3D CT data is usually done in an interactive way [8] but automated solution has been recently introduced [12]. More specifically, in such a system, the sizes of the implants and their 3D positions and orientations with respect to the host bones can be automatically computed, based on 3D surface models segmented from the CT data. A randomised comparison between the 3D CT-based planning and the 2D plain radiograph based templating for THA [10] showed that the prediction rate for the stem and the cup sizes for the 3D CT-based planning is two times more accurate than that for the 2D plain radiograph based templating. The concern on 3D CT-based planning of THA, however, lies in the increase of radiation dosage to the patient and the associated CT acquisition cost.

The concern on the increase of the radiation dosage with CT-based solution is addressed by the introduction of the low-dose EOS 2D-3D imaging system (EOS Imaging, Paris, France) to the market [14]. The EOS 2D-3D imaging system is based on the Nobel prize-winning work of French physicist Georges Charpak on multiwire proportional chamber, which is placed between the X-rays emerging from the to-be-imaged object and the distal detectors. Each of the emerging X-rays generates a secondary flow of photons within the chamber, which in turn stimulate the distal detectors that give rise to the digital image. This electronic avalanche effect explains why a low dose of primary X-ray beam is sufficient to generate a high-quality 2D digital radiograph, making it possible to cover a field of view of 175 cm by 45 cm in a single acquisition of about 20 s duration [15]. With an orthogonally co-linked, vertically movable slot-scanning X-ray tube/detector pairs, EOS has the benefit that it can take a pair of calibrated posteroanterior (PA) and lateral (LAT) images simultaneously [16]. EOS allows the acquisition of images while the patient is in an upright, weight-bearing (standing, seated or squatting) position, and can image the full length of the body, removing the need for digital stitching/manual joining of multiple images [17]. The quality and nature of the image generated by EOS system is comparable or even better than computed radiography (CR) and digital radiography (DR) but with much lower radiation dosage [16]. It was reported by Illes et al. [16] that absorbed radiation dose by various organs during a full-body EOS 2D/3D examination required to perform a surface 3D reconstruction was 800–1000 times less than the amount of radiation during a typical CT scan required for a volumetric 3D reconstruction. When compared with conventional or digitalized radiographs [18], EOS system allows a reduction of the X-ray dose of an order 80–90 %. The unique feature of simultaneously capturing a pair of calibrated PA and LAT images of the patient allows a full 3D reconstruction of the subjects skeleton. This in turn provides over 100 clinical parameters for pre- and post-operative surgical planning [16]. With a phantom study, Glaser et al. [19] assessed the accuracy of EOS 3D reconstruction by comparing it with 3D CT. They reported a mean shape reconstruction accuracy of 1.1 ± 0.2 mm (maximum 4.7 mm) with 95 % confidence interval of 1.7 mm. They also found that there was no significant difference in each of their analyzed parameters ($p > 0.05$) when the phantom was placed in different orientations in the EOS machine. The reconstruction of 3D bone models allows analysis of subject-specific morphology in a weight-bearing situation for different applications to a level of accuracy which was not previously possible. For example, Lazennec et al. [20] used the EOS system to measure pelvis and THA acetabular component orientations in sitting and standing positions. Further applications of EOS system in planning THA include accurate evaluation of femoral offset [21] and rotational alignment [22]. Though accurate, the EOS system at this moment is only available in a few big clinical centers and is not widely available due to the high acquisition and maintenance costs.

The situation for the post-operative measurements remains the same. 2D anteroposterior (AP) pelvic radiograph is the standard imaging means for measuring the post-operative cup position. Although it has an inferior accuracy in comparison to 3D techniques based on CT, it is used routinely because of its simplicity, availability, and minimal expense associated with its acquisition. While plain pelvic radiographs are easily obtained, their accurate interpretations are subject to substantial errors if the individual pelvic orientation with respect to the X-ray plate is not taken into consideration.

When a CT study of the patient is available at some point during treatment, CT-based 2D-3D rigid image registration methods [23–25] have been developed to measure the post-operative cup orientation with respect to an anatomical reference extracted from the CT images, which is a plane called the Anterior Pelvic Plane (APP) defined by the Anterior Superior Iliac Spines (ASIS) and the pubic tubercles. In such methods, a rigid transformation between the CT data coordinate system and the X-ray image coordinate system is estimated first by performing an intensity-based rigid 2D-3D registration, which then allows for computing the orientation of the acetabular cup with respect to the APP extracted from the CT images. While accurate, the extensive usage of CT-based 2D-3D image registration methods in clinical routine is still limited. This may be explained by their requirement of having a CT study of the patient at some point during treatment, which is usually not available for vast majority of THA procedures performed nowadays.

By using X-ray radiographs that will be acquired from a conventional X-ray machine, which is usually available in all types of clinics, we are aiming to provide a personalized X-ray reconstruction-based planning and post-operative treatment evaluation framework called iJoint for advancing modern THA (see Fig. 1 for an overview of how the iJoint framework works). Based on a mobile X-ray image calibration phantom and a unique 2D-3D reconstruction technique, iJoint can generate patient-specific models of hip joint by non-rigidly matching statistical shape models (SSMs) to the X-ray radiographs. Such a reconstruction enables a true 3D planning and treatment evaluation of hip arthroplasty from just 2D X-ray radiographs whose acquisition is part of the standard diagnostic and treatment loop. As part of the system, a 3D model-based planning environment provides surgeons with hip arthroplasty related parameters such as implant type, size, position, offset and leg length equalization. With this newly developed system, we

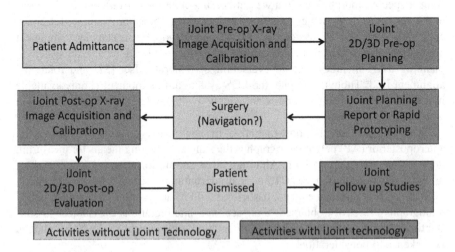

Fig. 1 The treatment protocol with the iJoint framework

are able to provide true 3D solutions for computer assisted planning of THA using only 2D X-ray radiographs, which is not only innovative but also cost-effective.

The chapter is organized as follows. In Sect. 2, we will present details about the calibration technique, followed by the description of statistical shape model construction in Sect. 3. Section 4 will give details on our hierarchical 2D-3D reconstruction techniques. Its applications for planning THA will be presented in Sect. 5. We will then present experimental design and preliminary results in Sect. 6, followed by discussions and conclusions in Sect. 7.

2 X-ray Image Calibration

The aim of the X-ray image calibration is to compute both the intrinsic and the extrinsic parameters of an acquired image. This is achieved by developing a mobile phantom as shown in Fig. 2 [26]. There are totally 16 sphere-shaped fiducials embedded in this phantom: 7 big fiducials with diameter of 8.0 mm and 9 small fiducials with diameter of 5.0 mm. The 16 fiducials are arranged in three different planes: all 7 big fiducials placed in one plane and the rest 9 small fiducials distributed in other two planes. Furthermore, the 7 big fiducials are arranged to form three line patterns as shown in Fig. 2, left. Every line pattern consists of three fiducials $\{M_i^1, M_i^2, M_i^3\}$, $i = 1, 2, 3$ with different ratios $\{r_i = |M_i^1 M_i^2|/|M_i^2 M_i^3|\}$. The exact ratio for each line is used below to identify which line pattern has been successfully detected.

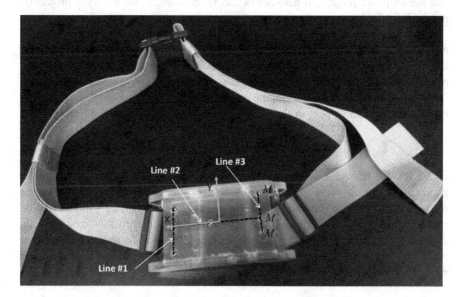

Fig. 2 X-ray calibration phantom and the belt-based fixation system

After an X-ray image is acquired, we first extract the sub-region containing the phantom projection. We then apply a sequence of image processing operations to the image. As those fiducials are made from steel, a simple threshold-based method is first used to segment the image. Connected-component labeling is then applied to the binary image to extract a set of separated regions. Morphology analysis is further applied to each label connected-component to extract two types of regions: candidate regions from big fiducial projections and candidate regions from small fiducial projections. The centers of these candidate regions are regarded as projections of the center of a potential fiducial. Due to background clutter, it is feasible that some of the candidate projections are outliers and that we may miss some of the true fiducial projections. Furthermore, to calculate both the intrinsic and the extrinsic parameters, we have to detect the phantom from the image. Here phantom detection means to establish the correspondences between the detected 2D fiducial projection centers and their associated 3D coordinates in the local coordinate system of the phantom. For this purpose, a robust simulation-based method as follows is proposed. The pre-condition to use this method to build the correspondences is that one of the three line patterns has been successfully detected. Due to the fact that these line patterns are defined by big fiducials, chance to missing all three line patterns is rare.

We model the X-ray projection using a pin-hole camera.

$$\alpha[I_x, I_y, 1]^T = \mathbf{K}(\mathbf{R}[x, y, z]^T + \mathbf{T}) = \mathbf{P}[x, y, z, 1]^T \tag{1}$$

where α is the scaling factor, $\mathbf{K}$ is the intrinsic calibration matrix, $\mathbf{R}$ and $\mathbf{T}$ are the extrinsic rotation matrix and translational vector, respectively. Both the intrinsic and the extrinsic projection parameters can be combined into a 3-by-4 projection matrix $\mathbf{P}$ in the local coordinate system established on the mobile phantom.

The idea behind the simulation-based method is to do a pre-calibration to compute both the intrinsic matrix $\mathbf{K}$ as well as the extrinsic parameters $\mathbf{R}_0$ and $\mathbf{T}_0$ of the X-ray image acquired in a reference position. Then, assuming that the intrinsic matrix $\mathbf{K}$ is not changed from one image to another (we only use this assumption for building the correspondences), the projection of an X-ray image acquired at any other position with respect to the phantom can be expressed as

$$\alpha[I_x, I_y, 1]^T = \mathbf{K}(\mathbf{R}_0(\mathbf{R}^x\mathbf{R}^y\mathbf{R}^z[x, y, z]^T + \mathbf{T}) + \mathbf{T}_0) \tag{2}$$

where $\mathbf{R}^x$, $\mathbf{R}^y$, $\mathbf{R}^z$ and $\mathbf{T}$ are the rotation matrices around three axes (assuming the z-axis is in parallel with the view direction of the calibration phantom at the reference position, see the middle column of Fig. 3 for details) and the translation vector from an arbitrary acquisition position to the reference position, respectively, expressed in the local coordinate of the mobile phantom. To detect the phantom projection when an image is acquired in a new position, the simulation-based method consists of two steps.

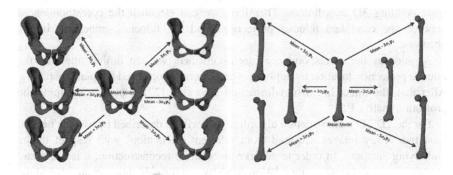

Fig. 3 Statistical shape models of the pelvis (*left*) and the femur (*right*). In each image, the mean model (in the *middle*) as well as the plusminus three times variations along the first three eigen-modes (from *top* to *down*) is shown

Image normalization The purpose of this step is to get rid of the influence of the parameters $\mathbf{R}^z$, α, and $\mathbf{T}$ on the phantom detection by normalizing the image acquired at the new position as follows. Assuming that we know the correspondences of fiducials on one line pattern, which is defined by 3 landmarks M_1, M_2, M_3 with their correspondent projections at IM_1, IM_2, IM_3, we can define a 2D coordinate system based on IM_1, IM_2, IM_3, whose origin O is located at $(IM_1 + IM_2)/2$ and the x-axis is defined along the direction $O \rightarrow IM_3$. Accordingly a 2D affine transformation $T_{normalize}$ can be computed to transform this line pattern based coordinate system to a standard 2D coordinate system with its origin at $(0, 0)$ and x-axis along direction $(1, 0)$ and at the same time to normalize the length of the vector $IM_1 \rightarrow IM_3$ to 1. By applying $T_{normalize}$ to all the fiducial projections, it can be observed that for a pair of fixed $\mathbf{R}^x$ and $\mathbf{R}^y$, we can get the same normalized image no matter how the other parameters $\mathbf{R}^z$, α, and $\mathbf{T}$ are changed because the influence of these parameters is just to translate, rotate, and scale the fiducial projections, which can be compensated by the normalization operation. Therefore, the fiducial projections after the normalization will only depend on the rotational matrices $\mathbf{R}^x$ and $\mathbf{R}^y$.

Normalized image based correspondence establishment Since the distribution of the fiducial projections in the normalized image only depends on the rotation matrices $\mathbf{R}^x$ and $\mathbf{R}^y$, it is natural to build a look-up table which up to a certain precision (e.g., 1°) contains all the normalized fiducial projections with different combination of $\mathbf{R}^x$ and $\mathbf{R}^y$. This is done off-line by simulating the projection operation using Eq. (1) based on the pre-calibrated projection model of the X-ray machine at the reference position. For an image acquired at position other than the reference, we apply the normalization operation as described above to all the detected candidate fiducial projections. The normalized candidate fiducial projections are then compared to those in the look-up table to find the best match. Since the items in the look-up table are generated by a simulation procedure, we know exactly the correspondence between the 2D fiducial projections and their corresponding 3D coordinates. Therefore, we can establish the correspondences between the candidate fiducial projections and the fiducials embedded in the phantom.

As soon as the correspondences are established, we can further fine-tune the fiducial projection location by applying a cross-correlation based template matching. After that, the direct linear transformation algorithm [27] is used to compute the projection matrix **P**.

For the 2D-3D reconstruction algorithm that will be described below, we need at least two X-ray images acquired from different orientations with respect to the underlying anatomy. In order to achieve an accurate reconstruction, it is important to maintain a fixed positioning relationship between the calibration phantom and the underlying anatomy. For this, we have developed a belt-based fixation system as shown in Fig. 2 to rigidly fix the calibration phantom to a patient.

3 Construction of Statistical Shape Models

The Point Distribution Model (PDM) [28] was chosen as the representation of the SSMs of both the pelvis and femur. The pelvic PDM used in this study was constructed from a training database of 114 segmented binary volumes with an equally distributed gender (57 male and 57 female) where the sacrum was removed from each dataset. After one of the binary volumes was chosen as the reference, diffeomorphic Demon's algorithm [29] was used to estimate the dense deformation fields between the reference binary volume and the other 113 binary volumes. Each estimated deformation field was then used to displace the positions of the vertices on the reference surface model, which was constructed from the reference binary volume, to the associated target volume, resulting in 114 surface models with established correspondences.

Following the alignment, the pelvic PDM was constructed as follows. Let $\mathbf{X}_i, i = 0, 1, \ldots, m - 1$ be m members in the aligned training population. Each member is described by a vector $\mathbf{X}_i$ containing N vertices:

$$\mathbf{X}_i = \{x_0, y_0, z_0, \ldots, x_{N-1}, y_{N-1}, z_{N-1}\} \tag{3}$$

The pelvic PDM is constructed by applying Principal Component Analysis (PCA) [28] on these aligned vectors:

$$\mathbf{D} = \frac{1}{(m-1)} \sum_{i=0}^{m-1} (\mathbf{x}_i - \bar{\mathbf{x}}) \cdot (\mathbf{x}_i - \bar{\mathbf{x}})^T$$

$$\mathbf{P} = (\mathbf{p}_0, \mathbf{p}_1, \ldots, \mathbf{p}_{m-2}); \mathbf{D} \cdot \mathbf{p}_i = \sigma_i^2 \cdot \mathbf{p}_i \tag{4}$$

where $\bar{\mathbf{x}}$ and $\mathbf{D}$ represent the mean vector and the covariance matrix, respectively; $\{\sigma_i^2\}$ are non-zero eigenvalues of the covariance matrix $\mathbf{D}$, and $\{\mathbf{p}_i\}$ are the corresponding eigenvectors. The descendingly sorted eigenvalues $\{\sigma_i^2\}$ and the corresponding eigenvector $\mathbf{P}_i$ of the covariance matrix are the principal directions spanning a shape space with $\bar{x}$ representing its origin.

A similar procedure was used to construct the femoral SSM from the 119 segmented binary volumes. Figure 3 shows the mean models as well as the plus-minus three times variations along the first three eigenmodes of the pelvic SSM (left) and the femoral SSM (right), respectively. Since usually an AP pelvic X-ray image only contains proximal femur part, we have accordingly derived a SSM of the proximal femur from the SSM of the complete femur by selecting only vertexes belonging to the proximal part.

4 Hierarchical 2D-3D Reconstruction

The existing feature-based 2D-3D reconstruction algorithms [30–32] have the difficulty in reconstructing concaving structures as they depend on the correspondences between the contours detected from the X-ray images and the silhouettes extracted from the PDMs. However, for THA, surgeons are interested not only in an accurate reconstruction of overall shape of the anatomical structures but also in an accurate reconstruction of the specific acetabular joint which consists of two surfaces: the acetabular surface and the proximal femur surface. The accuracy in reconstructing the acetabular joint will determine the accuracy of the pre-operative planning. Although the 2D-3D reconstruction scheme that we developed before can be used to reconstruct a patient-specific model of the proximal femur surface [33], its direct application to reconstruction of the acetabulum surface may lead to less accurate results.

To explain why we need to develop a new 2D-3D reconstruction scheme, the current 2D-3D correspondence establishment and thus the generation of the silhouette needs to be analyzed. Figure 4 shows a calculated silhouette of the pelvis. From this image it seems clear, that everything which does not generate a silhouette, will not be contributing to the finally found solution because it does not build any correspondence. To obtain more correspondences, more contours need to be drawn on the X-ray and accordingly identified on the SSM. This is realized with the so-called features [34] and patches [35] as explained below.

The features which build correspondences and should contribute to the reconstruction were selected in the SSM with an in-house developed SSM-Construction application. The acetabular rim was split into an anterior and posterior part. The boundary where the anterior part starts and ends was defined. Figure 5 shows the selected feature points for the anterior and posterior part of the acetabular rim respectively. The rim points were chosen as feature because they are almost located at the same position for all view angles, due to a more or less sharp edge.

The left and right hemi-pelvis as well as the acetabular fossa were introduced as patches, whereas a patch is a subregion of a surface. A patch is handled differently

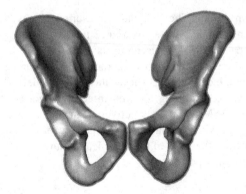

Fig. 4 Limitations of the existing 2D-3D reconstruction scheme. This image shows an AP view of the 3D pelvic surface model and the calculated silhouettes. Certain parts, i.e., the anterior acetabulum rims do not contribute to the silhouette generation. Thus, they will not participate in the 2D-3D reconstruction process

Fig. 5 Anterior (*grey points*) and posterior (*yellow points*) acetabular rims defined on the statistical shape model as features (Color figure online)

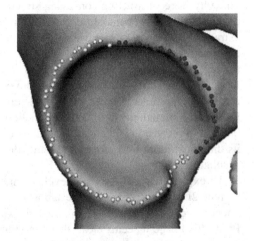

in the developed application compared to a feature. As the patch describes a sub-region of the surface model, a silhouette can be calculated, depending on the viewing direction.

All additional features and patches are listed in Table 1. There is no differentiation on the X-ray images between contours building correspondences with a patch, a feature or just the model. However, the contours for the acetabular anterior and posterior rims are not drawn on all the X-ray images. Although the rims are on an edge, the contour cannot be well identified on the X-ray images as it is not necessarily the most outer contour. Therefore the decision was made to draw only the inner (more medial) rim-contours, except for the AP image. On the AP image all contours are assigned, because the most outer contour can be clearly assigned to the rim. An example of the AP- and the outlet-image with the assigned contours is shown in Fig. 6.

Table 1 List of features and patches used in our hierarchical 2D-3D reconstruction

Anatomical name	Feature	Patch
Left hemi-pelvis		X
Right hemi-pelvis		X
Left anterior rim	X	
Left posterior rim	X	
Right anterior rim	X	
Right posterior rim	X	
Left acetabular fossa		X
Right acetabular fossa		

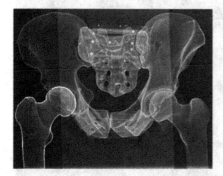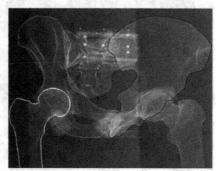

Fig. 6 Contours assigned on an AP (*left*) and an oblique (*right*) X-ray images for the hierarchical 2D-3D reconstruction. *Red* left hemi-pelvis contour, *orange* right hemi-pelvis contour, *light brown* left right acetabular fossa, *blue* left right anterior acetabular rim, *dark purple* left right posterior acetabular rims, *cyan* femur contours (Color figure online)

As the feature points do not generally coincide with the silhouette points, they are projected individually onto the image plane. To ensure that no wrong correspondences are found, the 2D-contours are assigned specifically to a feature or a patch. Instead of trying to find a matching point pair in the whole contour dataset, it only considers the contour assigned with the feature by their name. As now all the related features and patches contribute to the 2D-3D reconstruction process, it is expected that more accurate reconstruction of the hip joint models will be obtained [34, 35]. Figure 7 shows a reconstruction example.

5   2D-3D Reconstruction-Based Implant Planning

Once the surface models of the pelvis and the proximal femur are obtained, we can use morphological parameters extracted from these models to automatically plan THA. More specifically, we need not only to estimate the best-fit implants, their sizes, and positions but also to reconstruct leg length and the position of the center of rotation. In order to automatically plan the cup implant, we predefine the

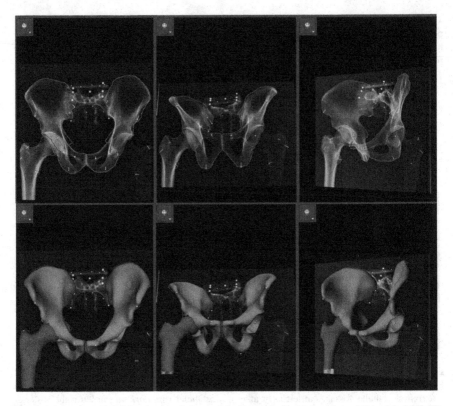

Fig. 7 The contour definition (*top*) and the models (*bottom*) obtained with our hierarchical 2D-3D reconstruction algorithm

acetabular rim using the mean model of the pelvic SSM (see Fig. 8 for details). Then, based on known vertex correspondences between the mean model and the reconstructed pelvic model, the acetabular rim from the reconstructed pelvic model can be automatically extracted. After that, we fit a 3D circle to the extracted rim points. The normal of the plane where the fitted 3D circle is located, the center and the diameter of the fitted circle and the fossa apex of the reconstructed acetabulum will then be used to plan the cup implant (see Fig. 9, left).

The cup implant is planned as follows. First the cup size can be determined by the diameter of the fitted 3D circle. Second, a Computer-Aided Design (CAD) model of the selected cup can be automatically positioned using the fitted 3D circle and the fossa apex (see Fig. 9, right). More specifically, orientation of the cup can be adjusted by aligning the axis of the cup CAD model with the normal to the plane where the acetabular rim is located.

The planning of the stem implant is mostly done semi-automatically. Again, we first predefine the center of the femoral head, the femoral neck axis as well as the femoral shaft axis from the mean model of the femoral SSM. Then, based on known vertex correspondences between the mean model and the reconstructed femoral

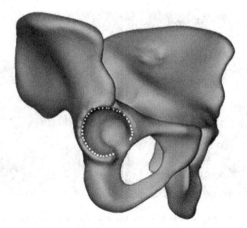

Fig. 8 Illustration of predefining acetabular rim on the mean model of the pelvic SSM

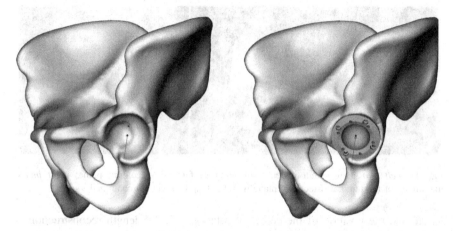

Fig. 9 *Left* estimate the normal (*red line*) to the plane where the acetabular rim is located, the acetabular center and the fossa apex (*red dots*). *Right* Using the extracted morphological parameters of the reconstructed acetabulum to plan the cup implant (Color figure online)

model, the center of the femoral head as well as the two axes can be automatically computed from the reconstructed femoral model. An initial position of a selected stem implant can then be achieved by aligning the CAD model of the implant to above mentioned morphological features. After that, the best fit stem implant can only be achieved by a manual fine-tuning (see Fig. 10 for an example).

The next step for the THA planning is to determine the femoral osteotomy plane. Based on the morphological features extracted from the reconstructed femoral model, the system automatically suggest an osteotomy plane (Fig. 11, the left two images). Its optimal location and orientation can then be fine-tuned interactively using a combined 2D-3D view (Fig. 11, the right image). After that, virtual femoral osteotomy will then be conducted. Final step for planning THA is to reconstruct leg

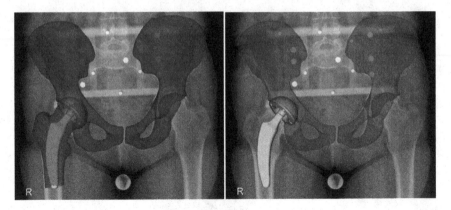

Fig. 10 A combined 2D-3D view was used to fine-tune the type, size and positioning of the stem implant

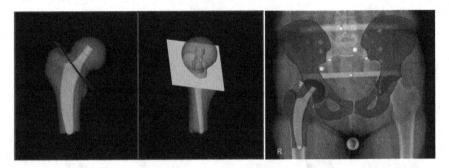

Fig. 11 *Left two images* automatically planning of femoral osteotomy plane; *right image* fine-tuning of the femoral osteotomy plane with the help of a 2D-3D combined view

length and the position of the center of rotation. The leg length reconstruction is achieved by interactively changing the position of the femoral bone after osteotomy along the femoral shaft axis until the leg length difference between two legs is eliminated. See Fig. 12 for an example.

6   Experimental Results

6.1   Preliminary Validation of 2D-3D Reconstruction Accuracy

To evaluate the accuracy of the reconstructions we conducted preliminary validation experiments based on calibrated X-ray radiographs. Three bones, i.e., two cadaveric hips (we named them as model #1 and #2, respectively) with each one cadaveric femur and one plastic hip containing two femurs with metallic coating

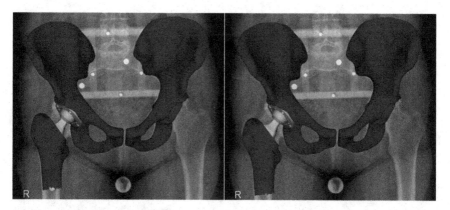

Fig. 12 The leg length reconstruction is achieved by interactively changing the position of the femoral bone after osteotomy along the femoral shaft axis

(we named these two as model #3 and #4), are used in our experiment. Three calibrated X-ray images (AP, Oblique, Outlet) were acquired for each of the four hip joints and used as the input for the reconstruction algorithms. For model #1 we reconstructed the right hip joint and for model #2 the left hip joint. For the plastic bone we did a reconstruction of both left and right hip joints.

The present hierarchical 2D-3D reconstruction algorithm was compared with the 2D-3D reconstruction algorithm introduced in [33]. Surface models segmented from CT scan of each bone were regarded as the ground truth. In order to evaluate the reconstruction accuracy, the surface models reconstructed from the X-ray images were transformed to the coordinate system of the associated ground truth models with a surface-based rigid registration before a surface-to-surface error can be computed.

When the 2D-3D reconstruction algorithm introduced in [33] was used, a mean surface reconstruction error of 1.1 ± 0.0 mm and 2.1 ± 0.3 mm was found for the femur and the pelvis, respectively. Using the hierarchical 2D-3D reconstruction algorithm led to a mean surface distance error of 0.8 ± 0.1 mm and 1.9 ± 0.2 mm for the femur and the pelvis, respectively.

6.2 Preliminary Clinical Study

From June 2014 to November 2014, X-ray radiographs of 18 patients (10 male, 8 female) were acquired. For each patient, two X-ray radiographs, one from the AP direction and the outer from the inlet direction (see Fig. 13 for details), were acquired. Out of these 18 image acquisitions, 13 were done post-operatively (control) and five were done pre-operatively. For two patients the acquisition of the inlet radiographs failed as the radiologic assistants were not yet experienced with the image calibration protocol. Thus, in total 16 reconstructions and prostheses plannings were done.

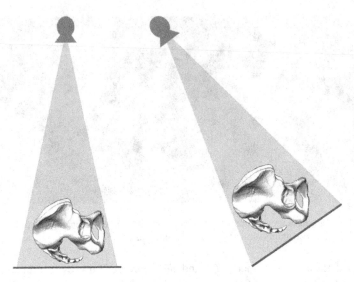

Fig. 13 Schematic drawing shows the two X-ray images acquired for each patient in our preliminary clinical study, one from the AP view (*left*) and the other from inlet view (*right*)

In order to validate the 2D-3D reconstruction-based planning of THA, we compared our planning results with the eventually implanted prostheses. In case of post-operative datasets, the implants were planned in 3D for the contralateral (healthy) patient side and compared to the already implanted prostheses of the ipsilateral (diseased) patient side.

Out of the 16 reconstructions and prostheses plannings, the information on 12 implanted cup implants was available. Out of these 12 datasets, two datasets were acquired pre-operatively and the remaining 10 were post-operative datasets. When the results achieved by our system were compared to the available implantation information, for five datasets, the cup size differed by two cup sizes, and for another four datasets by a single cup size (step width is 2 mm). While three cup sizes were correctly estimated, the cup was overestimated by in six cases and underestimated for three cases.

7 Discussions and Conclusions

An accurate pre-operative planning of hip arthroplasty depends not only on the overall model reconstruction accuracy but also on the reconstruction accuracy of the acetabular region. The hierarchical 2D-3D reconstruction scheme as proposed in this chapter allows to integrating more information for the acetabular region reconstruction, which will lead to more accurate pre-operative planning. Our preliminary experiment on 2D-3D reconstruction accuracy validation demonstrated that the hierarchical 2D-3D reconstruction achieved more accurate reconstruction results than the algorithm introduced in [33].

It is worth to discuss the preliminary clinical results. Due to the fact that we cannot do a post-operative CT scan of each patient involved in our preliminary clinical study, we don't have a solid ground truth to validate the results achieved by our system. Instead, we compared the results achieved by our system with those achieved with a conventional way based on 2D templating. This may partially explain why there is a relative large discrepancy between the sizes of the implants determined by our system and those determined by the conventional method.

It is interesting to compare the 2D-3D reconstruction based solution for planning THA with the current state of the art solutions. 2D plain X-ray radiograph-based solutions do not offer the capability of spatially imaging underlying anatomical structures, leading to inaccurate determination of implant size and positioning by using only standardized X-ray technology without considering the influence of the patient's individual anatomy and positioning on the X-ray magnification factor. Furthermore, with 2D plain X-ray radiograph-based solutions, there will be no coupling to navigation and no coupling of pre-operative planning to post-operative measurements. In contrast, the 2D-3D reconstruction-based solution as presented in this chapter provides the capability of reconstructing accurate 3D models of anatomical structures such that accurate determination of implant size and positioning will be possible by using the patient-specific 3D models that are reconstructed from the 2D X-rays. The influence of the patient's anatomy and positioning on the planning will be minimized. The 2D-3D reconstruction-based solution also offers the possibility in coupling with navigation and in coupling pre-operative planning with post-operative measurement.

In comparison with the 3D volumetric data-base solutions, the 2D-3D reconstruction-based solution offers several advantages. For example, no additional cost is needed as X-ray acquisition is part of the standard diagnosis and treatment loop. Furthermore, no time-consuming segmentation is needed. When compared with CT-based solutions, the 2D-3D reconstruction-based solution does not bring extra radiation, and thus is desirable for routine primary hip arthroplasty with an improved cost-benefit relation.

In summary, this chapter presented a personalized X-ray reconstruction-based planning and post-operative treatment evaluation framework called iJoint for advancing modern THA. Based on a mobile X-ray image calibration phantom and a unique 2D-3D reconstruction technique, iJoint can generate patient-specific models of hip joint by non-rigidly matching statistical shape models to the X-ray radiographs. Such a reconstruction enables a true 3D planning and treatment evaluation of hip arthroplasty from just 2D X-ray radiographs whose acquisition is part of the standard diagnostic and treatment loop. As part of the system, a 3D model-based planning environment provides surgeons with hip arthroplasty related parameters such as implant type, size, position, offset and leg length equalization. With this newly developed system, we are able to provide true 3D solutions for computer assisted planning of THA using only 2D X-ray radiographs, which is not only innovative but also cost-effective.

References

1. Kiefer H (2007) Differences and opportunities of THA in the USA, Asia and Europe. In: Chang J-D, Billau K (eds) Ceramics in orthopaedics. Proceedings of the 12th BIOLOXÂ® symposium on bioceramics and alternative bearings in joint arthroplasty. Springer, Berlin, pp 3–8
2. Charnley J (1979) Low friction arthroplasty of the hip. Springer, Berlin
3. Müller ME (1992) Lessons of 30 years of total hip arthroplasty. Clin Orthop Relat Res 274:12–21
4. Gonzalez Della Valle A, Slullitel G, Piccaluga F, Salvati EA (2005) The precision and usefulness of pre-operative planning for cemented and hybrid primary total hip arthroplasty. J Arthroplasty 20:51–58
5. Eggli S, Pisan M, Mueller M (1998) The value of preoperative planning for total hip arthroplasty. J Bone Joint Surg Br 80:382–390
6. Carter LW, Stovall DO, Young TR (1995) Determination of accuracy of preoperative templating of noncemented femoral prostheses. J Arthroplasty 10:507–513
7. Unnanuntana A, Wagner D, Goodman SB (2009) The accuracy of peroperative templating in cementless total hip arthroplasty. J Arthroplasty 24:180–186
8. Viceconti M, Lattanzi R, Antonietti B, Paderni S, Olmi R, Sudanese A, Toni A (2003) CT-based surgical planning software improves the accuracy of total hip replacement preoperative planning. Med Eng Phys 25:371–377
9. Huppertz A, Radmer S, Wagner M, Roessler T, Hamm B, Sparmann M (2014) Computed tomography for preoperative planning in total hip arthroplasty: what radiologists need to know. Skeletal Radiol 43:1041–1051
10. Sariali E, Mauprivez R, Khiami H, Pascal-Mousselard H, Catonne Y (2012) Accuracy of the preoperative planning for cementless total hip arthroplasty. A randomised comparison between three-dimensional computerised planning and conventional templating. Orthop Traumatol Surg Res 98:151–158
11. Huppertz A, Radmer S, Asbach P, Juran R, Schwenke C, Diederichs G, Hamm B, Sparmann M (2011) Computed tomography for preoperative planning in minimal-invasive total hip arthroplasty: radiation exposure and cost analysis. Eur J Radiol 78:406–413
12. Otomaru I, Nakamoto M, Kagiyama Y, Takao M, Sugano N, Tomiyama N, Tada Y, Sato Y (2012) Automated preoperative planning of femoral stem in total hip arthroplasty from 3D CT data: Atlas-based approach and comparative study. Med Image Anal 16:415–426
13. Hassani H, CheriX S, Ek ET, Ruediger HA (2014) Comparisons of preoperative three-dimensional planning and surgical reconstruction in primary cementless total hip arthroplasty. J Arthroplasty 29:1273–1277
14. Dubousset J, Charpak G, Skalli W, Deguise J, Kalifa G (2010) EOS: a new imaging system with low dose radiation in standing position for spine and bone & joint disorders. J Musculoskelet Res 13:1–12
15. Wybier M, Bossard P (2013) Musculoskeletal imaging in progress: the EOS imaging system. Joint Bone Spine 80:238–243
16. Illes T, Somoskeoy S (2012) The EOS imaging system and its use in daily orthopaedic practice. Int. Orthop 36:1325–1331
17. Wade R, Yang H, McKenna C, Faria R, Gummerson N, Woolacott N (2013) A systematic review of the clinical effectiveness of EOS 2D/3D X-ray imaging system. Eur Spine J 22:296–304
18. Deschenes S, Charron G, Beaudoin G, Labelle H, Dubois J, Miron MC, Parent S (2010) Diagnostic imaging of spinal deformities—reducing patients radiation dose with a new slot-scanning X-ray imager. Spine 35:989–994
19. Glaser DA, Doan J, Newton PO (2012) Comparison of 3-dimensional spinal reconstruction accuracy. Spine 37:1391–1397

20. Lazennec JV, Rousseau MA, Rangel A, Gorin M, Belicourt C, Brusson A, Catonne Y (2011) Pelvis and total hip arthroplasty acetabular component orientations in sitting and standing positions: Measurements reproductibility with EOS imaging system versus conventional radiographies. Orthop Traumatol Surg Res 97:373–380

21. Lazennec JV, Brusson A, Dominique F, Rousseau MA, Pour AE (2014) Offset and anteversion reconstruction after cemented and uncemented total hip arthroplasty: an evaluation with the low-cost EOS system comparing two- and three-dimensional imaging. Int Orthop. 39(7):1259–1267 Jul 2015, 10.1007/s00264-014-2616-3

22. Folinais D, Thelen P, Delin C, Radier C, Catonne Y, Lazennec JY (2011) Measuring femoral and rotational alignment: EOS system versus computed tomography. Orthop Traumatol Surg Res 99:509–516

23. Blendea S, Eckman K, Jaramaz B, Levison TJ, DiGioia III AM (2005) Measurements of acetabular cup position and pelvic spatial orientation after total hip arthroplasty using computed tomography/radiography matching. Comput Aided Surg 10:37–43

24. Penney GP, Edwards PJ, Hipwell JH, Slomczykowski M, Revie I, Hawkes DJ (2007) Postoperative calculation of acetabular cup position using 2D–3D registration. IEEE Trans Biomed Eng 54:1342–1348

25. Zheng G, Zhang X, Steppacher S, Murphy SB, Siebenrock KA, Tannast M (2009) HipMatch: an object-oriented cross-platform program for accurate determination of cup orientation using 2D-3D registration of single standard X-ray radiograph and a CT volume. Comput Methods Programs Biomed 95:236–248

26. Steffen S, Thelen B, Ballestra S, Nolte L-P, Buechler P, Zheng G (2014) X-ray image calibration and its application to clinical orthopedics. Med Eng Phys 36(7):968–974

27. Hartley R, Zisserman A (2004) Multiple view geometry in computer vision, 2nd edn. Cambridge University Press, Cambridge

28. Cootes TF, Taylor CJ, Cooper DH, Graham J (1995) Active shape models-their training and application. Comput Vis Image Underst 61:38–59

29. Vercauteren T, Pennec X, Perchant A, Ayache N (2009) Diffeomorphic demons: efficient non-parametric image registration. Neuroimage 45(1 Suppl):S61–S72

30. Lamecker H, Wenckebach TH, Hege H-C (2006) Atlas-based 3D-shape reconstruction from X-ray images. In: Proceedings of the 2006 international conference on pattern recognition 2006 (ICPR 2006). IEEE Computer Society, pp 371–374

31. Mitton D, Deschenes S, Laporte S, Godbout B, Bertrand S, de Guise JA, Skalli W (2006) 3D reconstruction of the pelvis from bi-planar radiography. Comput Methods Biomech Biomed Eng 9:1–5

32. Kadoury S, Cheriet F, Dansereau J, Labelle H (2007) Three-dimensional reconstruction of the scoliotic spine and pelvis from uncalibrated biplanar X-ray images. J Spinal Disord Tech 20:160–167

33. Zheng G, Gollmer S, Schumann S, Dong X, Feilkas T, and Gonzalez Ballester MA (2009) A 2D/3D correspondence building method for reconstruction of a patient-specific 3D bone surface model using point distribution models and calibrated X-ray images. Med. Image Anal. 13:883–899

34. Schumann S, Liu L, Tannast M, Bergmann M, Nolte LP, Zheng G (2013) An integrated system for 3D hip joint reconstruction from 2D X-rays: a preliminary validation study. Annals of Biomecial Engineering 41:2077–2087

35. Balestra S (2013) Statistical shape model-based articulated 2D-3D reconstruction. Master thesis, Institute for Surgical Technology and Biomechanics, University of Bern, Switzerland

Image Fusion for Computer-Assisted Bone Tumor Surgery

Kwok Chuen Wong

Abstract Conventionally, orthopaedic tumor surgeons have to mentally integrate all preoperative images and formulate a surgical plan. This preoperative planning is particularly difficult in pelvic or sacral tumors due to complex anatomy and nearby vital neurovascular structures. At the surgery, the implementation of these bone resections are more demanding if the resections not only are clear of the tumor but also match with custom implants or allograft for bony reconstruction. CT and MRI are both essential preoperative imaging studies before complex bone tumor surgery. CT shows good bony anatomy, whereas MRI is better at indicating tumor extent and surrounding soft tissue details. Overlaying MRI over CT images with the same spatial coordinates generates fusion images and 3D models that provide the characteristics of each imaging modality and the new dimensions for planning of bone resections. This accurate planning may then be reproduced when it is executed with computer-assisted surgery. It may offer clinical benefits. Although current navigation systems can integrate all the preoperative images for resection planning, they do not support the advanced surgical planning that medical engineering CAD software can provide, such as virtual bone resections and assessment of the resection defects due to system incompatibility. Translating the virtual surgical planning to computer navigation by manual measurements may be prone to processing errors. Currently, the integration of CAD data into navigation system is made possible by converting the virtual plan in CAD format into navigation acceptable DICOM format. The fusion of the modified image datasets that contain the virtual planning with the original image datasets allow easy incorporation of virtual planning for navigation execution in bone tumor surgery. Image fusion technique has also been used in bone allograft selection from a 3D virtual bone bank for reconstruction of a massive bone defect. This facilitates and expedites in finding a suitable allograft that matches with the skeletal defect after bone tumor resection. The technique is also utilized in image-to-patient registration of preoperative CT/MR images. It takes the forms of 2D/3D or 3D/3D image registration.

K.C. Wong (✉)
Orthopaedic Oncology, Department of Orthopaedics and Traumatology,
Prince of Wales Hospital, Hong Kong, China
e-mail: skcwong@ort.cuhk.edu.hk

© Springer International Publishing Switzerland 2016   217
G. Zheng and S. Li (eds.), *Computational Radiology*
for Orthopaedic Interventions, Lecture Notes in Computational
Vision and Biomechanics 23, DOI 10.1007/978-3-319-23482-3_11

It eliminates the step of surface registration of the operated bones that normally require large surgical exposure. The registration method has great potential and has been applied for minimally invasive surgery in benign bone tumors. This article provides an up-to-date review of the recent developments and technical features in image fusion for computer-assisted tumor surgery (CATS), its current status in clinical practice, and future directions in its development.

1 Introduction

Orthopaedic tumor surgeons have to mentally integrate two-dimensional (2D) preoperative images and formulate three-dimensional (3D) surgical plans of resection and reconstruction in orthopaedic bone tumors. Accurate implementation of the surgical plans at operating room (OR) is important not only to remove tumors with clear oncological margin, but also maximally preserve normal critical structures for reconstruction with a better limb function. This preoperative planning and its translation at OR is particularly difficult at some anatomical sites such as pelvic or sacral tumors due to complex anatomy and nearby vital neurovascular structures, or when some limited tumor resection such as multiplanar or joint-preserving intercalated tumor resection is contemplated. The implementation of bone tumor resections is even more technically demanding when the resections also have to match with custom-made implants or allograft for bony reconstruction.

Although primarily developed for neurosurgical procedures, computer-assisted navigation has been utilized for improving precision in various orthopaedic applications, such as spinal pedicle screws insertion, arthroplasty and fracture fixation [1–3]. It is logical to apply the well-established principles in computer-assisted orthopaedic surgery to facilitate the surgical planning and execution of the intended resections in bone tumor surgery, so to improve the accuracy and precision of the surgical procedures that may offer clinical benefits. The growing appeal in computer-assisted tumor surgery (CATS) is reflected by the increasing scientific papers published in the journals after the technique is first reported in 2004 [4]. The early and intermediate results are encouraging and suggest that the computer-assisted approach facilitate and help reproduce the complex surgical planning that may lead to a better oncological or functional outcome in orthopaedic tumor surgery [5–11].

Medical image fusion is the process of registering and overlaying or combining images from multiple imaging modalities to provide additional information in order to increase its clinical applications. The fusion images have been effectively used for oncologic indications. CT and MRI are both essential preoperative imaging studies before complex bone tumor surgery. CT shows good bony anatomy, whereas MRI is better at delineating both intraosseous and extraosseous tumor extent and surrounding soft tissue details. Fusion of CT and MR images yields hybrid images that combine the key characteristics of both imaging modality, thus enabling better interpretation of each and accurate localization of the lesion. The fused images can then be utilized for real-time applications in image-guided navigational procedures. This image manipulation technique has been commonly used in craniomaxillofacial tumor surgery [12, 13] but there are only few reports in orthopaedic tumor surgery [14, 15].

This article provides an up-to-date review of the developments and technical features in image fusion for CATS, its current status in clinical practice, and future directions in its development.

2 Image Fusion in Orthopaedic Tumor Surgery

Preoperative computer-assisted planning is as important as intraoperative computer navigation that is used as an intraoperative tool to locate surgical anatomy and guide subsequent surgical procedures. The more detailed the planning is, the greater the chance the surgical goals can be achieved when the surgeons execute the intended resections at OR. The role of image fusion in the workflow of CATS is summarized in Fig. 1.

CT image is the commonest imaging modality used for image-guided computer navigation in bone tumor surgery [7, 8, 10, 14] though MR-based computer navigation was also reported [16]. CT-based navigation can provide a reliable image-to-patient registration [10, 11, 14] that is a prerequisite for any accurate image-guided navigational procedures. Also, bone information can be easily extracted from CT image dataset that is essential for custom-made computer-aided design (CAD) implants and allograft selection in bony reconstruction after tumor

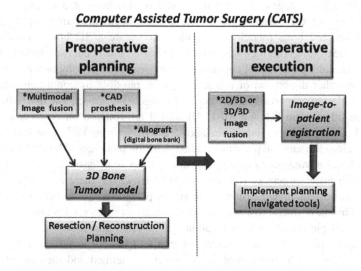

Fig. 1 The applications of image fusion (marked with *) in the workflow of CATS

resections. Therefore, CT images are normally used as the base layer of imaging on which other image modality (MRI/PET) or CAD data (prosthesis/allograft) are overlaid.

2.1 CT/MR Image Fusion for Surgical Planning

In the fusion of two or more imaging modalities, the image co-registration can be performed by "rigid" or "non-rigid" registration. As bone structures are not shifted or deformed by surgical manipulation as for abdominal soft tissue organs, rigid registration of imaging modalities is sufficiently accurate and the most ideal for surgical and interventional procedures for bone tumors. The technique of image fusion for CATS was described in 2008 [15]. Overlaying MRI over CT images with the same spatial coordinates generates fusion images (Fig. 2a–d). The multimodal images can be overlaid automatically in the navigation software (OrthoMap 3D module, version 2.0, Stryker, Hong Kong) that is the only commercially available navigation software dedicated to bone tumor surgery. The process can also be manually adjusted as different imaging datasets may be acquired at different scanning positions. The MR image dataset can be shifted manually to match the corresponding axial, coronal and sagittal views with that of CT images. The fusion image is considered to be acceptable for subsequent navigation planning if the bony contours on the CT/MR images at the region of interest matched within a 1-mm margin of error as visually assessed by surgeons. In general, the process of image takes less than few minutes and will improve with practice. Tumor extent is outlined from MRI which best shows the pathological details of bone tumors. The tumor edge is determined by looking at the transition of marrow signal from abnormal to normal in T1-weighted MR images. A 3D bone model is also generated by adjusting the contrast level of CT images. The tumor volume segmented from MRI and the CT-reconstructed bone model are combined to create a 3D bone–tumor model (Fig. 3a–c). Surgeons can then scrutinize all the fused image data sets in three spatial dimensions and the 3D model concurrently on one screen of computer navigational display. Another dimension of image analysis is possible by continuous blending between different proportions of CT and MR images (Fig. 2a–c). The image fusion is not restricted to plain CT and MRI. Additional information on vascular anatomy from CT angiography or tumor metabolic activities from PET scan can also be integrated into the surgical planning via image fusion technique. The PET fusion has the additional advantage to distinguish tumor from scar tissue in recurrent tumor patient with prior surgery or radiotherapy [15]. However, there is no report investigating the role of image fusion in diagnosis or assessment in orthopaedic bone tumors. This new interactive way of image analysis allows surgeons to obtain a better mental picture of tumor's location and regional anatomy. The best surgical access can be planned. Vital structures in relation to tumor location can be visualized. Resection levels and planes can be precisely defined and then executed via intraoperative image-guided computer navigation.

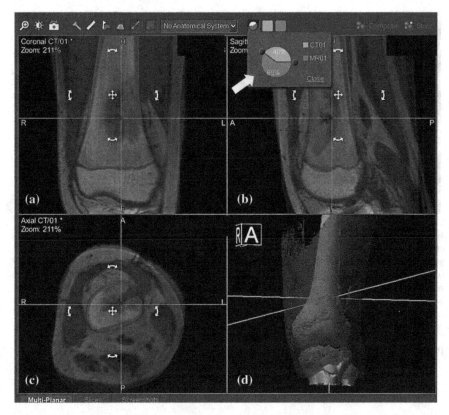

Fig. 2 CT/MR fusion images are shown in the navigation display in a 11-year-old patient with osteosarcoma at left distal femur metaphysis. MR images (*orange*) are overlaid and realigned on CT images (*green*) so that the region of interest (*femur bone*) has the same spatial coordinates on both images. The process can be done automatically or manually (by moving MR images with cross or *curved arrows in white*). The accuracy of image fusion is verified in reformatted coronal (**a**) and sagittal (**b**), axial (**c**) views and 3D bone models (**d**) of both images. The fusion is acceptable only when the femur bone surfaces of both images exactly coincided. Different proportions of individual image dataset can be blended (*big white arrow*) to examine the characteristic features of each image dataset (Color figure online)

Wong et al. [15, 17] reported the results of image fusion in 22 cases with CATS. The results showed all tumor resections could be carried out as planned under navigation guidance. Navigation software enabled surgeons to examine all fused image datasets (CT/MRI) together in reformatted 2D images and 3D models. It enabled surgeons to understand the tumor extent and its anatomical relationship with nearby structures. The mean time for preoperative navigation planning was 1.85 h (1–3.8). Intraoperatively, image guidance with fusion images could provide precise orientation of surgical anatomy, easy identification of tumor extent, neural structures and intended resection planes in all cases. The bone resection could be

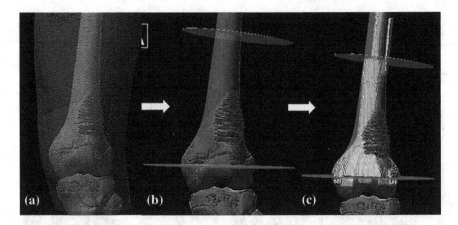

Fig. 3 3D models of the patient with left distal femur osteosarcoma. After fusing CT/MR images and outlining the tumor extent from MRI, a 3D model that includes skin, femur bone and tumor **a** is generated for resection planning. **b** The intended resection planes of joint-preserving tumor surgery (*pink and green*) can be defined precisely in the navigation software after examining all the reformatted views of the fused images and 3D models. **c** A CAD prosthesis is designed that matches with the bone defect and the remaining distal femur epiphysis. It is performed by integrating the data of CAD prosthesis into the navigation planning by image fusion technique (Color figure online)

precisely implemented in terms of the resection level and orientation, according to the pre-defined tumor volume and data of custom prosthesis.

2.2 Integrating Virtual CAD Planning into CATS

Surgical planning in orthopaedic bone tumors require precise definition of bone resections with planes that are away from the tumor edge and also in correct orientation to accommodate a custom implant if it is chosen for bony reconstruction. Virtual planning with manipulation of the resection bone models, incorporation of CAD implant models or allograft bone models for reconstruction are not possible in surgical navigation systems. They only accept medical imaging data in Digital Imaging and Communications in Medicine (DICOM) format and do not offer complex surgical simulation on these data. They offer relatively simple planning and serve more as intraoperative guidance tools. On the other hand, CAD engineering software allows surgical simulation with complex tumor resections with preoperative image datasets, design of custom tumor prostheses and selection of bone allografts from a 3D digital bone bank [18]. However, the planning in its proprietary format of the engineering software is incompatible for direct use in surgical navigation system. Therefore, successful implementation of this virtual surgical plan at OR still relies on the experience of surgeons if no intraoperative guiding tools are available. The difficulty increases with case complexity. The technique of integrating virtual CAD planning into CATS was first described in

2010 [19]. For surgical planning of musculoskeletal tumors, the virtual surgical plan datasets contain the original CT images, intended resection planes and CAD custom prostheses. This modified CT datasets in CAD format are back converted and exported in DICOM format. The virtual surgical plan datasets are then imported into the navigation machine and co-registered with the original CT datasets that will be used for navigation-guided surgery. The exact locations of intended resection and custom CAD implants that are obtained from the engineering software can be accurately transferred to the navigation planning (Fig. 4a–d). Therefore, this type of

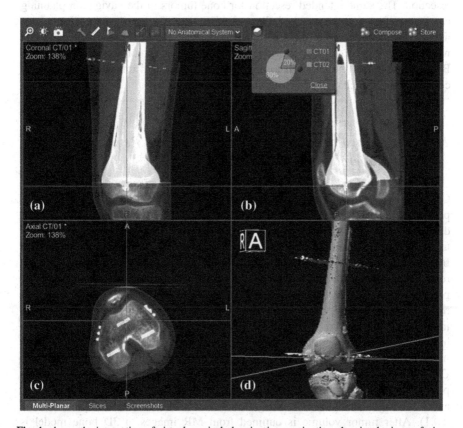

Fig. 4 shows the integration of virtual surgical planning into navigation planning by image fusion technique in the patient with left distal femur osteosarcoma. Virtual planning with femur bone resection and reconstruction with a CAD prosthesis was first performed in the CT image datasets in a medical engineering software. The virtual planning was exported in DICOM format and then imported into the navigation system. This modified image datasets (CT02) that contain the resection planes and the CAD prosthesis was fused with original CT image datasets (CT01). Reformatted coronal (**a**) and sagittal (**b**), axial (**c**) views and 3D bone models (**d**) of both image datasets shows the femur bones of both CT01 and CT02 image datasets are aligned with same spatial coordinates. This technique of CAD to DICOM conversion of virtual surgical planning and image fusion with modified CT datasets enables accurate translation of virtual CAD planning to subsequent navigation-assisted tumor surgery

image fusion can enhance the capacity of surgical navigation and may enable surgeons to precisely execute complex virtual surgical simulation with any CT-based surgical navigation system at OR.

Image fusion technique has also been used in selection of the most optimal bone allograft for reconstruction of a massive bone defect after tumor resection [18]. CT scan of both patients' bones and allograft are performed. The 3D bone model of the patient is then matched with that of allograft in digital bone bank by 3D image registration in an engineering software. This facilitates and expedites in finding a suitable allograft that matches with the planned skeletal defect after bone tumor resection. The same intended resection for bone tumors in the navigation planning can be transferred to the matched allograft when the CT images of the allograft are imported and fused with the CT images of bone tumors. Same image-guided navigation resection can be performed on both the bone tumor and allograft at OR. Better matching of the allograft to the resection defect will improve host-allograft contact, with less chance of nonunion.

2.3 Image Fusion by 2D/3D or 3D/3D Image Registration of Preoperative CT Images

CT-2D-fluoroscopy matching was originally designed for use in spine surgery performed under CT-based navigation guidance. Two fluoroscopic images with two different views of the bone at the region of interest are acquired after a navigation tracker is inserted into the patient's operated bone. The 3D bone model generated from preoperative CT images is overlaid on the 2D fluoroscopic images. It can be regarded as a form of 2D/3D image fusion by allowing an indirect registration of preoperative CT images without the need of open surgical exposure for surface registration. Therefore, this less invasive registration technique can facilitate minimal access surgery for navigation-guided procedures. 2D/3D image registration has been successfully applied in vertebro-pelvic fixation [20] and femoro-acetabular impingement surgery [21]. Wong et al. [22] reported a novel technique of using CT-2D-fluoroscopy matching to register preoperative CT image (Fig. 5) and then perform navigation and endoscopic guided curettage in benign bone tumors. CT and MR images are fused in a navigation system (BrainLAB iPlan Spine, version 2.0.1). After tumor volume is outlined from MR images, a 3D bone model is generated and then the most suitable surgical approach and cortical window can be precisely planned. Intraoperatively, CT-2D-fluoroscopy matching allowed less invasive registration of the preoperative CT images that contain the surgical planning. The amount of bone curettage can be checked by real-time visual feedback from navigated CT images and the actual burring procedures can be visualized on endoscopic images at the same setting. The technique is particularly useful when the benign bone tumors contain bony septae that have to be removed. Further studies are required to investigate the registration accuracy of CT-2D-fluoroscopy

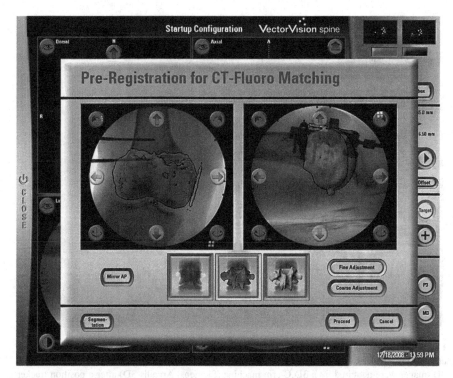

Fig. 5 A 2D/3D image-to-patient registration was performed by CT-2D-fluoroscopy matching for minimally invasive curettage in a patient with right distal femur benign bone tumor (chondromyxoid fibroma). Two fluoroscopic 2D images were acquired after a position tracker was inserted into the operated femur bone. A segmented 3D bone volume from preoperative CT images was fused with the intraoperatively acquired fluoroscopic 2D images. It allows for a minimally invasive registration of preoperative CT images without surface registration that requires a large surgical exposure

matching in other bone regions as the registration method is originally designed for spine procedures only.

3D/3D Image fusion by CT-3D-fluoroscopy matching was reported in percutaneous iliosacral screw insertion with navigation system [23]. Intraoperatively acquired 3D-fluoroscopic images were co-registered with preoperative CT images in a navigation system (Stryker Navigation System II-Cart, Stryker, Kalamazoo, MI, USA). The authors reported that in both experimental cadaver studies and a clinical series of six patients, the CT-3D-fluoroscopy matching navigation system was accurate and robust with mean target registration error of about 2 mm regardless of pelvic ring fracture type and fragment displacement. Our group also used this 3D/3D image fusion for minimally invasive, navigation-assisted surgery in benign bone tumors (Fig. 6a–d). Navigation-assisted tumor surgery with 3D/3D image registration has advantages over 2D/3D image registration. 50–100 2D images are acquired intraoperatively by 3D fluoroscopic machine. The images are

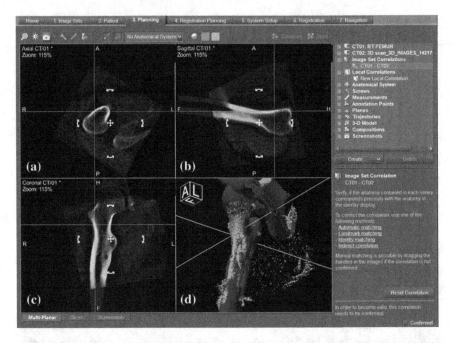

Fig. 6 A 3D/3D image-to-patient registration was performed by CT-3D-fluoroscopy matching for curettage in a patient with a suspected small osteoid osteoma in right proximal femur. Fluoroscopic 3D images were acquired by a 3D C-arm machine (Siemens, Arcadis 3D) after a position tracker was inserted into the distal femur. The fluoroscopic 3D images were transferred to the navigation machine in which the preoperative CT images (CT01) with surgical plan was fused with the intraoperatively acquired fluoroscopic 3D images (CT02). The two image datasets were matched and verified on reformatted axial (**a**), sagittal (**b**) and coronal (**c**) views and also 3D bone models (**d**). The accuracy of the 3D/3D registration is expected to be greater than that in 2D/3D registration as 50–100 2Dimages and their reformatted views are available for the matching

then reformatted into different views (axial, sagittal and coronal) and a 3D bone model can be generated. As more images are available for images coregistration, it not only renders the process of image fusion with preoperative CT images more accurate, but also enables image fusion with preoperative MR images that contain unique tumor information not available in CT images. Although the scanning volume of intraoperative 3D fluoroscopy is limited and about 12 cm^3, the fused images enable surgeon to utilize preoperative CT/MR images that have greater image resolution, cover larger anatomical area and all the surgical planning can be done prior to the actual surgery. Even if image fusion is not performed intraoperatively, the image-guided navigation procedure can still be performed with the registered 3D fluoroscopic images. This may not be possible with 2D/3D image registration as only two 2D fluoroscopic images are available for navigation. A recent study on registration accuracy of CATS in eight bone sarcomas showed a large mean target registration error of 12.21 ± 6.52 mm when performing manual registration by paired-points and surface matching on the exposed bone surface

[24]. The accurate CT-3D-fluoroscopy matching has the potential of improving the registration accuracy to navigate tumor boundaries and ensure clear resection margins in CATS.

2.4 Surgical Indications and Limitations in CATS with Image Fusion

Currently, CATS with image fusion may be beneficial in malignant bone tumors surgery (1) if there are anticipated difficulties in achieving an accurate bone tumor resection, (2) in obtaining a satisfactory resection plane to accommodate a custom prosthesis, or (3) in shaping an allograft to fit a resection defect [9, 10, 25]. The CATS technique has been successfully applied in pelvic or sacral tumor resection, joint-preserving tumor resection, multiplanar tumor resection and tumor reconstruction with CAD prosthesis or allograft [10, 25]. Current evidence suggests that the technique is safe without increase in operative complications. Also, it helps surgeons reproduce surgical planning accurately and precisely and it may offer clinical benefits [7, 8, 10, 14]. In a recent study of 31 patients with pelvic or sacral tumors, the technique could reduce intralesional resection from 29 to 8.7 % and the local recurrence rate was 13 % at a mean follow-up of 13.1 months [8]. As the technique enable surgeons to perform more accurate resection, more conservative bone resection around tumors may be achieved and it can maximally preserve normal structures for reconstruction and restoration of limb function [5, 11]. The technique reduced the rate of nonunion at host-allograft junction in bone tumor resections due to better allograft shaping [6].

However, as surgeons work on virtual images in a navigation console, surgeons have to know and understand the potential errors of CATS technique, so to avoid misinterpretation of navigational information and adversely affect the clinical outcome. CATS errors were described in details before [11]. CATS technique only improves the accuracy of bone resection but not soft tissue resection that still requires conventional surgical technique. Soft tissue deforms after surgical exposure and the spatial coordinates of soft tissue change are different from that of preoperative imaging. The performance of a navigation system and CATS technique is only as good as raw imaging data. The better the quality of each imaging modality and the larger the number of imaging modality for image fusion are, the more information is available for navigation planning and there will be a higher chance of achieving an accurate resection as planned.

2.5 Future Developments in CATS with Image Fusion

Image fusion in CATS is mainly focused on morphological imaging (CT/MRI) for surgical planning of musculoskeletal tumors. In addition to excellent soft tissue contrast for anatomical delineation, MRI can provide functional information from diffusion or perfusion MRI scans that have been used in neuroimaging [26]. With advent in imaging technology, multiple good quality image datasets with ana- tomical (CT/MR) and functional information (MR/PET) can be co-registered. The fused images with integrated tumor information can facilitate the accurate planning for navigation-assisted surgery in musculoskeletal tumors.

Real-time Ultrasound (US) has been used to fuse with MRI for soft tissue navigation-assisted procedures. US-MR fusion targeted biopsy was shown to enhance the diagnostic capabilities of unguided or single-modality image-guided biopsy to detect clinically significant prostate cancer [27]. Intraoperative US-MR fusion was also used to correct brain shift problem during neuronavigation in brain tumor surgery [28]. CATS with image fusion is only accurate with regards to bone resections. The real-time US-MR/CT fusion may have potentials to facilitate the soft tissue resection, soft tissue sarcoma resection and image-guided musculo- skeletal tumor biopsy.

Currently, only one commercially available navigation system (OrthoMap 3D module, version 2.0, Stryker, Hong Kong) is dedicated to surgical planning for musculoskeletal tumors. It allows for multimodal image fusion, tumor segmentation and definition of the resection planes. However, due to system incompatibility, it does not support the advanced surgical planning such as virtual tumor resection, manipulation of resection parts or prosthetic or allograft reconstruction that are possible in medical engineering CAD software. An integrated system that combines both the capacity of navigation planning and advanced surgical planning in engi- neering software is an important area for future development in CATS with image fusion.

3   Conclusions

There have been tremendous advances in computer-assisted technology for mus- culoskeletal tumors surgery in the last decade. Current literature suggests that CATS with image fusion can help reproduce intended bone resections of bone tumors in an accurate and precise manner. It may lead to better clinical outcomes. The technique is particularly useful in difficult resection and reconstruction such as pelvic or sacral tumors, limited resection in joint-preserving tumor surgery or reconstruction with custom CAD prostheses or allograft. Image fusion technique facilitates the incorporation of complex CAD planning into CATS and enables an indirect registration of preoperative CT/MR images for minimally invasive surgery in benign bone tumors. With advances in imaging technology, future images will achieve better anatomical and functional visualization of musculoskeletal tumors in CATS.

Conflict of Interest Kwok-Chuen WONG declares that he has no conflict of interest. The Stryker, Materialise, Stanmore Implants Limited and BrainLab companies did not fund or sponsor this research in any way.

References

1. Laine T, Lund T, Ylikoski M, Lohikoshi J, Schlenzja D (2000) Accuracy of pedicle screw insertion with and without computer assistance: a randomized controlled clinical study in 100 consecutive patients. Eur Spine J 9:235–240
2. Anderson KC, Buehler KC, Markel DC (2005) Computer assisted navigation in total knee arthroplasty: comparison with conventional methods. J Arthroplasty 20(Suppl 3):132–138
3. Grutzner PA, Suhm N (2004) Computer aided long bone fracture treatment. Injury 35(Suppl 1):S-A57–64
4. Hüfner T, Kfuri MJr, Galanski M, Bastian L, Loss M, Pohlemann T, Krettek C (2004) New indications for computer-assisted surgery: tumor resection in the pelvis. Clin Orthop Relat Res 426:219–225
5. Aponte-Tinao LA, Ritacco LE, Ayerza MA, Muscolo DL, Farfalli GL (2013) Multiplanar osteotomies guided by navigation in chondrosarcoma of the knee. Orthopedics 36(3):e325–e330
6. Aponte-Tinao L, Ritacco LE, Ayerza MA, Luis Muscolo D, Albergo JI, Farfalli GL (2015) Does intraoperative navigation assistance improve bone tumor resection and allograft reconstruction results? Clin Orthop Relat Res 473(3):796–804. doi:10.1007/s11999-014-3604-z
7. Cho HS, Oh JH, Han I, Kim HS (2012) The outcomes of navigation assisted bone tumour surgery: minimum three-year follow-up. J Bone Joint Surg Br 94(10):1414–1420
8. Jeys L, Matharu GS, Nandra RS, Grimer RJ (2013) Can computer navigation-assisted surgery reduce the risk of an intralesional margin and reduce the rate of local recurrence in patients with a tumour of the pelvis or sacrum? Bone Joint J 95-B(10):1417–1424
9. Wong KC, Kumta SM, Chiu KH, Antonio GE, Unwin P, Leung KS (2007) Precision tumour resection and reconstruction using image-guided computer navigation. J Bone Joint Surg Br 89:943–947
10. Wong KC, Kumta SM (2013) Computer-assisted tumor surgery in malignant bone tumors. Clin Orthop Relat Res 471(3):750–761
11. Wong KC, Kumta SM (2013) Joint-preserving tumor resection and reconstruction using image-guided computer navigation. Clin Orthop Relat Res 471(3):762–773
12. Leong JL, Batra PS, Citardi MJ (2006) CT-MR image fusion for the management of skull base lesions. Otolaryngol Head Neck Surg 134:868–876
13. Nemec SF, Donat MA, Mehrain S, Friedrich K, Krestan C, Matula C, Imhof H, Czerny C (2007) CT-MR image fusion for computer assisted navigated neurosurgery of temporal bone tumors. Eur J Radiol 62:192–198
14. Ritacco LE, Milano FE, Farfalli GL, Ayerza MA, Muscolo DL, Aponte-Tinao LA (2013) Accuracy of 3-D planning and navigation in bone tumor resection. Orthopedics 36(7):e942–e950
15. Wong KC, Kumta SM, Antonio GE, Tse LF (2008) Image fusion for computer-assisted bone tumor surgery. Clin Orthop Relat Res 466(10):2533–2541
16. Cho HS, Park IH, Jeon IH, Kim YG, Han I, Kim HS (2011) Direct application of MR images to computer-assisted bone tumor surgery. J Orthop Sci 16(2):190–195
17. Wong KC, Kumta SM, Tse LF, Ng EWK, Lee KS (2011) Image fusion for computer assisted tumor surgery (CATS), In: Ukimura, O (ed) Image fusion. InTech, Available from: http://www.intechopen.com/books/image-fusion/image-fusion-for-computer-assisted-tumor-surgery-cats. ISBN: 978-953-307-679-9

18. Ritacco LE, Farfalli GL, Milano FE, Ayerza MA, Muscolo DL, Aponte-Tinao L (2013) Three-dimensional virtual bone bank system workflow for structural bone allograft selection: a technical report. Sarcoma 2013:524395
19. Wong KC, Kumta SM, Leung KS, Ng KW, Ng EW, Lee KS (2010) Integration of CAD/CAM planning into computer assisted orthopaedic surgery. Comput Aided Surg. 15(4–6):65–74
20. Marintschev I, Gras F, Klos K, Wilharm A, Mückley T, Hofmann GO (2010) Navigation of vertebro-pelvic fixations based on CT-fluoro matching. Eur Spine J 19(11):1921–1927
21. Kendoff D, Citak M, Stueber V, Nelson L, Pearle AD, Boettner F (2011) Feasibility of a navigated registration technique in FAI surgery. Arch Orthop Trauma Surg 131(2):167–172. doi:10.1007/s00402-010-1114-3 (Epub 2010 May 20)
22. Wong KC, Kumta SM, Tse LF, Ng EW, Lee KS (2010) Navigation Endoscopic Assisted Tumor (NEAT) surgery for benign bone tumors of the extremities. Comput Aided Surg 15(1–3):32–39
23. Takao M, Nishii T, Sakai T, Yoshikawa H, Sugano N (2014) Iliosacral screw insertion using CT-3D-fluoroscopy matching navigation. Injury 45(6):988–994
24. Stoll KE, Miles JD, White JK, Punt SE, Conrad EU 3rd, Ching RP (2015) Assessment of registration accuracy during computer-aided oncologic limb-salvage surgery. Int J Comput Assist Radiol Surg (Epub 2015 Jan 13)
25. Wong KC, Kumta SM (2014) Use of computer navigation in orthopedic oncology. Curr Surg Rep. 2:47 (eCollection 2014, Review)
26. Lee IS, Jin YH, Hong SH, Yang SO (2014) Musculoskeletal applications of PET/MR. Semin Musculoskelet Radiol 18(2):203–216. doi:10.1055/s-0034-1371021 (Epub 2014 Apr 8)
27. Valerio M, Donaldson I, Emberton M, Ehdaie B, Hadaschik BA, Marks LS, Mozer P, Rastinehad AR, Ahmed HU (2015) Detection of clinically significant prostate cancer using magnetic resonance imaging-ultrasound fusion targeted biopsy: a systematic review. Eur Urol. 68(1):8–19. doi:10.1016/j.eururo.2014.10.026 (Epub 2014 Nov 1)
28. Prada F, Del Bene M, Mattei L, Lodigiani L, DeBeni S, Kolev V, Vetrano I, Solbiati L, Sakas G, DiMeco F (2015) Preoperative magnetic resonance and intraoperative ultrasound fusion imaging for real-time neuronavigation in brain tumor surgery. Ultraschall Med 36 (2):174–186. doi:10.1055/s-0034-1385347 (Epub 2014 Nov 27)

Intraoperative Three-Dimensional Imaging in Fracture Treatment with a Mobile C-Arm

Jochen Franke and Nils Beisemann

Abstract Intraoperative 3D imaging offers benefits in the treatment of fractures in many anatomical regions. Using this procedure, inadequate reduction results and implant malpositions that were not identified by two-dimensional fluoroscopy can be detected in all the regions described. In addition, there is the possibility of connecting to a 3D navigation system. This allows a precise placement of the osteosynthesis material in anatomically narrow corridors. This chapter will provide an overview of intraoperative 3D imaging in fracture treatment with a mobile C-Arm.

1 Introduction

The use of mobile C-arms for intraoperative visualisation of structures that are not directly visible at the operative site has been standard practice for many decades. Modern digital devices have for a long time now offered increasingly better image quality. This enables most questions that arise intraoperatively about the reduction result of fractures and the implant position in the treatment of long bones to be answered satisfactorily. In complex anatomical regions, such as the spinal column and pelvis, but also in joints and joint surfaces, such as the ankle joint and the tibial head, the interpretation of two-dimensional fluoroscopic images can be difficult. The summation of image information in a limited number of viewing planes makes misinterpretations possible. For this reason, the gold standard for preoperative and postoperative diagnostic procedures in these areas is computed tomography (CT) However, information about reduction and implant placement from postoperative CT scans is too late to influence the intraoperative procedure and might only give reasons for revision surgery, which should be avoided in the patient's interest.

J. Franke (✉) · N. Beisemann
MINTOS—Medical Imaging and Navigation in Trauma and Orthopaedic Surgery,
BG Trauma Centre Ludwigshafen at Heidelberg University Hospital,
Ludwig-Guttmann-Str. 13, 67071 Ludwigshafen, Germany
e-mail: jochen.franke@bgu-ludwigshafen.de

© Springer International Publishing Switzerland 2016
G. Zheng and S. Li (eds.), *Computational Radiology*
for Orthopaedic Interventions, Lecture Notes in Computational
Vision and Biomechanics 23, DOI 10.1007/978-3-319-23482-3_12

Nevertheless, the method is of limited intraoperative use as only a few centres possess the possibility of intraoperative CT. The costs and staffing demands of a device of this kind are substantial and its distribution correspondingly limited.

As a result of the development of mobile C-arms with the option of 3D imaging, a considerable improvement in intraoperative visualisation has been achieved for trauma surgeons. While the image quality of C-arms is slightly poorer than that of computed tomography, it is very good to adequate for assessing the relevant findings, depending on the region and the implant inserted.

C-arms with a 3D capability have been in clinical use since the beginning of 2000 and several devices from various manufacturers have become established on the market since then.

Among these, mention may be made of the Ziehm Vision Vario 3D and the Ziehm Vision FD Vario 3D from Ziehm (Nuremberg/Germany), the Pulsera 3D from Philips (Amsterdam/Netherlands) and the Siremobile Iso-C 3D and Arcadis Orbic 3D from Siemens (Erlangen/Germany).

The various devices differ in their mode of operation.

Since Siemens devices are used in the authors' hospital, the mode of operation of a 3D C-arm of the Arcadis Orbic 3D type is described in the following section by way of example.

The radiation source and the detector of the Arcadis Orbic, unlike that of conventional devices, is arranged in such a way that the central point of a line connecting these two units on rotation of the C-arm always remains focused on one spot, the isocentre. This is the only way that 3D reconstructions can be generated from the individual images. This is not the case with conventional image converters, where the isocentre follows a similar path to the rotation of the C-arm.

To produce the 3D dataset, a continuous orbital rotation around 190° then occurs on a motor-controlled rail. A defined number of fluoroscopic images are taken at fixed angular distances (referred to below as scans). Depending on the desired quality, 50 or 100 images are taken.

A three-dimensional reconstruction of the area being examined is calculated from the two-dimensional fluoroscopic images on an in-line computer (Fig. 1).

In this way a 3D data cube is obtained from the isocentre of the C-arm with a side length of about 12 cm. The maximum resolution in the centre of the cube is approximately 0.47 mm. Within this data cube, any number of planes can be viewed at an in-line workstation, this technique is also called cone-beam CT. It has become established practice to set the standard planes known from CT diagnostic procedures, including for documentation. This allows a continuous standardised assessment, the three-dimensional orientation is easier, and comparison with any existing preoperative CT series is readily possible. In contrast to the conventional computer tomography, the position between the scanned object and the 3D C-arm is unknown, therefore the adjustment of the standard planes is an important step. For special questions, any number of free adjustable planes can be set, such as axial imaging of one single screw in order to be able to assess its exact course. Therefore the adjustment of the views are oriented by the position of the screw. The data are

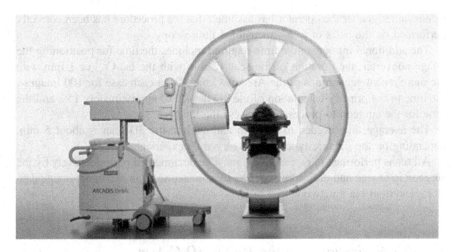

Fig. 1 Arcadis Orbic from Siemens

portrayed as high-contrast imaging and strong density differences, and the definition of bony structures, in particular, are optimally portrayed in this way.

Artefacts due to inserted implant material cannot be avoided. The less metal there is in the scan field, the fewer the number of artefact formations as well (Fig. 2).

In the area of the lower extremities, superimpositions from the opposite extremity can be avoided by excluding it from the scan field, for example by bending the leg. By positioning the patient on a metal-free radiolucent carbon leg plate or on a combined carbon and metal leg plate, artefacts from the operating table can to a large extent be excluded. The patient's own movements must be avoided during the scan. The time of the scan is specified by the operator and typically occurs following the completion of the essential parts of the reduction and

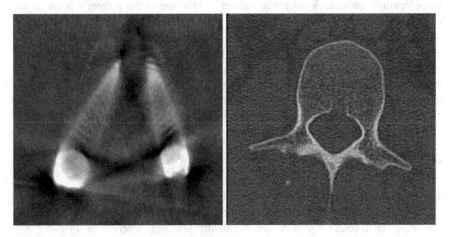

Fig. 2 Image quality with and without implants

osteosynthesis, after the operator has assumed that the procedure has been correctly performed on the basis of two-dimensional fluoroscopy.

The additional intraoperative time required includes the time for positioning the image converter, the duration of the scan (2 min with the Iso-C^{3D} or 1 min with the more recent generations of the Arcadis Orbic 3D, in each case for 100 images), the time to the primary calculation of the 3D dataset of approximately 15 s and the time for the surgeon to process and assess the images.

The average time needed to perform and analyse the 3D scan is about 5 min, depending on the fracture type and the surgeon's experience.

All scans performed intraoperatively must be documented postoperatively by the surgeon in charge and duly archived, e.g. by connecting to an electronic image data storage system via a network.

1.1 3D Navigation with the Mobile 3D C-Arm

In addition to the multislice imaging of cross-sectional planes, an additional area of use of the 3D image converter is 3D navigation.

Navigation is particularly used in complex fractures in the area of the spinal column and pelvis, but also in challenging procedures in the area of the foot and ankle.

By visualising instruments in the three-dimensional dataset, its use allows surgical procedures in complex anatomical regions without a direct view of the operative region. The general benefits over conventional procedures include reduced radiation exposure for patients and surgical team, as well as the possibility of being able to operate more precisely using a minimally invasive procedure. This results in increased safety and quality of outcome for patient and surgeon. The disadvantage may be a longer operative time depending on the region [1, 2].

3D navigation requires a 3D image intensifier and a navigation system with an optoelectronic camera. A reference marker is attached to the patient's bone in the vicinity of the region to be operated upon. After registering the marker, a 3D scan is performed. This 3D dataset is transferred to the navigation system. Surgical navigation is then performed using calibrated instruments in the corresponding section of the image dataset. After implant positioning, another 3D scan is performed to verify the placement. Where it is useable in the anatomical region concerned, the indications for the use of 3D navigation are discussed in the following sections.

2 Spinal Column

The incidence of injuries of the spinal column is 64/100,000 in Germany.

Although overall there is an equal sex distribution, two frequency peaks may be identified: at a young age men are frequently affected, whereas at an older age women tend to be affected more.

Differences in the causes of trauma are also apparent according to the age structure: in younger patients, high-speed trauma such as traffic accidents involving cars and motorbikes, as well as sporting injuries in extreme sports such as climbing and paragliding, are frequently the causes of fractures.

In elderly patients, osteoporosis resulting in a fracture of the vertebral body is considerably more common.

Overall, the cervical spine and the thoracolumbar junction are the sections most commonly affected.

Common to all injuries is the fact that there is a risk of injury to neurological structures because of the vicinity of the spinal cord and nerve roots in an unstable situation.

In 40 % of cervical spine injuries and 20 % of fractures in the area of the thoracic and lumbar spine there is an accompanying neurological impairment, which can range from paraesthesia to symptoms of a transverse lesion.

Fractures in the area of the upper cervical spine and their treatment concept are described below first of all and this is then followed by an overview of fractures in the lower cervical, thoracic and lumbar spine due to their similar anatomical structure.

3 Upper Cervical Spine

The upper cervical spine is composed of the atlas and axis.

The atlas consists of an anterior and a posterior arch and allows sliding and extension/flexion movements over its joint surfaces in the head region.

The axis is seated below the atlas and is connected to it via a peg, known as the dens. Rotational movements in the area of the head are undertaken through this jointed connection.

Laterally to the cervical spine, the vertebral arteries pass normally from the sixth cervical vertebra through the vertebral foramina in the direction of the head.

Typical injuries to the upper cervical spine are dens fractures and atlas arch fractures. If there is a risk of respiratory paralysis due to the vicinity of the dens to the brainstem in the event of dislocation, there is a strong indication for surgery. But many dens fractures in old patients are rather stable and will result in a stable pseudarthrosis, with no need of an require upfront fusion.

If there is an indication for fixation, screws are usually used to stabilise the fracture.

The challenging aspect of fracture care is the exact placement of the osteosynthesis without injuring the spinal cord or perforating the vertebral artery running through the vertebral foramen (Fig. 3).

In treatment of dens fractures, the stabilization is possible in different ways, depending of the fracture type and the age of the patient. Using the anterior approach, reduction is usually obtained by the insertion of two traction screws. In order to be able to check the position of the screws, Kirschner wires are first

Fig. 3 C1 vertebral foramen
of the vertebral arteries

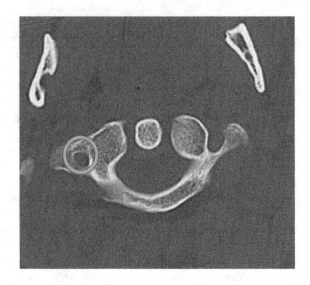

inserted temporarily in the desired direction. The position of the wires is then
checked by fluoroscopy. One problematic aspect of this procedure is that a slight
malposition can result in injuries to the spinal marrow.

In this case, 3D imaging with the C-arm is particularly valuable.

Following insertion of the Kirschner wires, a scan can be performed first of all to
check their position and secondly to correct this if necessary.

After placement of the screws, a further scan is often also performed, the reason
for this being that the diameter of the screws exceeds that of the K wires, so that
perforation of the bone can be excluded by the scan (Fig. 4).

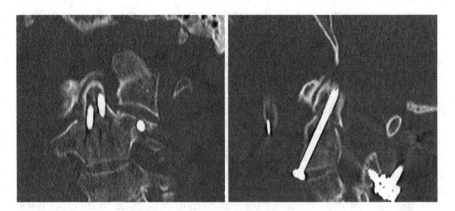

Fig. 4 Osteosynthesis of a dens fracture with screws

4 Lower Cervical Spine, Thoracic Spine and Lumbar Spine

The fractures predominantly found in the lower cervical spine and in the thoracic and lumbar spine are those in the area of the vertebral bodies.

Surgical treatment is frequently necessary in these cases, which is partly due to the vicinity of the spinal marrow and hence the threat or actual presence of neurological symptoms.

To relieve the pressure on the fractured vertebral body, the motion segments lying above and below the affected vertebra must be rigidified.

To this end, two screws are inserted into each of the unaffected vertebral bodies through the vertebral pedicles. The screw heads are then connected by a rod and the transmission of force is diverted away from the fractured vertebra.

The challenge of the procedure here is the exact placement of the above mentioned pedicle screws. These must be positioned from the back of the vertebra through the vertebral arch. There is the risk here of damaging the spinal column running through the vertebral arches or the nerve roots running above or below with the screw.

Similarly to the wire described in the upper cervical spine, a wire-like Jamshidi needle is inserted into the pedicle in this operation as well. This needle traces the planned route of the pedicle screw and the entry point and the course in the pedicle can be altered under fluoroscopy. A guide wire is then placed through this needle for the subsequent screw and the needle is removed.

A 3D scan can also be performed here to check the exact position of the wires.

Following the insertion of the pedicle screw and completion of the instrumentation by the assembly of the rod, a further scan can be taken. This can be used to check firstly the position of the osteosynthesis material, but also secondly the quality of the fracture reduction.

On a critical note, it should be pointed out that intraoperative 3D imaging can in some cases come up against its limits in the thoracic spine. The chest cage structures surrounding the cervical spine and in some cases those of the upper extremities can reduce the image quality in this area.

Nevertheless, despite these problems, the 3D scan is superior to traditional fluoroscopy where imaging of the anatomy is concerned.

5 3D Navigation in the Area of the Spine

As well as the "simple" multislice verification of the position of the osteosynthesis material and the quality of the reduction, 3D imaging by navigation can be used.

The aims are increased accuracy in the placement of the pedicle screws and a reduction in the radiation burden as a result of a shorter fluoroscopy time.

Navigation in the area of the spine was introduced because of the extremely high number of pedicle screw malpositions. There were reports of up to 30 %

malpositions in the cervical spine and up to 55 % malpositions in the area of the lumbar spine [3, 4].

In the subsequent studies, the malposition rate was reduced to 0–8 % following the introduction of navigation procedures [1, 5].

In one study, which compared navigated screw placement with conventional screw placement, it was shown that in addition to the greater accuracy of the pedicle screw position, a marked reduction in fluoroscopy time was achieved [5].

Navigation in the area of the cervical spine has not been a standard procedure so far. Dens screws and transpedicular screw placements in the area of the axis are described. However, robust comparative studies demonstrating a benefit of the navigated group over the non-navigated group are lacking.

6 Pelvis

6.1 Posterior Pelvic Ring

The placement of screws in the area of the posterior pelvic ring in the case of fractures of the sacrum as well as dislocations in the area of the iliosacral joint is challenging because of the complex anatomy.

The posterior pelvic ring is surrounded by nerves and blood vessels and the bony corridors for inserting a screw are narrow.

As a result, extensive fluoroscopy is frequently required to check the conventional percutaneous insertion of screws. Exposure times of up to 10 min are reported. This is complicated by the difficulties that occur in obtaining the appropriate projections for assessing the correct implant position depending on the patient's stature. Using 3D scans, both the position of the initial wire and screws, as well as the reduction result obtained, can be checked considerably better.

6.2 Acetabulum

The gold standard in the treatment of acetabular fractures is an open reduction and plate osteosynthesis. Surgical accesses are often very extensive and are associated with corresponding soft tissue trauma.

Depending on the fracture morphology, percutaneous screw osteosynthesis is also possible for fractures in the area of the acetabulum. The prerequisite for this is that the fractures are not dislocated or only slightly so. However, because of the complex anatomy of the acetabulum, here too screw placement is extremely challenging for the surgeon. An intra-articular screw position causes impairment of movement in the hip, while perforation of the pelvis can lead to vascular and nervous damage. For this reason, the benefits of the use of intraoperative 3D imaging described for the posterior pelvic ring apply here as well.

6.2.1 3D Navigation of the Pelvis

Using 3D-based navigation, exposure times can be minimised. A Kirschner wire is introduced with a navigated drill guide. After the position of the wire has been verified, a cannulated screw can be inserted over it. On a note of criticism, it should be pointed out that there might be slight deviations here because of the flexibility of the Kirschner wire. If this is the case, the position can be checked once again by means of a further scan. Where necessary, the standard planes can be omitted and the position of the wire can be followed accurately in free mode by imaging in the longitudinal axis. As also described for the spine, a conclusive assessment of the reduction result is possible at the same time.

7 Tibial Head

Fractures in the area of the tibial head are fairly rare. Overall, they account for about 1 % of all fractures. A peak frequency between the ages of 50 and 70 is found in the age distribution.

The causes of trauma are traffic accidents, falls and sports accidents.

There are different forms of fracture depending on the trauma and bone quality: axial compression, for example as a result of a jump from a great height, causes a cleavage fracture in young patients. The good bone quality transmits the impact fully and lets the condyles fracture as a whole.

Due to the frequently reduced bone quality in elderly people, there is a greater tendency to impression fractures.

In order to prevent subsequent osteoarthritis and the associated limitations of movement and pain, the indication for surgical therapy is very broad. A good clinical outcome can only be achieved with the greatest possible anatomical restoration of the joint surface. A conventional C-arm fluoroscope is usually used for intraoperative follow-up of the reduction result and for follow-up of the implant position. The problem with this procedure is that the whole tibial head joint surface cannot be seen.

Other bony structures in the beam path are often superimposed on the concave joint surface, which is referred to as a summation effect.

A complete assessment is therefore frequently not possible. Intraoperative 3D imaging provides the opportunity here to detect reduction errors and implant malpositions during surgery that would not be visible with the traditional C-arm.

In order to verify both the short-term and long-term benefit of intraoperative 3D imaging in the treatment of tibial head fractures, the author's research group undertook a retrospective clinical cohort study:

In our own procedure, we first of all checked the reduction result and the implant position in the surgical treatment of tibial head fractures by two-dimensional fluoroscopy in the standard planes.

If the outcome was satisfactory, a 3D scan was then performed routinely. If this revealed reduction results that required improvement or implant malpositions, these were corrected. As part of the documentation obligation since 2001, the scans and the recorded findings were documented immediately postoperatively in a register in hard copy and were afterwards transferred into an excel-sheet.

The register was analysed and evaluated for the analysis of the short-term benefit.

This showed that out of a total of 264 operated fractures an immediate revision of the original operative outcome was performed in 21.6 % of cases as a result of 3D imaging. Correction of the reduction accounted for by far the greatest proportion of all revisions with 68.3 %.

In most cases, gaps or step-offs that had not previously been identified on fluoroscopy were detected by the 3D scan.

An example is given below. A restored joint surface was seen in both anterior-posterior and lateral fluoroscopic planes. In the subsequent 3D scan, a gap and step-offs of almost 4 mm were then visible (Figs. 5 and 6).

A change of screw was the second most common type of revision with 14.3 %, followed by the removal of an intra-articular screw with 12.7 %.

On a consideration of the consequences in relation to fracture classification, it was found that B3 and C3 fractures according to the AO classification, with severe damage to the joint surface, particularly benefited [6].

The particularly high rate of intraoperative revisions in C3 and B3 fractures may be explained by the major damage to the tibial head in these types of fractures.

Traditional fluoroscopy by C-arm rapidly comes up against its limits at this point since the joint surface is markedly altered because of the severity of the injury and the reduction result can only be assessed to a limited extent under traditional fluoroscopy because of superimposition effects.

Thus, because of the multislice imaging of the tibial head and the free choice of viewing angle, 3D imaging offers a substantial advantage in assessing the reduction result.

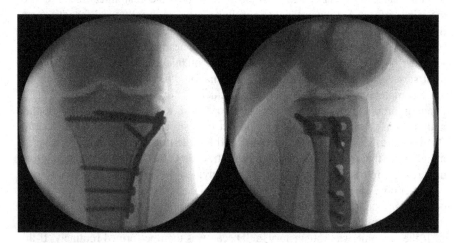

Fig. 5 Apparently restored joint surface

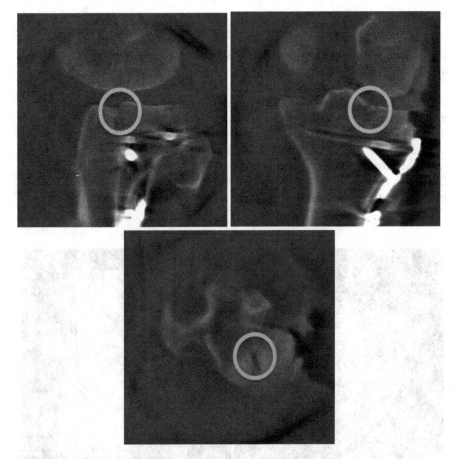

Fig. 6 3.6-mm large step-off visible on the 3D scan

A clinical follow-up study of the patient cohort was performed to ascertain the long-term benefit. Two patient groups were constituted on the basis of their reduction result. Group 1 had a maximum gap/step-off in the joint surface of 2 mm, while Group 2 included all patients with gap/step-off >=2 mm. The clinical result was then determined on the basis of the Lysholm score and the measurement of the range of movement.

There was a significantly better result in Group 1 on both the Lysholm score and the range of movement.

It can be deduced from these results that patients also gain long-term benefit from intraoperative 3D imaging. This is because the possibility for correction only exists if the residual step-offs and gap formation as well as screw malpositions are also detected.

8 Pilon

Dislocated fractures in the area of the distal tibia, known as the tibial pilon, are also a rare but surgically challenging injury.

Only about 5 % of all fractures in the area of the tibia concern the pilon.

The causes of pilon fractures are predominantly high-speed traumas such as motorbike and car accidents and falls from a great height. The particular problem of these fractures is the associated, in most cases severe, soft tissue injuries and the often extensive displacement and dislocation of bone fragments (Fig. 7).

In the absence of an adequate, anatomically correct reduction of the fracture fragments, there is a markedly increased risk of subsequent osteoarthritis with the restoration of the joint surface and hence loss of function in the ankle joint.

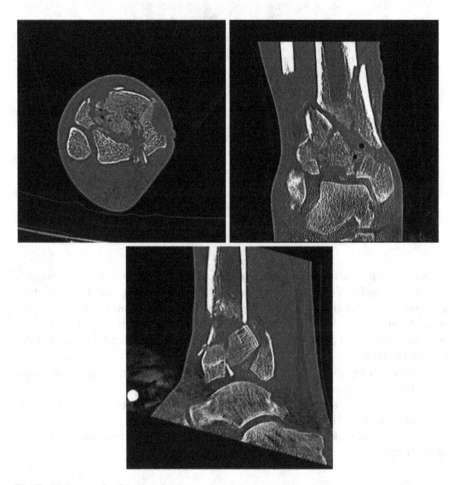

Fig. 7 CT image of a C3 pilon fracture

For this reason surgical treatment is necessary in most cases to achieve a good long-term functional outcome.

In assessing the joint surface and placement of the osteosynthesis intraoperatively, there are limits to standard dual plane fluoroscopy. It has been shown in cadaveric studies that adequate assessment of the joint surface is frequently not possible in spite of optimal conditions.

The gold standard for imaging of the joint surface is computed tomography. This is used preoperatively for planning. Intraoperatively, the technique is available only in a few highly specialised centres.

Postoperatively, computed tomography can again be used to detect implant malposition and any residual step-offs. If CT reveals findings that require improvement, a further correction by a second operation is necessary.

Therefore intraoperative 3D imaging can also be used in this case and further surgery can often be avoided.

In our own study, the benefit of intraoperative 3D imaging in the treatment of pilon fractures was documented in a large number of patients.

In total, the data from 161 patients treated by osteosynthesis for a pilon fracture and undergoing intraoperative 3D imaging were analysed.

The scans and the recorded findings were documented immediately postoperatively in a register in hard copy and were afterwards transferred into an excel-sheet.

The most frequent causes of trauma were motorcycle accidents and falls from a great height.

With a proportion of 72 % male patients, this was also consistent with the proportion described in the literature. The proportion of fractures with joint involvement was 89 %, of which 35 % were B injuries and 54 % C injuries.

In the analysis of the above mentioned X-ray register, a consequence was found in 29 % of cases on the basis of 3D imaging.

In 38 patients (81 %) the reduction was improved, while in 8 patients an intra-articular screw position was detected and corrected.

As in the previously described anatomical regions, the benefit of intraoperative 3D imaging is once again demonstrated here. The surgeon can check his surgical results immediately and, if necessary, improve the reduction or change the implant position. Only in this way that the best possible outcome can be achieved and further revision surgery be avoided.

This is of particular benefit because of the frequently taut soft tissue conditions in pilon fractures.

9 Ankle Joint/Syndesmosis

The ankle joint is composed of the tibia, fibula and talus and allows extension and flexion in the foot.

There is a ligamentous connection proximally between the fibula and tibia via the interosseous membrane that merges distally into the syndesmosis, composed of four ligaments.

Injuries to the syndesmosis are in most cases associated with a fracture of the fibula.

Fractures in this area are the most common fractures of the lower extremity.

The incidence is 174 fractures per 100,000 adults per year.

Ankle joint fractures usually result from indirect impact forces resulting from falls and twisting in the area of the foot. This frequently occurs in ball sports and sports with high intensity running.

As well as the AO classification for grading fractures, Weber's classification has become particularly well established [7].

This is based on the level of the fracture in the lateral malleolus.

Weber A fractures can be found far distally in the area of the lateral malleolar tip, Weber B fractures are found in the area of the syndesmosis and Weber C are defined as fractures above the syndesmosis.

In Weber A fractures the syndesmosis is for the most part uninjured, in Weber B it is damaged in 20–30 % of cases and in Weber C a tear of the syndesmosis is almost always present.

In the case of acute unstable injuries of the tibiofibular syndesmosis of the ankle joint, surgical stabilisation with prior anatomical reduction of the fibula in the tibiofibular incisura is necessary.

In the event of inadequate treatment of the syndesmosis rupture, there is the risk of premature osteoarthritis, accompanied by pain and restricted movement.

If a syndesmosis rupture is present, the syndesmosis is stabilised according to the accompanying bone injury and the surgeon's usual procedure.

Adjustment screws are most widely used in this case.

Following stabilisation of the syndesmosis, the position is inspected under fluoroscopy.

Usually an inspection is performed with a conventional image converter in the three standard planes, i.e. anterior-posterior, lateral and with the foot internally rotated 15°–20°.

As well as the tibiofibular overlap, attention is paid among other aspects to the width of the gap between the medial talar shoulder and the medial malleolus, known as the medial clear space. The most reliable parameter for assessing a syndesmosis injury with conventional X-ray images appears to be the width of the syndesmosis itself. A conclusion about a possible lateral malleolar malrotation cannot be drawn and hence a syndesmosis malposition with the previously described methods.

In general, a definitive assessment of the joint situation is not possible with X-ray fluoroscopy alone.

The correct positioning of the ankle mortise is best checked by postoperative CT.

Again the 3D C-arm technique can be used here. By means of multiplanar reconstruction, the ankle mortise can be viewed in all planes and thus the position of the fibula as well as the position of the implants can be assessed in all cross-sectional planes. Furthermore, in the event of an inadequate reduction or an implant malposition a revision can be performed immediately and in most cases further surgery avoided.

Initial publications reported direct intraoperative consequences with 3D imaging in 22–42 % of cases following surgical treatment in the area of the ankle joint, despite the fact that the reduction had previously been found to be good on 2D fluoroscopy. This was confirmed by our own study in a large patient population with a revision rate of 32.7 % [8].

Patients from 2001 to 2011 who had been operated upon for a simple or bimalleolar ankle joint fracture and an accompanying injury of the syndesmosis and treated with an adjustment screw were studied.

Following stabilisation of the syndesmosis with an adjustment screw, the reduction was first checked in the standard planes. If an insufficient reduction was seen here, this was corrected immediately.

If the result was adequate under fluoroscopy, a 3D scan was performed.

In a similar way to CT, a 3D dataset was then checked in coronal, axial and sagittal planes for a satisfactory reduction.

The following criteria were used as a basis for an anatomical reduction result of the ankle mortise:

Axial plane:

- Ending of the anterior border of the fibula in a harmonious elliptical line with the tibial pilon.
- Correct positioning of the fibula in the incisura in respect of the topography and width of the syndesmosis, with due allowance for interindividual differences.
- Equal width of the joint clefts between talus and malleoli.

Correct rotation of the lateral malleolus, reflected in the congruent position of the malleoli in relation to the talus

Coronal plane:

- Same width of joint clefts between talus and malleoli
- Correct length of fibula

Axial, coronal and sagittal plane:

- Correct implant position

In the analysis of a total of 251 patients, it was found that reduction in the area of the syndesmosis or the fracture was unsuccessful in 32.7 % of cases despite previous verification in the standard two-dimensional planes. In 30.7 % (based on all patients) the syndesmosis was not adequately reduced. The most common reduction error was a defective localisation of the fibula in the tibiofibular incisura in 77 cases (Fig. 8).

In order to validate the established criteria, a clinical study then followed.

In this study, patients generated from the previously mentioned patient population following the application of several exclusion criteria were followed up clinically and radiologically. Two patient groups were then constituted to validate the criteria.

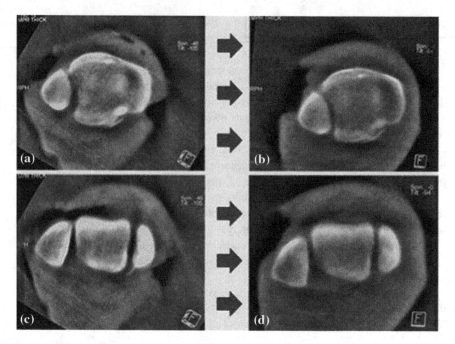

Fig. 8 Axial reconstruction of the 3D scan at the level of the syndesmosis (**a**, **b**) and the malleoli (**c**, **d**) before and after correction of an insufficient reduction with ventral subluxation and internal rotation of the distal fibula in a Maisonneuve injury following stabilisation

Group 1 contained 41 patients in whom all the above-mentioned criteria were obtained. In the 32 patients in Group 2 this was not possible on at least one criterion.

The Olerud and Molander score was used for the clinical outcome and the Kellgren and Lawrence score for the radiological outcome.

The statistical analysis showed significantly better clinical and radiological outcomes for Group 1 in which the criteria mentioned were met.

On the Olerud and Molander score, Group 1 obtained a mean of 92.44 points (SD 10.73; range 50–100) and Group 2 a mean of 65.47 points (ST 28.77; range 5–100). With a mean difference of 26.97 points, this was a highly significantly better result. This was also confirmed in the multivariate analysis.

In the comparison of the groups with respect to the grade of osteoarthritis according to Kellgren and Lawrence, there was again a significantly better result in Group 1 with an average grade of osteoarthritis of 1.24 points compared with Group 2 with 1.79 points.

In summary, therefore, it was shown that, if the described criteria can be achieved, a better clinical and radiological outcome may be expected.

The conclusion of this study is that if intraoperative 3D imaging is not available, postoperative computed tomography is advisable to detect any reduction errors and to enable them to be corrected where necessary.

10   Calcaneus

Calcaneal fractures account in total for a proportion of 2 % of all fractures in humans. However, they are the most common fracture of the tarsal bone. Typical causes of trauma are a fall from a great height or a traffic accident. Involvement of the joint surface is seen in 75 % of all cases. In 18 % of patients both calcanei are involved.

In most cases, calcaneal fractures should be treated by osteosynthesis to achieve the best possible clinical outcome.

Due to the complex anatomy of the calcaneus, computed tomography should be done preoperatively to analyse the fracture morphology and to be able to decide on the surgical tactics.

This is usually then followed by intraoperative monitoring of the surgical outcome by means of axial and lateral fluoroscopy and Broden's view.

As with the tibial head and pilon, here again customary two-dimensional fluoroscopy does not always do justice to the complex anatomy of the calcaneus with its four joint surfaces so that joint step-offs or intra-articular screw positions may be overlooked.

In the past, studies have revealed an immediate revision rate of up to 41 % on the basis of intraoperative 3D imaging.

In our own study, we investigated the intraoperative revision rate after 3D imaging in calcaneal fractures treated by osteosynthesis in 377 calcanei [9].

This study confirmed the previously mentioned very high revision rates with 40.3 %.

The reasons for the revisions are listed in the following table (Fig. 9).

Cause	n	%
Total	152	40.3
Fracture reduction (total)	74	19.6
Isolated fracture reduction	59	15.6
Fracture reduction and intra-articular screw	13	3.4
Fracture reduction and other screw	2	0.5
Intra-articular screw	65	17.2
Isolated intra-articular screw	46	12.2
Intra-articular screw and other screw	6	1.6
Other screw	31	8.2
Isolated other screw	23	6.1
Spongiosaplasty	1	0.3
Plate changed	1	0.3

Other screw means that the screw does not reach the targeted fragment or projects >4 mm on the other side

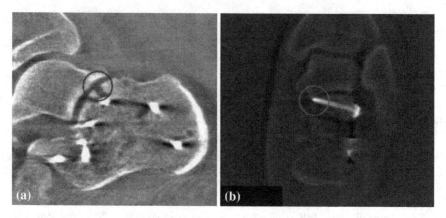

Fig. 9 Gap formation **a** in the dorsal joint facet and **b** intra-articular screw placement not visible by conventional fluoroscopy but detectable on intraoperative 3D imaging

The analysis of the intraoperative revision rate was followed by a clinical study. Similarly to the clinical study in tibial head fractures, patients were again divided into two groups on the basis of their reduction result.

Group 1 had a successful anatomical reconstruction of the joint surfaces or slight step-off or gap formation (<2 mm). Patients with residual step-offs of >=2 mm were assigned to Group 2.

A total of 89 patients then underwent clinical and radiological follow-up according to AOFAS (American Foot and Ankle Society) and Kellgren and Lawrence.

In the bivariate analysis of the AOFAS score, it was apparent that joint surface congruency significantly affects the outcome, i.e. a step-off or gap >2 mm is associated with significantly poorer outcomes ($p < 0.001$).

Analysis of the grade of osteoarthritis also revealed significantly poorer results in Group 2 ($p = 0.0239$).

Both the clinically and radiologically significantly better results in this study were also confirmed by multivariate analysis.

This shows that intra-articular incongruences that are not apparent on two-dimensional fluoroscopic follow-up can be detected by intraoperative 3D reduction follow-up in a considerable number of cases.

As a result of the possibility of better joint surface reconstruction that this offers, the clinical results can be improved on the assumption of a successful reduction and post-traumatic osteoarthritis changes can be reduced.

11   Summary and Conclusion

Intraoperative 3D imaging offers benefits in the treatment of fractures in many anatomical regions. Using this procedure, inadequate reduction results and implant malpositions that were not identified by two-dimensional fluoroscopy can be detected in all the regions described. In addition, there is the possibility of connecting to a 3D navigation system. This allows a precise placement of the osteosynthesis material in anatomically narrow corridors.

References

1. Grützner PA, Beutler T, Wendl K, von Recum J, Wentzensen A, Nolte LP (2004) Navigation an der Brust- und Lendenwirbelsäule mit dem 3D-Bildwandler. Chirurg 75(10):967–975
2. Stockle U, Schaser K, Konig B (2007) Image guidance in pelvic and acetabular surgery–expectations, success and limitations. Injury 38(4):450–462
3. Gertzbein SD, Robbins SE (1990) Accuracy of pedicular screw placement in vivo. Spine 15(1):11–14
4. Tjardes T, Shafizadeh S, Rixen D, Paffrath T, Bouillon B, Steinhausen ES et al (2010) Image-guided spine surgery: state of the art and future directions. Eur Spine J 19(1):25–45 Official publication of the European Spine Society, the European Spinal Deformity Society, and the European Section of the Cervical Spine Research Society
5. Wendl K, von Recum J, Wentzensen A, Grutzner PA (2003) Iso-C(3D0-assisted) navigated implantation of pedicle screws in thoracic lumbar vertebrae. Der Unfallchirurg 106 (11):907–913
6. Müller ME (1990) The Comprehensive classification of fractures of long bones, vol xiii, 1st edn. Springer, Berlin, 210 p
7. Weber B (1972) Die Verletzungen des oberen Sprunggelenkes. Huber, editor. Bern, Stuttgart, Wien, 102–7 p
8. Franke J, von Recum J, Suda AJ, Grutzner PA, Wendl K (2012) Intraoperative three-dimensional imaging in the treatment of acute unstable syndesmotic injuries. J Bone Joint Surg 94(15):1386–1390
9. Franke J, Wendl K, Suda AJ, Giese T, Grutzner PA, von Recum J (2014) Intraoperative three-dimensional imaging in the treatment of calcaneal fractures. J Bone Joint Surg 96(9):e72

Augmented Reality in Orthopaedic Interventions and Education

Pascal Fallavollita, Lejing Wang, Simon Weidert and Nassir Navab

Abstract During surgery, the surgeon's view of the patient is complemented with imaging that provides an indirect visual feedback of the treated anatomy. For an advanced visualization, existing image-guided surgical systems employ tracking and registration methods to fuse medical images. Lately, medical augmented reality has enabled in situ visualization by registering three-dimensional preoperative data (e.g. CT, MRI, and PET) with two-dimensional intraoperative images (e.g. X-ray, Ultrasound, and Optics). Including the virtual locations of the surgical instruments within the visualization provides the most intuitive way to facilitate surgical navigation since the surgeon mental mapping between medical images, instruments, and patient is not necessary anymore. This book chapter investigates medical augmented reality solutions within orthopaedics and highlights the innovations that lead surgeons in surgical planning, navigation, and education in an efficient and timely manner.

P. Fallavollita · N. Navab
Fakultät Für Informatik, Technische Universität München, Garching bei München, Germany
e-mail: fallavol@in.tum.de

N. Navab
e-mail: navab@in.tum.de

L. Wang (✉)
Metaio GmbH, Munich, Germany
e-mail: lejingwang@msn.com

S. Weidert
Klinik für Allgemeine Unfall-, Hand- und Plastische Chirurgie, Klinikum der Universität München, Munich, Germany
e-mail: simon.weidert@med.uni-muenchen.de

© Springer International Publishing Switzerland 2016
G. Zheng and S. Li (eds.), *Computational Radiology for Orthopaedic Interventions*, Lecture Notes in Computational Vision and Biomechanics 23, DOI 10.1007/978-3-319-23482-3_13

1 Introduction

Orthopaedic surgeons rely greatly on intraoperative X-ray images to visualize bones, implants and surgical instruments to complete patient procedures especially during minimally invasive surgery. Within the past two decades, mobile C-arm fluoroscopes have become the imaging modality of choice to acquire X-rays though the images are not directly aligned with the real scene as viewed by the surgeon. Surgeons perform a mental mapping to understand the spatial relationship between images and patient that absorbs considerable mental effort and time which may introduces complications and potential human mistakes impairing the quality of surgery.

Augmented Reality (AR) supplements the real scene with a virtual scene. It has been widely used in areas such as manufacturing and repair, annotation and visualization, robot path planning, entertainment, and military aircraft guidance [1]. Lately, surgeons have welcomed AR to assist them only in specific workflow tasks and in surgical planning. Consequently, it was discussed that augmented reality is a promising solution to improve the accuracy of surgical procedures, decrease the variability of surgical outcomes, reduce trauma to the critical anatomical structures, increase the reproducibility of a surgeons' performance, and reduce radiation exposure [2].

Medical augmented reality has been successfully applied in various disciplines of surgery, such as neurosurgery, orthopaedic surgery, and maxillofacial surgery. Stetten et al. [3] presents a Real-Time Tomographic Reflection (RTTR) as an image guidance technique for needle biopsy. RTTR is a new method of in situ visualization, which merges the visual outer surface of a patient with a simultaneous ultrasound scan of the patient's interior using a half-silvered mirror. The ultrasound image is visually merged with the patient, along with the operator's hands and the invasive tool in the operator's natural field of view. Fichtinger et al. [4] proposes an intraoperative CT-based medical AR system for visualizing one CT slice onto the patient in situ using a specific arrangement of a half transparent mirror and a monitor rigidly attached to a CT scanner. A similar technique was proposed for the in situ visualization of a single MRI slice [5]. Feuerstein et al. [6] augments laparoscopic video images with intraoperative 3D cone-beam CT by tracking both C-arm and laparoscope using the same external optical tracking system. Wendler et al. [7] fuses real-time ultrasound images with synchronized real-time functional nuclear information from a gamma probe based on optical tracking the ultrasound and nuclear probes in a common coordinate system. Wendler et al. [8] also propose freehand SPECT to augment the 3D reconstruction of radioactive distributions via live video by using a calibrated optical tracking and video camera system. A clinical study of the freehand SPECT system for sentinel lymph node biopsy on over 50 patients is reported in [9].

In general, AR has not been widely accepted in clinical practice due to its cumbersome system setups, the requirement of a line-of-sight for tracking, on-site calibration, and a lack of real-time accurate registration between medical data, patient and surgical instruments. Additionally, real-time non-rigid registration methods, display techniques and user-interfaces must be addressed and delivered for surgeon acceptance inside the operating room.

2 Camera Augmented Mobile C-Arm (CamC) for Orthopaedic Surgery

The first medical augmented reality system worldwide to be deployed in the operating room was the Camera augmented mobile C-arm (CamC). The CamC technology is attractive from both clinical and economical aspects, as no additional separated devices and calibration are required during surgery.

CamC extends a standard mobile C-arm fluoroscope by a video camera and mirror construction (Fig. 1). Their locations are optimized such that the camera optical center and the X-ray source virtually coincide, allowing the camera to have the same view of the patient as the X-ray [10, 11]. This enables CamC to deliver an intuitive real-time intraoperative visualization of X-ray images co-registered with live video (Fig. 2). Since the spatial relation between bones, implants, surgical instruments, and patient surface can be quickly and intuitively perceived in the X-ray and video overlay, surgeons can perform operations more confidently with less radiation exposure, reduced rate of surgical mistakes, and increased reproducibility.

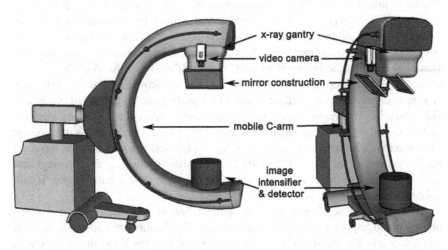

Fig. 1 The camera augmented mobile C-arm. The mobile C-arm is extended by an optical camera and mirror construction

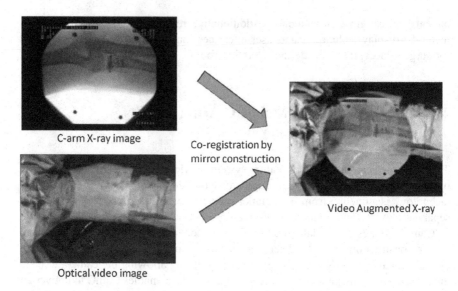

C-arm X-ray image

Co-registration by
mirror construction

Video Augmented X-ray

Optical video image

Fig. 2 CamC implicitly registers X-ray images (*upper-left*) and video images (*lower-left*) to provide video augmented X-ray imaging (*right*). The picture depicts an elbow fracture reduction surgery

2.1 CamC System Components

The first clinical CamC (Fig. 3) was built by affixing a *Flea2* camera (*Point Grey Research Inc., Vancouver, BC, Canada*) and mirror construction to a mobile *Powermobile* isocentric C-arm (*Siemens Healthcare, Erlangen, Germany*). The system comprises three monitors: the common C-arm monitor showing the conventional fluoroscopic image, the CamC monitor displaying the video image augmented by fluoroscopy, and a touch screen display providing the user-interface. A simple alpha blending method is used to visualize the co-registered X-ray and video images. The surgical crew can operate the CamC system like any standard mobile C-arm. A two-level phantom is used to align the camera and X-ray source. The phantom's bottom level has five X-ray visible xed markers and the top level has 5 movable small rings. The calibration consists of aligning the top rings and bottom markers in both X-ray and video images and then a homography is calculated for image overlay [11].

Misalignments between the treated patient anatomy in the X-ray and video images could happen due to patient movement. To visually inform surgeons about misalignments, the initial positions of a marker visible in the X-ray image are drawn as green quadrilaterals and their positions in the current video image are drawn as red quadrilaterals. Moreover, a gradient color bar is shown on the right side of the video images, whose length indicates the pixel-difference between the marker initial and current positions (Fig. 4).

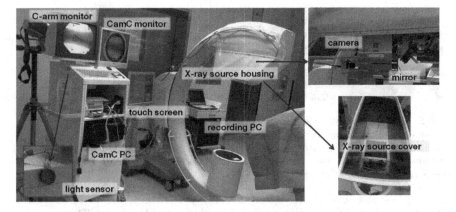

Fig. 3 The system components of CamC used during patient surgeries

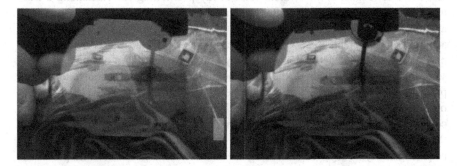

Fig. 4 Visual square marker tracking to inform surgeons about a possible misalignment between the X-ray and video image

2.2 CamC Patient Study

Forty-three patients were treated successfully with CamC support between July 2009 and March 2010 at the Klinikum der Universität München, Klinik für Allgemeine, Unfall-, Hand- und Plastische Chirurgie, Germany. This is the first time that a medical augmented reality technology was used consistently in real orthopaedic surgeries worldwide. Once an X-ray image is acquired, the surgical procedure may be continued solely under the visualization of video without additional radiation exposure. The patient study resulted in identifying the following surgical tasks which directly benefit from the medical AR imaging of CamC: X-ray positioning, incision, and instrument guidance.

First, X-ray positioning consists in iteratively moving the C-arm over the patient and is completed once the C-arm is correctly positioned to show the treated anatomy. Although this task seems intuitive, it may take several C-arm positions to acquire the correct X-ray image. The CamC overlay resolves this task by

Fig. 5 *Left* The CamC video augmentation can play the role of an aiming circle to position the C-arm optimally to acquire the desired X-ray image. *Center* Incision above the implant hole is facilitated using the CamC overlay between X-ray and video. *Right* Positioning of surgical instruments, such as a K-wire, relative to distal radius fractures becomes intuitive by visualizing the X-ray and video overlay

showcasing a semi-transparent grey circle within the video image. This augmentation allows the medical staff to efficiently position the C-arm above the anatomy to be treated. Figure 5-left demonstrates X-ray positioning for visualizing a fractured distal radius. Secondly, after acquiring an X-ray image showing the bone anatomy or implants, it is often required to open the skin in order to access bone anatomies or implants. The CamC overlay is used to plan the correct incision, placing the scalpel exactly above the entry point of interest. An example of this is seen in Fig. 5-center, where the entry point localization during an interlocking procedure is facilitated by positioning the scalpel on the nail hole. Lastly, many orthopaedic surgeries require the precise alignment of surgical instruments with the correct axis. This is usually achieved by first aligning the tip of the instrument with the entry point and then orienting the instrument to be aligned with the axis by acquiring many X-ray images. Using the CamC video guidance, this process is simplified. As an example, K-wires are often used for temporary fixation of bone fragments or for temporary immobilization of a joint during surgery. With the CamC system, the direction of a linear K-wire relative to the bone can be intuitively anticipated from the overlay image showing the projected direction of the K-wire and bone in a common image frame (Fig. 5-right).

2.3 Example CamC Orthopaedic Applications

Multi-view AR: The CamC provides AR visualization and guidance in two-dimensional space and no depth control is possible. Thus, the system is limited to applications where depth is not a factor such as in the interlocking of intramedullary nailing procedure. To resolve this limitation, Traub et al. [12] developed a multi-view opto-Xray system that provides visualization capable of depth control during surgery (Fig. 6). In addition to the original video camera mounted to the C-arm gantry, a second camera is attached orthogonal to the gantry such that its view is aligned with the X-ray image taken at a 90° orbital angle to the current C-arm position. After a one time calibration of the newly attached video camera, the

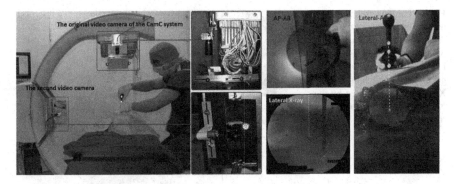

Fig. 6 *Left* The multi-view opto-X-ray system. *Center* The AP view of the video augmented X-ray image for axis control with the lateral X-ray image. *Right* The lateral view of the video augmented X-ray image for depth control

surgical instrument tip position can be augmented within a lateral X-ray image while the surgeon navigates the procedure using an anterior–posterior view. The feasibility of the system has been validated through cadaver studies [12].

Panoramic X-ray imaging: X-ray images generated by mobile C-arms have a limited field of view. Hence, one X-ray image may not capture the entire bone anatomy. Acquiring several individual X-rays to capture the entire bone in separate images only gives a vague impression of the relative position and orientation of the bone segments. This often compromises the quality of surgeries. Panoramic X-ray images that could show the entire bone as a whole would be helpful particularly for long bone surgeries. Existing methods in state-of-art have a common limitation which is parallax errors [13, 14]. A new method based on CamC was developed to generate parallax-free panoramic X-ray images during surgery by enabling the mobile C-arm to rotate around its X-ray source center (Fig. 7) relative to the patient table [15]. Rotating the mobile C-arm around its X-ray source center is impractical and sometimes impossible due to the mechanical design of mobile C-arms. To ensure that the C-arm motion is a relative pure rotation around its X-ray source center, the table is moved to compensate for the translational part of the motion based on C-arm pose estimation. C-arm pose estimation methods, like the one proposed by [16–20], use visible markers or radiographic fiducials during X-ray imaging. Such methods are not suitable for positioning of the table since continuous X-ray exposure is required, and therefore a large amount of radiation is inevitable. The CamC is a logical alternative for pose estimation since it contains a video camera. Moving the table and maneuvering the C-arm to reach a relative pure rotation of the X-ray source suffers from the complexity of the user interaction. Therefore, it is preferable for surgeons to specify a target position of X-ray images in a panorama frame and to be guided by the system on how to move the C-arm and the table as a kinematic unit. An integrated kinematic chain of the C-arm and the table with its closed-form inverse kinematics was developed by Wang et al. [21]. The technique could be employed to determine C-arm joint movements and table

Fig. 7 Parallax-free X-ray image stitching method. A parallax-free panoramic X-ray image of a plastic lumbar and sacrum is generated by stitching three X-ray images acquired by the X-ray source undergoing pure rotations

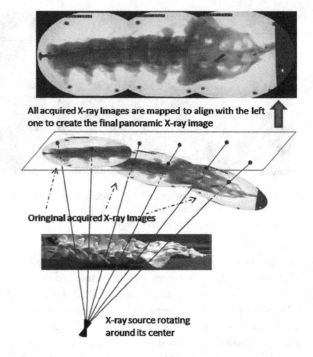

translations needed for acquiring an optimal X-ray image defined by its image position in a panorama frame. Given an X-ray image position within the panorama, the required C-arm pose is first automatically computed, and then the necessary joint movements and table translations from the closed-form inverse kinematics are solved. The surgeon keeps defining targets within the panorama frame until he/she are content with the final parallax-free panoramic image. The final image was used to determine, for example, the mechanical axis deviation in a cadaver leg study [22].

Distal locking of intramedullary nails: Intramedullary nailing is the surgical procedure mostly used in fracture reduction of the tibial and femoral shafts. Following successful insertion of the nail into the medullary canal, it must be fixed by inserting screws through its proximal and distal locking holes. Prior to distal locking of the nail, surgeons must position the C-arm device and patient leg in such a way that the nail holes appear as circles in the X-ray image. This is considered a 'trial and error' process, is time consuming and requires many X-ray shots. Londei et al. [23] proposed an augmented reality application that visually depicts to the surgeon two 'augmented' circles, their centers lying on the axis of the nail hole, making it visible in space. After an initial X-ray image acquisition, real-time video guidance using CamC allows the surgeon to superimpose the 'augmented' circles by moving the patient leg; the resulting X-ray showing nail holes appearing as circles. Following this, distal locking of the nail hole can be completed. Authors in [24, 25] designed a radiation-free guide to enable surgeons to complete the

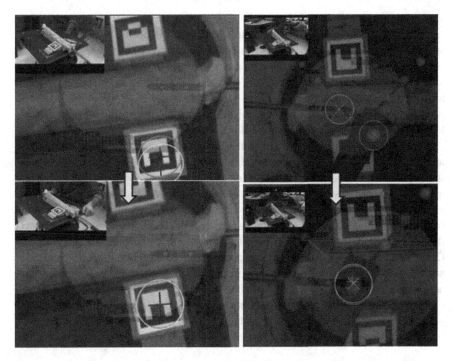

Fig. 8 *Left* Augmented circles visualization for the down-the-beam positioning of the nail, and *right* the augmented drill and target location for distal locking

interlocking of intramedullary nailing procedure without a significant number of X-ray acquisitions. The radiation-free guide is detected by the CamC camera and its tip is measured in real-time and displayed to the surgeon in both the X-ray and video images. A cannula consisting of two branches with markers are affixed to the radiolucent drill. Computer vision algorithms are developed that exploit cross ratio properties in order to estimate the tip-position of the novel radiation-free instrument allowing the surgeon to visualize and guide distal locking (Fig. 8). A recent study presented the complete pipeline of distal locking using augmented reality [26].

Multi-modal visualization: It is increasingly difficult to rapidly recognize and differentiate anatomy and objects in alpha-blended images as is the case with the CamC system. The surgeon's depth perception is altered since: (i) the X-ray anatomy appears floating on top of the scene in the optical image, (ii) surgeon hands and surgical instruments occlude the visualization, and (iii) there is no correct ordering between anatomy and objects in the fused images. With these issues in mind, we observe that all pixels in X-ray and optical images do not have the same importance and contribution to the final blending (e.g. the background is not important compared to the surgical instruments). This reflection suggests extracting only relevant-based data according to pixels belonging to background, instruments, and surgeon hands [27]. The labeling of the surgical scene by a precise segmentation and differentiation of its different parts allows a relevant blending respecting the desired ordering of structures. A few attempts have been endeavored, such as in [28]. In these early works, a Naive Bayes classification approach based on color and radiodensity is applied to recognize the different objects in X-ray and video images. Depending on the pair of pixels it belongs to, each pixel is associated to a mixing value to create a relevant-based fused image. While authors showed promising results, recognizing each object on their color distribution is very challenging and not robust to changes in illumination.

Pauly et al. [29] introduced a surgical scene labeling paradigm based on machine learning using the CamC system. In their application, the depth is a useful hint for the segmentation and ordering of hands and instruments with respect to patient anatomy since the surgeon performs the procedure on the patient. Thus, their visualization paradigm is founded on segmentation consisting in modeling the background via depth data. They perform in parallel color image segmentation via the state-of-art Random Forests. To refine the segmentation method, they use the GrabCut algorithm. Lastly, the authors combine the background modeling and color segmentation in order to identify the objects of interests in the color images and achieve successfully ordering of structures (Fig. 9).

C-arm positioning via DRRs: Dressel et al. [30] propose to guide C-arm positioning using artificial fluoroscopy based on the CamC system. This is achieved by computing digitally reconstructed radiographs (DRRs) from preoperative CT

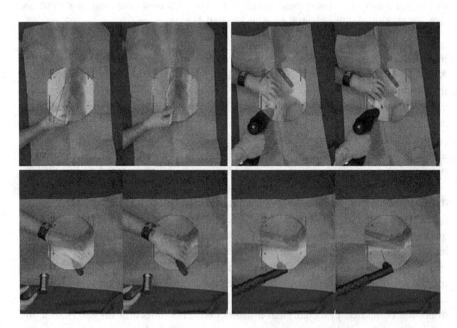

Fig. 9 Four clinical examples comparing the traditional alpha-blending visualization technique to the relevance-based fused images

data or cone-beam CT. An initial pose between the patient and the C-arm is computed by rigid 2D/3D registration. After the initial pose estimation, a spatial transformation between the patient and the C-arm is obtained from the video camera affixed to the CamC. Using this information, it is able to generate DRRs and simulate fluoroscopic images. For positioning tasks, this solution appears to match conventional fluoroscopy; however simulating the images from the CT data in real time as the C-arm is moved is performed radiation-free.

3 AR Solutions for Orthopaedic Surgery, Perception, and Medical Education

In the previous section, the Camera augmented mobile C-arm was presented as a viable technology that facilitates many tasks within surgery. We now present other works based on augmented reality and how these have solved difficulties associated with orthopaedic procedures and surgeon perception, and conclude the section by presenting applications with respect to resident trainee education.

3.1 AR Solutions for Orthopaedic Surgery

Zheng et al. [31] propose a novel technique to create a reality-augmented virtual fluoroscopy for computer-assisted diaphyseal long bone fracture osteosynthesis. With this solution, repositioning of bone fragments during closed fracture reduction and osteosynthesis can lead to image updates in the virtual imaging planes of all acquired images without any radiation. After acquiring two calibrated X-ray images and prior to fracture reduction, a data preparation phase interactively identifies and segments the bone fragments from the background in each image. After that, a second phase repositions the fragment projection onto each virtual imaging plane in real-time during using an OpenGL-based texture warping. Combined with a photorealistic virtual implant model rendering technique, the presented solution allows the control of closed indirect fracture osteosynthesis in the real world through direct insight into the virtual world. The first clinical study results showed a reduction in X-ray radiation to the patient and surgical team, and improved operative precision guaranteeing more safety for the patient.

Abe et al. [32] present a study introducing a novel AR guidance system called virtual protractor with augmented reality (VIPAR) to visualize a needle trajectory in 3D space during percutaneous vertebroplasty. The AR system used for this study comprised a head-mounted display (HMD) with a tracking camera and a marker sheet. An augmented scene was created by overlaying the preoperatively generated needle trajectory path onto a marker detected on the patient using AR software, thereby providing the surgeon with augmented views in real time through the

HMD. Early clinical results and experimental accuracy in spine phantoms demonstrated that the virtual protractor with augmented reality can successfully assist the surgeon as a virtual protractor by indicating the trajectory of the needle and preventing neurovascular complications during percutaneous vertebroplasty of the thoracolumbar spine.

Wu et al. [33] present an advanced augmented reality system for spinal surgery assistance, and develop entry-point guidance prior to vertebroplasty spinal surgery. Based on image-based marker detection and tracking, the proposed camera-projector system superimposes preoperative 3D images onto patients. The patients' preoperative 3D image model is registered by projecting it onto the patient such that the synthetic 3D model merges with the real patient image, enabling the surgeon to see through the patients' anatomy. The proposed method is much simpler than heavy and computationally challenging navigation systems, and also reduces radiation exposure. The results of the clinical trials are extremely promising, with surgeons reporting favorably on the reduced time of finding a suitable entry point coupled with reduced radiation dose to patients.

In traditional remote instruction, clinicians at the remote site may select a region of interest (ROI) in medical images which is captured and sent by a camera at the local site. After the ROI selection of the remote-site clinician, the 2D information within the ROI will be sent back to the local site. Because the clinician viewpoint at the local site may change, the 2D information of the ROI can't be projected on the camera image in the correct position. To deal with this issue, Chang et al. [34] propose a method to calculate the 3D position of the selected ROI and apply it for AR display. This method utilizes the line-of-sight of the local-site camera and pre-registered medical images. In addition, a marker-less AR visualization was also implemented to integrate the preoperative medical image(s) with the real patient position. With this solution, the 3D information of the selected ROI can be displayed on the local camera image even if the viewpoint of the local-site surgeon changes.

Fritz et al. [35] developed a low-cost augmented reality system that can be used for MRI guidance with almost any conventional MRI system. Virtual MRI guidance is accomplished by projecting cross-sectional MR images into the patient space, generating a hybrid view of reality and MR images. With this system, a variety of spinal injection procedures appear possible. The augmented reality image overlay system facilitated accurate MRI guidance for successful spinal procedures in a lumbar spine model. It exhibited potential for simplifying the current practice of MRI-guided lumbar spinal injection procedures.

Hu et al. [36] present a convenient method for video see-through augmented reality on the basis of an existing image-guided surgery system. Authors describe the prototype of the system, the registration of virtual objects in a marker coordinate system, and the motion tracking of the Head Mounted Display (HMD). The experiment results show that the proposed AR technology can provide augmented information for the surgeons and works well with an existing image-guided surgery system.

Wang et al. [37] present a framework for speeding up workflow of orthopaedic surgery helping surgeons to simply transformations between several imaging displays. Their solution is based on epipolar geometry and simple feature landmarks. A 3D superimposed imaging approach is developed via the construction of the camera-projector system. The superimposed approach is to rectify X-ray and projected image pair using the perspective projection model. The proposed method not only simplifies the computation between surgical instruments and patient for surgeons, but also reduces the radiation exposure. The experimental results for both the synthetic spinal image on dummy and real patient testing have demonstrated the feasibility of the technique.

3.2 AR Solutions for Perception

Some augmented reality applications use polygonal models to augment a real scene. However, most of the medical applications have volumetric data to be visualized. Therefore, it is desirable for the medical AR systems to provide the visualization of these volumes into the real scene in real-time. Macedo et al. [38] introduce a real-time semi-automatic approach for on-patient medical volume data visualization. This solution is possible in a marker-less augmented reality (MAR) environment, whereas the medical data consists of a volume reconstructed from 3D computed tomography (CT) image data. A 3D reference model of the region of interest in the patient is generated and tracked from a Kinect depth stream. From the estimated camera pose, volumetric medical data can be displayed inside the patient's anatomy at the location of the real anatomy. The authors evaluated the use of standard volume rendering techniques in the context of a MAR environment and demonstrated that these techniques can be applied in this scenario in real-time. In another work, Maceo et al. [39], introduced an on-patient focus + context medical data visualization based on volume clipping. From the estimated camera pose, the volumetric medical data can be displayed inside the patient's anatomy at the location of the real anatomy. To improve the visual quality of the final scene, three methods based on volume clipping are proposed to allow new focus + context visualizations. Moreover, the whole solution supports occlusion handling. From the evaluation of the proposed techniques, the results demonstrate that these methods improve the visual quality of the final rendering.

Dixon et al. [40] assess whether perceptual blindness is significant in a surgical context and evaluate the impact of on-screen navigational cuing with augmented reality. Surgeons and trainees performed an endoscopic navigation exercise on a cadaveric specimen. The subjects were randomized to either a standard endoscopic view (control) or an AR view consisting of an endoscopic video fused with anatomic contours. Two unexpected findings were presented in close proximity to the target point: one critical complication and one foreign body (i.e. a screw). Task completion time, accuracy, and recognition of findings were recorded. Authors demonstrated that perceptual blindness was evident in both groups. Although more accurate, the AR group was less likely to identify unexpected findings clearly within view. Advanced navigational displays may increase precision, but strategies to mitigate perceptual costs need further investigation to allow safe implementation.

Blum et al. [41] describes first steps towards a Superman-like X-ray vision where a brain-computer interface (BCI) device and a gaze-tracker are used to allow the user to control the augmented reality (AR) visualization. A BCI device is integrated into two medical AR systems. To assess the potential of this technology feedback from medical clinicians was gathered. While in their pilot study only electromyographic signals are used, the clinicians provided very positive feedback on the use of BCI for medical AR.

3.3 AR Solutions for Medical Education

Understanding human anatomy is essential for practicing medicine since anatomical knowledge supports the formulation of a diagnosis and communication of that diagnosis to patient and colleagues [42]. Anatomy education is traditionally performed by the dissection of cadavers. The value of dissection classes as a teaching format lies in the fact that it provides a 3D view on human anatomy including tactile learning experiences. It enables elaboration of knowledge already acquired in lectures and study books and it provides an overall perspective of anatomical structures and their mutual relations in a whole organism [43]. This training format is, however, quite costly. And so far, no objective empirical evidence exists concerning the effectiveness of dissection classes for learning anatomy [42]. AR technology could offer an additional teaching method for anatomy education, depending on how it is implemented. Strong points are the visualization capabilities including the 3D rendering of anatomical imagery. Other sensory experiences could be implemented as well, such as tactile feedback. AR provides real-time manipulation of these visualizations and direct feedback to students. With that, AR technology could comply with some of the affordances of traditional dissection classes.

Several AR systems have already been developed specifically for anatomy education [44–46]. Blum et al. [46] describes the magic mirror ('Miracle') which is an AR system that can be used for undergraduate anatomy education (Fig. 10). The set-up of that system is as follows. The trainee stands in front of a TV screen that has a camera and the Kinect attached to it. The camera image of the trainee is flipped horizontally and is shown on the TV screen, mimicking a mirror function. Part of an anonymous patient CT dataset is augmented on the user's body and shown on the TV screen. This creates the illusion that the trainee can look inside their body. A gesture-based user interface allows real-time manipulation and visualization of the CT data. The trainee can scroll through the dataset in sagittal, transverse and coronal slice mode, by using different hand gestures [47].

Consistent with modern learning theories advocating for active participation with immediate application of knowledge [48], virtual interactive presence (VIP) technology enables learners to immerse themselves in a real surgical

Fig. 10 The magic mirror system for anatomy learning

environment. The free-form feedback from the participants in a study organized by Ponce et al. [48] highlights the potential for an improved educational experience through greater involvement of a resident surgeon. Resident surgeons expressed greater comfort with having the attending surgeon outside the operative room because they experienced a feeling of autonomy while still having sufficient oversight. The attending surgeon commented that instruction was improved since virtually "touching" anatomy helped to better communicate degrees of motion. The use of the system to convey hand gestures or point to a particular anatomy was thought to be more effective. Both the resident and attending surgeons thought that the identification of anatomy was better with use of this technology. This pilot study revealed that the VIP technology was efficient, safe, and effective as a teaching tool. The attending and resident surgeons agreed that training was enhanced, and this occurred without increasing operative times. Furthermore, the attending surgeon believed that this technology improved teaching effectiveness [48].

4 Conclusions

Different aspects have emerged that may be important when developing future AR user interfaces to be used in the surgical workflow of orthopaedic surgery. Some of those aspects may also be valid for other application domains or other technology. Bichlemeier [49] proposes four aspects to investigate for the successful development and deployment of AR technology inside the operating rooms. these are:

- *Workflow-driven application of AR*: the AR system won't be used during the whole procedure. The AR system is considered as a surgical tool like a scalpel. It is provided when needed and removed when dispensable. Situations during the workflow that might benefit from AR views can be automatically determined by a major workflow analysis of a particular intervention.
- *Inter-disciplinarity*: It is also important that a common physical, interdisciplinary platform is provided such as a real world lab to get to know and

understand each other. In the case of collaboration between surgeons and engineers or computer scientists, it is essential that the latter group physically moves their workspace close to the surgeon's workspace. By the same token physicians have to get to know what is technically possible.

- *Importance of the User Interface*: It is important to understand that only a sophisticated user interface will create an impact in the medical workflow. A proposed technical method that might improve a certain working step has to be provided in a way so that it can be easily controlled and used by the end-user. In addition, it has to fulfill the infrastructural requirements such as sterility rules. Otherwise the probability that the technology or features of it are really used would extremely decrease.

- *Stepping into the OR*: The indication of an additional device may have an impact on many of the subtasks of the surgical team during surgery. It may affect not only the work of the surgeon, but also the job of the anesthetist or the assisting nurse. For this reason it is important that the whole surgical team gets to know the objectives of a project that influences their jobs in order to achieve tolerance and acceptance. The transfer of the AR system into its designated intraoperative workspace has to be a smooth, iterative procedure.

References

1. Van Krevelen DWF, Poelman R (2010) A survey of augmented reality technologies, applications and limitations. Int J Virtual Reality 9(2):1
2. Navab N, Blum T, Wang L, Okur A, Wendler T (2012) First deployments of augmented reality in operating rooms. Computer 7:48–55
3. Stetten G, Cois A, Chang W, Shelton D, Tamburo R, Castellucci J, von Ramm O (2005) C-mode real-time tomographic reflection for a matrix array ultrasound sonic flashlight 1. Acad Radiol 12(5):535–543
4. Fichtinger G, Deguet A, Masamune K, Balogh E, Fischer GS, Mathieu H, Taylor RH, James Zinreich S, Fayad LM (2005) Image overlay guidance for needle insertion in CT scanner. IEEE Trans Biomed Eng 52(8):1415–1424
5. Fischer GS, Deguet A, Csoma C, Taylor RH, Fayad L, Carrino JA, James Zinreich S, Fichtinger G (2007) MRI image overlay: application to arthrography needle insertion. Comput Aided Surg 12(1):2–14
6. Feuerstein M, Mussack T, Heining SM, Navab Nassir (2008) Intraoperative laparoscope augmentation for port placement and resection planning in minimally invasive liver resection. IEEE Trans Med Imaging 27(3):355–369
7. Wendler T, Feuerstein M, Traub J, Lasser T, Vogel J, Daghighian F, Ziegler SI, Navab N (2007) Real-time fusion of ultrasound and gamma probe for navigated localization of liver metastases. In: Medical image computing and computer-assisted intervention–MICCAI 2007. Springer, Berlin, pp 252–260
8. Wendler T, Hartl A, Lasser T, Traub J, Daghighian F, Ziegler SI, Navab N (2007) Towards intra-operative 3D nuclear imaging: reconstruction of 3D radioactive distributions using tracked gamma probes. In: Medical image computing and computer-assisted intervention–MICCAI 2007. Springer, Berlin, pp 909–917

9. Okur A, Ahmadi SA, Bigdelou A, Wendler T, Navab N (2011) MR in OR: first analysis of AR/VR visualization in 100 intra-operative freehand SPECT acquisitions. In: 10th IEEE international symposium on mixed and augmented reality (ISMAR). IEEE, pp 211–218

10. Navab N, Bani-Kashemi A, Mitschke M (1999) Merging visible and invisible: two camera-augmented mobile C-arm (CAMC) applications. In: 2nd IEEE and ACM international workshop on augmented reality (IWAR'99) proceedings. IEEE, pp 134–141

11. Navab N, Heining SM, Traub J (2010) Camera augmented mobile C-arm (CAMC): calibration, accuracy study, and clinical applications. IEEE Trans Med Imaging 29(7): 1412–1423

12. Traub J, Heibel TH, Dressel P, Heining SM, Graumann R, Navab N (2007) A multi-view opto-Xray imaging system. In: Medical image computing and computer-assisted intervention–MICCAI 2007. Springer, Berlin, pp 18–25

13. Yaniv Z, Joskowicz L (2004) Long bone panoramas from fluoroscopic X-ray images. IEEE Trans Med Imaging 23(1):26–35

14. Messmer P, Matthews F, Wullschleger C, Hügli R, Regazzoni P, Jacob AL (2006) Image fusion for intraoperative control of axis in long bone fracture treatment. Eur J Trauma 32(6):555–561

15. Wang L, Traub J, Weidert S, Heining SM, Euler E, Navab N (2010) Parallax-free intra-operative X-ray image stitching. Med Image Anal 14(5):674–686

16. Kainz B, Grabner M, Rüther M (2008) Fast marker based C-arm pose estimation. In: Medical image computing and computer-assisted intervention–MICCAI 2008. Springer, Berlin, pp 652–659

17. Pati S, Erat O, Wang L, Weidert S, Euler E, Navab N, Fallavollita P (2013) Accurate pose estimation using single marker single camera calibration system. In: SPIE medical imaging. International Society for Optics and Photonics, pp 867126–867126

18. Fallavollita P, Burdette EC, Song DY, Abolmaesumi P, Fichtinger G (2011) Technical note: unsupervised c-arm pose tracking with radiographic fiducial. Med Phys 38(4):2241–2245

19. Moult E, Burdette EC, Song DY, Abolmaesumi P, Fichtinger G, Fallavollita P (2011) Automatic C-arm pose estimation via 2D/3D hybrid registration of a radiographic fiducial. In: SPIE medical imaging. International Society for Optics and Photonics, pp 79642S–79642S

20. Fallavollita P, Burdette C, Song DY, Abolmaesumi P, Fichtinger G (2010) C-arm tracking by intensity-based registration of a fiducial in prostate brachytherapy. In: information processing in computer-assisted interventions. Springer, Berlin, pp 45–55

21. Wang L, Fallavollita P, Zou R, Chen X, Weidert S, Navab N (2012) Closed-form inverse kinematics for interventional C-arm X-ray imaging with six degrees of freedom: modeling and application. IEEE Trans Med Imaging 31(5):1086–1099

22. Wang L, Fallavollita P, Brand A, Erat O, Weidert S, Thaller PH, Euler E, Navab N (2012) Intra-op measurement of the mechanical axis deviation: an evaluation study on 19 human cadaver legs. In: Medical image computing and computer-assisted intervention–MICCAI 2012. Springer, Berlin, pp 609–616

23. Londei R, Esposito M, Diotte B, Weidert S, Euler E, Thaller P, Navab N, Fallavollita P (2014) The 'augmented' circles: a video-guided solution for the down-the-beam positioning of IM nail holes. In: Information processing in computer-assisted interventions. Springer International Publishing, pp 100–107

24. Diotte B, Fallavollita P, Wang L, Weidert S, Euler E, Thaller P, Navab N (2014) Multi-modal intra-operative navigation during distal locking of intramedullary nails. IEEE Trans Med Imaging 34(2):487–495

25. Thaller PH, Diotte B, Fallavollita P, Wang L, Weidert S, Euler E, Mutschler W, Navab N (2013) Sa4. 8 video-assisted interlocking of intramedullary nails. Injury 44:S25

26. Londei R, Esposito M, Diotte B, Weidert S, Euler E, Thaller P, Navab N, Fallavollita P (2015) Intra-operative augmented reality in distal locking. Int J Comput Assisted Radiol Surg 1–9

27. Pauly O, Katouzian A, Eslami A, Fallavollita P, Navab N (2012) Supervised classification for customized intraoperative augmented reality visualization. In: IEEE International symposium on mixed and augmented reality (ISMAR), 2012. IEEE, pp 311–312

28. Erat O, Pauly O, Weidert S, Thaller P, Euler E, Mutschler W, Navab N, Fallavollita P (2013) How a surgeon becomes superman by visualization of intelligently fused multi-modalities. SPIE medical imaging. International Society for Optics and Photonics, pp 86710L-86710L

29. Pauly O, Diotte B, Habert S, Weidert S, Euler E, Fallavollita P, Navab N (2014) Relevance-based visualization to improve surgeon perception. In: Information processing in computer-assisted interventions. Springer International Publishing, pp 178–185

30. Dressel P, Wang L, Kutter O, Traub J, Heining SM, Navab N (2010) Intraoperative positioning of mobile C-arms using artificial fluoroscopy. In: SPIE medical imaging. International Society for Optics and Photonics, pp 762506–762506

31. Zheng G, Dong X, Gruetzner PA (2008) Reality-augmented virtual fluoroscopy for computer-assisted diaphyseal long bone fracture osteosynthesis: a novel technique and feasibility study results. Proc Inst Mech Eng [H] 222(1):101–115

32. Abe Y, Sato S, Kato K, Hyakumachi T, Yanagibashi Y, Ito M, Abumi K (2013) A novel 3D guidance system using augmented reality for percutaneous vertebroplasty: technical note. J Neurosurg Spine 19(4):492–501

33. Wu JR, Wang ML, Liu KC, Hu MH, Lee PY (2014) Real-time advanced spinal surgery via visible patient model and augmented reality system. Comput Methods Programs Biomed 113 (3):869–881

34. Chang TC, Hsieh CH, Huang CH, Yang JW, Lee ST, Wu CT, Lee JD (2015) Interactive medical augmented reality system for remote surgical assistance. Appl Math 9(1L):97–104

35. Fritz J, U-Thainual P, Ungi T, Flammang AJ, Cho NB, Fichtinger G, Iordachita II, Carrino JA (2012) Augmented reality visualization with image overlay for MRI-guided intervention: accuracy for lumbar spinal procedures with a 1.5-T MRI system. Am J Roentgenol 198(3): W266–W273

36. Hu L, Wang M, Song Z (2013) A convenient method of video see-through augmented reality based on image-guided surgery system. In: Seventh international conference on internet computing for engineering and science (ICICSE), 2013. IEEE, pp 100–103

37. Wang ML, Wu JR, Liu KC, Lee PY, Chiang YY, Lin HY (2012) Innovative 3D augmented reality techniques for spinal surgery applications. In: International symposium on intelligent signal processing and communications systems (ISPACS), 2012. IEEE, pp 16–20

38. Macedo MCF, Lopes Apolinário A, Souza ACS, Giraldi GA (2014) A semi-automatic markerless augmented reality approach for on-patient volumetric medical data visualization. In: 2014 XVI symposium on virtual and augmented reality (SVR). IEEE, pp 63–70

39. Macedo MCF, Apolinário AL Jr (2014) Improving on-patient medical data visualization in a markerless augmented reality environment by volume clipping. In: Proceedings of the 2014 27th SIBGRAPI conference on graphics, patterns and images. IEEE Computer Society, pp 149–156

40. Dixon BJ, Daly MJ, Chan H, Vescan AD, Witterick IJ, Irish JC (2013) Surgeons blinded by enhanced navigation: the effect of augmented reality on attention. Surg Endosc 27(2):454–461

41. Blum T, Stauder R, Euler E, Navab N (2012) Superman-like X-ray vision: towards brain-computer interfaces for medical augmented reality. In: IEEE international symposium on mixed and augmented reality (ISMAR), 2012. IEEE, pp 271–272

42. Frank JR (ed) (2005) The CanMEDS 2005 physician competency framework: better standards, better physicians, better care. Royal College of Physicians and Surgeons of Canada

43. McLachlan JC, Bligh J, Bradley P, Searle J (2004) Teaching anatomy without cadavers. Med Educ 38(4):418–424

44. Thomas RG, William John N, Delieu JM (2010) Augmented reality for anatomical education. J Vis Commun Med 33(1):6–15

45. Chien CH, Chen CH, Jeng TS (2010) An Interactive augmented reality system for learning anatomy structure. In: Proceedings of the international multiconference of engineers and computer scientists, IMECS

46. Blum T, Kleeberger V, Bichlmeier C, Navab N (2012) Miracle: an augmented reality magic mirror system for anatomy education. In: Virtual reality short papers and posters (VRW), 2012. IEEE, pp 115–116

47. Kamphuis C, Barsom E, Schijven M, Christoph N (2014) Augmented reality in medical education? Perspect Med Edu 3(4):300–311
48. Ponce BA, Jennings JK, Clay TB, May MB, Huisingh C, Sheppard ED (2014) Telementoring: use of augmented reality in orthopaedic education. J Bone Joint Surg 96(10):e84
49. Bichlmeier CP (2010) Immersive, interactive and contextual in-situ visualization for medical applications. PhD dissertation, München, Technische Universität Dissertation

State of the Art of Ultrasound-Based Registration in Computer Assisted Orthopedic Interventions

Steffen Schumann

Abstract The preferred modality for intraoperative imaging in orthopedic interventions is fluoroscopic imaging. But the exposure of the patient and the surgical team to high dose of x-ray radiation and the inappropriate handling of the fluoroscopic c-arm limit its application. In contrast to fluoroscopy, ultrasound imaging does not expose the patient to harmful ionizing radiation. Due to its excellent spatial resolution, B-mode ultrasound imaging is commonly used in the clinical routine to examine soft tissue such as muscles and organs. Thus, it is perfectly suited for preoperative diagnosis and is an essential tool in prenatal care. However, ultrasound images are subject to various types of artifacts, degrading the quality of the data and making the perception and interpretation rather difficult. Despite its known drawbacks, ultrasound imaging has the potential to become an efficient modality for intraoperative imaging. But the integration of ultrasound into the clinical workflow of a computer-assisted orthopedic surgery gives rise to new challenges. In this chapter the advantages and disadvantages of using ultrasound imaging in image-guided orthopedic interventions are pointed out. Moreover, an overview of the latest clinical applications and the current research is given. A special focus will be on the application of B-mode ultrasound imaging for intraoperative registration in image-guided interventions.

1 Introduction

Since its introduction into orthopedic operating rooms in the mid-nineties [2, 38, 48], surgical navigation systems have significantly improved. Surgical navigation systems enhance the reproducibility and improve the accuracy of the performed surgical action and are thus superior to conventionally performed interventions [4, 32, 60].

S. Schumann (✉)
Institute for Surgical Technology and Biomechanics, University of Bern,
Stauffacherstrasse 78, 3014 Bern, Switzerland
e-mail: steffen.schumann@istb.unibe.ch

© Springer International Publishing Switzerland 2016   271
G. Zheng and S. Li (eds.), *Computational Radiology*
for Orthopaedic Interventions, Lecture Notes in Computational
Vision and Biomechanics 23, DOI 10.1007/978-3-319-23482-3_14

In the meantime, these systems have been developed for various surgical interventions and are well accepted in the field of orthopedics. Even though various concepts and techniques have been developed, all surgical navigation systems share a common idea and basic components. Each navigated procedure consists of a therapeutic object, a virtual object and a navigator. The therapeutic object corresponds to the anatomy that is being operated and is supplemented by a virtual representation, for example in terms of medical imaging data. The navigator has the central role in such a navigation system and links both therapeutic and virtual object. A more detailed explanation of these major components is given by Langlotz and Nolte [37].

The navigator of a conventional navigation system is a sensing device, which is able to measure the position and orientation of a sensor in space at any point in time. Although, many different physical principles exist to accomplish this task, most navigators rely on optical tracking using infrared light. Such tracking cameras emit infrared light, which is reflected by the tracked sensors. Thus, the tracked sensors consist of specifically arranged infrared light reflecting spheres, which are then in turn recognized by the navigator. For a navigated surgery, the anatomy of the patient and all surgical tools need to be equipped with such sensors. A major limitation of optical tracking is the need of a direct line-of-sight between the navigator and the tracked objects (surgical tools and patient's anatomy). As the space is limited around the operating table, such restriction can be a critical issue. Thus, one popular alternative to optical tracking relies on the concept of electromagnetic tracking. It generates a local magnetic field next to the operation site with a special emitter coil. The tracked sensors are accordingly equipped with receiver coils. But magnetic tracking is subject to certain distortions, if highly conductive metals are in close proximity, decreasing the tracking accuracy and limiting its application in the clinical routine.

The process of tracking the sensors by the navigator is denoted as referencing. As the sensors are attached to the surgical tools and the anatomy of the patient (therapeutic object), the relative motion of both objects is recorded in real-time. In order to link the virtual representation of the therapeutic object to its real entity via the navigator, a registration step has to be performed. By this step a transformation is calculated, which aligns the virtual object with the intraoperative situation of the therapeutic object. The virtual object could be generated either preoperatively or intraoperatively. Preoperatively, the virtual object could be generated by acquisition of a medical imaging dataset (e.g. computed tomography (CT), magnetic resonance imaging (MRI), positron emission tomography (PET)). In addition to the image dataset, a preoperative planning (e.g. trajectory, tumor location) could be assigned to the virtual object. After calculating the registration transformation, this information is merged with the intraoperative situation. This enables the guidance of the surgical tools to accomplish the desired preoperative planning. The registration step is commonly achieved by the intraoperative acquisition of landmarks from the tracked patient's anatomy using a tracked pointer tool. In a first step, corresponding points need to be defined on the virtual object (software-based). Based on the corresponding points, an initial registration is computed using paired-point matching technique. In a second step, the initial registration is refined by matching the virtual object to the intraoperatively acquired cloud of points using surface matching technique [5, 40].

Intraoperatively, the virtual object could be generated by means of tracked fluoroscopic imaging or a tracked surgical tool to collect anatomical landmarks from the patient's anatomy. In order to visualize a complete three-dimensional (3D) model of the patient's anatomy, this intraoperatively collected sparse information could be extrapolated using a statistical shape model [15]. Such a statistical shape model is constructed from a database of aligned 3D bone models and is non-rigidly deformed to optimally match the collected sparse information (e.g. 3D points collected from bony anatomy) as described by Stindel et al. [59] or Rajamani et al. [53]. An overview of existing registration methods for the purpose of surgical navigation is presented by Zheng et al. [69]. As the intraoperative virtual object is generated based on tracked data, an explicit registration to the therapeutic object is not required.

For the registration of the preoperative virtual objects and the generation of intraoperative virtual objects, patient-specific information needs to be gathered from the tracked bony anatomy. This step is normally done using a tracked pointer tool. This pointer tool could either have a round or a sharp tip and is specifically designed for the digitization of anatomical landmarks. In order to use such a pointer tool, its tip needs to be determined in the coordinate system of the attached sensor. Therefore, a calibration step needs to be performed to identify the tip in the sensor's coordinate system before starting the acquisition of the landmarks. Intraoperatively, specific landmarks are selected with the pointer tip directly on the bony surface. Thereby, the accuracy of this registration step is very important for the overall clinical outcome of a navigated surgery. In order to achieve a precise registration, points from a large bony area need to be acquired. The larger the distribution of the points, the more robust and accurate will be the registration step. This in turn means that large parts of the bone might need to be exposed in order to get a well-distributed cloud of points. If certain regions of a bone are not exposed during surgery (e.g. the contralateral side of the pelvis during total hip replacement), points could also be acquired percutaneously. This percutaneous landmark acquisition can only be precisely performed for prominent bony regions, which are conventionally palpable. But particularly for obese patients, the percutaneous landmark acquisition can lead to certain errors in the registration step. In addition, the trend towards minimal invasive surgeries is contradicting with the need of large exposures of the bone for intraoperative landmark acquisition. While minimal invasive surgeries normally reduce the operation time and allow for a faster rehabilitation of the patient, open surgeries involve an increased damage of soft tissue and impose a higher risk of infection.

In order to overcome the limitations of pointer-based digitization of bony landmarks, ultrasound imaging was identified as a potential alternative. Ultrasound is well suited for the visualization of inner organs and is commonly used in the clinical routine for general diagnosis and obstetrics. Ultrasound imaging has distinct advantages making it a perfect candidate to replace pointer-based landmark acquisition:

- inexpensive
- generally available
- harmless, no radiation
- capable of real-time acquisition
- compact, portable

On the contrary, the use of ultrasound imaging imposes also certain challenges to the operator. As the field of view is reasonably small and the interpretation of ultrasound images is rather difficult, the application of ultrasound imaging requires a lot of training [47]. Moreover, the acquisition of ultrasound images is strongly operator-dependent. Hence, the quality of the images is depending on the experience and skills of the respective operator. Another challenge for the operator is that the imaging is restricted to only shallow depths and does not facilitate the imaging of structures behind bones. For a flawless integration of ultrasound imaging for intraoperative registration, these drawbacks need to be efficiently resolved by suitable software algorithms. A sketch of the main components of such a surgical navigation system is exemplary shown for total hip replacement in Fig. 1.

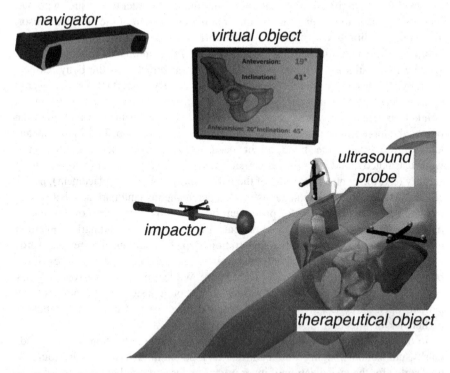

Fig. 1 Typical example of involved components in an ultrasound-based surgical navigation system: B-mode ultrasound probe, infrared tracking camera (=navigator), surgical tool (here: impactor), the therapeutic object (here: pelvis) to be operated and the virtual object (representation on a computer screen)

In the next section, the basic principles of ultrasound imaging will be explained, before ultrasound-based registration methods will be presented in detail. This chapter concludes with a summary on the clinical state of the art and an outlook on the future work of making ultrasound-based registration a valid alternative to pointer-based registration methods.

2 Basic Principles of Ultrasound Imaging

This section deals with the principles of ultrasound image formation and the basic physical concepts, which are relevant for the application of ultrasound imaging for intraoperative registration. An explanation of the basic principles is of particular importance for the understanding of the section on ultrasound-based registration algorithms.

Ultrasound is generally referring to a high-energy acoustic sound wave. Due to its high frequency, ultrasound waves can propagate within a matter medium and thus are extremely suitable to penetrate body tissue. Ultrasound waves, which are sent into an inhomogeneous medium such as the human body, are reflected and scattered at the boundaries of different tissue types. These reflections can be measured in terms of echoes. In medical ultrasound, these echoes are utilized to visualize inner structures of the human body. Typically used ultrasound frequencies for medical purpose are in the range of 2–30 MHz. An ultrasonic transducer (probe) is placed in direct contact to the skin or organ and short ultrasound pulses are continuously sent into the body. These pulses are accordingly reflected back towards the transducer at the different tissue boundaries and detected as echoes. By means of the time it takes for the echoes to return back to the transducer, the depth of these originating reflections can be determined.

Ultrasonic imaging can be operated in different modes. Depending on the medical purpose, various imaging modes exist. The simplest mode is the A-mode (amplitude mode). A single transducer element sends pulses of ultrasound waves and records the returning echo intensity (amplitude). Because only a single scan line exists, the resulting echoes are plotted as a curve in terms of the travel time. The travel time is directly related to the depth of the recorded echo and thus the A-mode is primarily used to measure the distance to certain structures. The most commonly used mode in the clinical routine is the B-mode (brightness mode). An array of transducer elements sends and receives ultrasound pulses and therefore represents an extension of the A-mode. As many transducer elements are simultaneously measuring the echo intensities, a cross-sectional image can be formed. Thereby, the pixel grayscale (brightness) of such a two-dimensional (2D) ultrasound image reflect the measured echo amplitude. Beside the commonly used A- and B-mode, Doppler ultrasonography and M-mode are frequently used in medical diagnostics. While the Doppler mode is particularly relevant to measure and visualize the blood flow, the M-mode (motion mode) is mainly used to determine the motion of organs such as the one of the heart. But mainly relevant for the intraoperative registration of bony structures are the A- and B-mode. While the A-mode can be used to measure the distance through the skin to a bony structure in a single direction, the B-mode is able to form a cross-sectional image and thus gives two-dimensional information of the anatomy treated with ultrasound.

While different A-mode transducer mainly differ in size, B-mode transducer differ in shape and size. Mainly relevant for the imaging of the musculoskeletal system is the linear array transducer. This transducer consists of a large number of linearly arranged elements and consequently generates a rectangular 2D ultrasound image. As conventional ultrasound imaging only provides two-dimensional information, two technologies can be used to retrieve 3D volumetric data. The first technology extends the B-mode imaging by attaching a trackable sensor (e.g. infrared light reflecting spheres) to the ultrasound transducer. It requires a tracking system, which measures the position and orientation of the trackable sensor in space. Thus, for each B-mode ultrasound image, a unique 3D transformation in space is recorded. As the transducer can be freely moved over the patient's anatomy, this technique is referred to as "3D free-hand". The second technique is based on a special 3D transducer and does not require a tracking system. This 3D transducer is able to sample a pyramidal volume by internally rotating the transducer array of up to 90°. As the acquisition angle of each single ultrasound image is internally encoded, the imaged 3D volume can be consequently assembled. For a detailed description of the functionality of such a 3D transducer I would like to refer to Hoskins et al. [29]. As the dimensions of the assembled image volume are restricted by the hardware, stitching techniques were developed to overcome this limitation [17, 52, 64]. But since the 3D free-hand technique allows for a more flexible image acquisition (volumes do not need to overlap), this technique is preferred over using a 3D transducer for ultrasound-based registration.

In order to correctly interpret a two-dimensional ultrasound image, it is important to understand the process of echo generation. As already indicated, echoes are generated when the ultrasound waves encounter a change in the medium properties (e.g. at tissue boundaries). The relevant material properties are the medium density ρ and the stiffness k, which are expressed as the acoustic impedance z:

$$z = \sqrt{\rho \cdot k} \tag{1}$$

Generally speaking, the acoustic impedance describes the response of a medium to an incoming ultrasound wave of certain pressure p. The medium density and stiffness are also essential for the propagating speed c of the ultrasound waves:

$$c = \sqrt{\frac{k}{\rho}} \tag{2}$$

Thus, the speed of sound is a material property and has been measured for all relevant tissue types of the human body. The speeds of sound for different tissue types are listed in Table 1. At a tissue boundary, the ultrasound wave is partially

Table 1 Speed of sound and acoustic impedances for different tissue types [28, 29]

Body tissue	Speed of sound c (m/s)	Acoustic impedance Z (Pa m^{-3} s)
Fat	1475	1.38×10^6
Water	1480	1.5×10^6
Blood	1570	1.61×10^6
Muscle	1580	1.70×10^6
Bone	3190–3406	6.47×10^6
Average tissue	1540	1.63×10^6
Air	333	0.00004×10^6

transmitted or reflected—depending on the involved acoustic impedances. This relationship can be expressed by the reflection coefficient R:

$$R = \frac{p_r}{p_i} = \frac{z_2 - z_1}{z_2 + z_1} \qquad (3)$$

whereas p_i and p_r are the pressures of the incident and the reflected ultrasound waves. The ultrasound wave is thereby travelling from the first medium with acoustic impedance z_1 into the second medium with acoustic impedance z_2. Consequently, a transition between tissue types of similar acoustic impedance ($z_1 \approx z_2$) will lead to a rather weak echo, while a transition between tissue types of great difference (e.g. $z_2 \gg z_1$) in acoustic impedance will result in a strong echo. Out of the reflection coefficient, the transmission coefficient T can be determined. As the pressure in the first medium and the second medium must be the same (see Fig. 2), the pressure of the incident wave and of the reflecting wave equals to the one of the transmitted wave p_t,

$$p_i + p_r = p_t \qquad (4)$$

Thereby the direction of the reflected wave is in the opposite direction to the incident wave. Out of this relationship, the transmission coefficient can be calculated as follows:

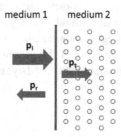

Fig. 2 The incident wave is transmitted and reflected at the transition of the first and the second medium

$$T = 1 + R = 1 + \frac{z_2 - z_1}{z_2 + z_1} = \frac{2z_2}{z_2 + z_1} \qquad (5)$$

Out of the acoustic impedances of the first and the second medium, the amount of reflection and transmission can be calculated. The acoustic impedances for different tissue types are listed in Table 1.

Since the described reflection process is generally applicable to smooth interfaces, it is referred to as specular reflection. This type of reflection is important for the interpretation of the ultrasound image. In diffuse reflection, the ultrasound wave is reflected uniformly in all directions. It also largely contributes to the image formation, as it generates echoes from surfaces, which are not perpendicular to the ultrasound beam. A similar effect degrading the image quality is speckle noise. When a wave strikes particles of much smaller size compared to the ultrasound wavelength, echoes are scattered randomly in many directions. This speckle pattern is normally less bright than specular reflection and is frequently observed in ultrasound images of the liver and lung [11]. As it is not directly related to anatomical information, speckle noise is considered to be an ultrasound artifact. Another important ultrasound artifact is related to the assumption of a constant propagation of sound. As the ultrasound system assumes a value for speed of sound of 1540 m/s (corresponds to the average speed of sound in human soft tissue), echoes of tissues with deviating speed of sound (see Table 1) appear at an incorrect depth. This artifact might also degrade the ultrasound-based registration accuracy: A certain thick layer of fat d_{fat} on top of the bone will lead to an echo delay ($c_{fat} < c_{avg}$) so that the depth of the bone surface would be displaced to a greater depth. The greater the penetration depth d_{tissue}, the larger is the depth localization error. We will come back to this issue in the section on the ultrasound-based registration strategies.

For more detailed information on the basics of ultrasound, I would like to refer to the relevant literature in this field [6, 11, 28, 29, 61].

3   Ultrasound-Based Registration

As indicated in the previous section, mostly relevant for the intraoperative registration step are A- and B-mode ultrasound imaging. Thus, this section focuses on the review of ultrasound-based registration techniques using A- or B-mode imaging published in the literature. While most of the presented methods are dealing with the registration of bones, some approaches are beyond orthopedics, but might get relevant for this field in the near future.

3.1 Registration Methods Using A-Mode Imaging

The direct replacement of pointer-based palpation by the acquisition of A-mode signals is quite evident. While the pointer-based digitization delivers a single 3D point at a time, A-mode imaging provides a one-dimensional signal in beam direction. Most commonly, this signal is processed using the Hilbert transform to determine the echo location related to the reflection of the wave from the bone surface [27, 42]. This however presumes that the beam axis of the A-mode transducer can be arranged perpendicular to the surface of the bone. In order to utilize a tracked A-mode transducer, a calibration step needs to be performed to relate the signal echo to a 3D point. Thus, the origin and the main axis of the ultrasound beam need to be determined with respect to the attached trackable sensor. As the case of the ultrasound transducer is normally of cylindrical shape, the symmetry axis of the case corresponds to the beam propagation axis. Therefore, the origin of the transducer and another point on the symmetry axis could be simply digitized with a tracked pointer to define the calibration parameters. A more advanced calibration method was presented by Maurer et al. [42]. They developed an ultrasound-based registration for the skull. A similar method was also presented by Amstutz et al. [3]. Both approaches used A-mode imaging to match the surface model derived from a CT-scan to the intraoperative situation. In a plastic bone study Maurer et al. [42] acquired 150 points to achieve a surface residual error of 0.20 mm. In addition, one patient trial was performed, obtaining a surface residual error of 0.17 mm for a set of 30 points. Sets of only 12–20 points were recorded by Amstutz et al. [3] in 12 patient trials, obtaining a mean surface registration error of 0.49 ± 0.20 mm. In order to overcome the difficulty of accurately aligning the A-mode transducer with respect to the bone surface, a man/machine interface was proposed by Heger et al. [27]. Instead of using an optical or magnetic tracking system, they proposed to use a mechanical localizer system. After an initial interactive registration step, the orientation of the transducer is adjusted, to get an optimal perpendicular alignment of the beam axis with respect to the bone surface. The approach was validated by repetitive registration of a femoral bone model, resulting in a mean root-mean-square (RMS) error of 0.59 mm. Similarly, A-mode registration of the pelvis was investigated by Oszwald et al. [49]. The surface matching accuracy was analyzed in an in vitro study using two synthetic pelvis models. The identified registration errors were in the range of 0.98–1.51 mm.

3.2 Registration Methods Using B-Mode Imaging

Even though, all presented approaches using A-mode ultrasound imaging for intraoperative registration showed promising results, its general application is restricted. A-mode imaging only allows the recording of single points at a time. As the angle of the beam axis to the bone surface has to be approximately 90°, the transducer cannot be easily swept over the area of interest. On the contrary, B-mode ultrasound allows a much more flexible image acquisition. Thus, the existing methods for B-mode ultrasound-based registration will be covered in greater detail.

3.2.1 B-Mode Calibration

As indicated in the previous section, 3D free-hand B-mode imaging requires the attachment of a trackable sensor to the transducer. Consequently, a calibration step needs to be performed to relate 2D image information (in pixels) to the tracked 3D space (in mm). Figure 3 gives in overview on the involved coordinate systems and transformations.

While the transformations of the patient {Pat} and the ultrasound transducer {US} are inherently known by the tracking system, the relationship between the tracked transducer coordinate-system {US} and the image coordinate-system {Im_{loc}} needs to be determined in a calibration step:

$$T_{calib} = {}^{Im_{glob}}_{US}T \cdot T_{scale} \qquad (6)$$

T_{scale} is a scaling matrix and describes the transformation between the local 2D image coordinate-system {Im_{loc}} and the 3D global image coordinate-system {Im_{glob}} (see Fig. 3), ${}^{Im_{glob}}_{US}T$ is the transformation from the 3D global image coordinate-system {Im_{glob}} to the 3D coordinate-system of the US probe {US} and T_{calib} is the resulting calibration transformation. This calibration transformation can be applied to transform 2D points in the ultrasound image ${}_{im}P$ to 3D points in the coordinate-system of the ultrasound transducer ${}_{US}P$:

$$_{US}P = T_{calib} \cdot {}_{im}P \qquad (7)$$

Fig. 3 Overview on different coordinate-systems (*in curly brackets*) involved during an ultrasound-based registration

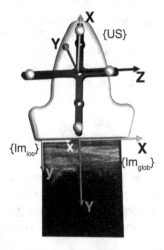

These 3D points can be further transformed to a point cloud $_{Pat}P$ in common 3D patient space via the transformations of the trackable patient sensor $^{Pat}_{Cam}T$ and the ultrasound transducer $^{US}_{Cam}T$ with respect to the tracking camera:

$$_{Pat}P = {}^{Pat}_{Cam}T^{-1} \cdot {}^{US}_{Cam}T \cdot {}_{US}P \tag{8}$$

The calibration transformation T_{calib} consists of nine parameters: While T_{scale} contains the two mm-to-pixel scaling factors in x- and y-direction and a translation, $^{Im_{glob}}_{US}T$ has three degrees of freedom for translation and three for rotation. In order to determine these calibration parameters the general concept of image calibration is employed. Therefore, a calibration phantom of known geometrical properties is imaged and its features are detected on the ultrasound image. As the positions of the phantom features are known in physical space, the spatial relationship to its imaged features can be estimated using a least-squares approach. A comprehensive review of existing B-mode calibration techniques was presented by Mercier et al. [43]. According to this review, all calibration phantoms have a common setup: They consist either of small spherical objects or of intersecting wires and are placed in a container of a coupling medium. During the calibration procedure, the probe is adjusted to image all relevant phantom features. The features are either automatically or interactively segmented on the image and used to find the unknown calibration parameters [43]. Even though, ultrasound calibration is a standard procedure, it is actually only valid for a specific speed of sound. Thus, most commonly water is used as its speed of sound corresponds to the one of the average speed of sound of soft tissue (1540 m/s). But for a medium with a speed of sound different than the one used for calibration, a depth localization error will occur. While the translational and rotational parameters of $^{Im_{glob}}_{US}T$ purely rely on the hardware configuration (see Fig. 3), only the scaling factor in scanning direction of T_{scale} is affected by a deviating speed of sound. Techniques to compensate for this deficiency will be presented later in this section.

The main goal of ultrasound-based registration is the fusion of a preoperative planning with the intraoperative situation. Many different approaches have been published in literature. For convenience the existing approaches have been categorized according to their employed strategy. Accordingly, three main categories were determined:

- landmark digitization
- surface-based registration
- volume-based registration

3.2.2 Landmark Digitization

Equivalent to the approaches using A-mode imaging, B-mode images could be used to digitize single bony landmarks. But instead of computing a registration transformation to preoperative image data, the digitized bony landmarks could be used to set up an intraoperative reference plane for safe implantation of acetabular cup implants [34, 50, 65]. This reference plane was employed by Jaramaz et al. [31] to guide the cup implantation during navigated total hip replacement. This plane— generally referred to as 'anterior pelvic plane' (APP)—requires the digitization of three pelvic landmarks and has been introduced by Lewinnek et al. [39]. They have investigated the relationship between the orientation of cup implants with respect to this APP and the probability of dislocation. They identified a 'safe zone', in which the dislocation rate was significantly low. As the cup orientation can be measured with respect to this APP in terms of two angles (anteversion and inclination), this safe zone defined the safe range for both angles. Thus, the common goal of a navigated total hip replacement is to place the cup implant within this safe zone. While conventionally, the corresponding APP landmarks were percutaneously digitized using a tracked pointer tool, some approaches proposed the use of B-mode imaging. Parratte et al. [50] compared the effect of percutaneous and ultrasound digitization of the APP landmarks. In an in vitro study, landmarks on two cadaveric specimen were digitized with both modalities. Higher reliability was found for the ultrasound modality. The same objective was investigated by Kiefer and Othman [34]. Comparing the data of 37 patient trials showed higher validity for the APP defined with ultrasound. Wassilew et al. [65] analyzed the accuracy of ultrasound digitization in a cadaver trial. In order to determine the ground truth APP, radio-opaque markers were placed into the cadaveric specimen before CT acquisition. The ground truth APP was defined in the CT-scan and transferred to the tracking space by locating the radio-opaque markers with a tracked pointer. Five observers repeated the ultrasound digitization five times for both cadaveric specimen, resulting in average errors for inclination of $-0.1 \pm 1.0°$ and for anteversion of $-0.4 \pm 2.7°$.

While the first category dealt with the landmark digitization for setting up a reference system, the next two categories deal with the registration of intraoperative ultrasound data to preoperative images or statistical shape models (SSM) for extrapolating the sparse information. An overview of the individual components and their particular transformations is shown in Fig. 4. Thus, the primary goal of the all the methods being presented is to determine the transformation between the virtual data (e.g. CT, SSM) and the patient.

3.2.3 Surface Based Registration

For most surgical interventions the digitization of three landmarks is not sufficient to set up a reference system or to provide a valuable guidance to the surgeon. Particularly if preoperative image data and planning need to be visualized intra-operatively, a sophisticated method is required. A convenient approach is based on a registration of point clouds extracted from ultrasound images to a surface model. While the surface model could be segmented from the image data (e.g. from CT or MRI) prior to the surgery, the extraction of the point data from the ultrasound

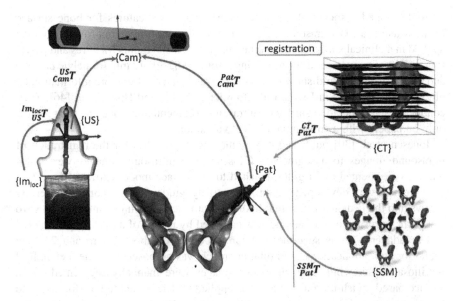

Fig. 4 Involved transformations during the image-guided intervention. The transformations highlighted in *magenta color* are either known by the trackable sensors $\left(^{US}_{Cam}T, ^{Pat}_{Cam}T\right)$ or due to the calibration step $\left(^{Im_{loc}}_{US}T\right)$. The registration transformation in cyan color (e.g. $^{CT}_{Pat}T, ^{SSM}_{Pat}T, \ldots$) needs to be determined in a registration step. {US}, {Pat} and {Cam} represent the coordinate systems of the ultrasound probe, the patient and the tracking camera

images followed by the actual registration step would need to take place on-line during the surgery. Thus, the first step to be solved is the segmentation of the ultrasound images. So far, many segmentation approaches have been published in literature. As most of them have been developed for a specific clinical application, I would like to highlight a few approaches with a focus on orthopedic applications. Thomas et al. [62] developed an automatic ultrasound segmentation method for estimating the femur length in fetal ultrasound images. They applied basic image processing algorithms such as morphological operators, contrast enhancement and thresholding to determine the bone surface. The method was validated by means of 24 ultrasound datasets, showing a good agreement to the manual measurement. An automatic segmentation based on a priori knowledge about the osseous interface and ultrasound physics was proposed by Daanen et al. [16]. This a priori knowledge was fused by the use of fuzzy logic to produce an accurate delineation of the sacrum. An extensive validation study with about 300 ultrasound images of cadavers and patients was conducted, showing a mean error of less than 1 mm. A more general segmentation approach was presented by Kowal et al. [36]. The first out of two steps determines a region of interest, which most likely contains the bone contour, while the second step tries to extract the bone contour. Both steps were based on general image processing algorithms and were tested with animal cadavers. A more sophisticated segmentation approach was proposed by Hacihaliloglu et al. [24].

They developed a special detector to extract ridge-like features for bone surface location using a 3D ultrasound probe. The work was further improved [26] and applied in a clinical study to support the imaging of distal radius fractures and pelvic ring injuries. On average, a surface fitting error of 0.62 ± 0.42 mm for pelvic patients and 0.21 ± 0.14 mm for distal radius patients was obtained. For more information on ultrasound segmentation I would like to refer to Noble and Boukerroui [46]. They conducted a very detailed survey on ultrasound segmentation, focusing their review only on papers with a substantial clinical validation.

Ionescu et al. [30] published one of the first approaches for the registration of ultrasound images to a segmented CT-scan. The ultrasound images were automatically segmented and rigidly matched to the surface model extracted from CT. The method was developed to intraoperatively guide the surgeon for inserting screws into the pedicle of vertebral bodies or into the sacro-iliac joint. In an in vitro study, the accuracy of the method was analyzed by means of a plastic spine model and a cadaveric pelvis specimen. Maximum errors of about 2 mm and 2° were found for both applications. A similar approach was proposed by Tonetti et al. [63] for iliosacral screwing. The ultrasound images were manually segmented and a surface-based registration algorithm was applied to determine the transformation to the preoperative CT-scan. The accuracy was analyzed by comparing it to the standard procedure of percutaneous iliosacral screwing using a fluoroscope. Thereby, the ultrasound-based technique showed higher precision and a lower complication rate. Amin et al. [1] combined the steps of ultrasound segmentation and CT registration for image-guided total hip replacement. After an initial landmark-based matching between the segmented CT-model and the intraoperative space, the aligned surface model of the pelvis is used as a shape prior to guide the ultrasound image segmentation. Thus, ultrasound segmentation and registration steps are solved simultaneously. The validity of the proposed approach was analyzed by means of a pelvic phantom model resulting in an average translational error of less than 0.5 mm and an average rotational error of less than 0.5°. In addition, 100 ultrasound images were recorded during a navigated surgery. In ten registration trials using a subset of 30 images each, the ultrasound-based registration was compared to conventional percutaneous pointer-based registration. A maximum difference of 2.07 mm in translation and of 1.58° in rotation was found. Barratt et al. [8] proposed a registration method to solve the error introduced by the assumption of a constant speed of sound. During the rigid registration between the ultrasound-derived points and the CT-segmented surface model not only the 3D transformation, but also the calibration matrix T_{calib} is optimized. Therefore, the scaling in scanning direction (see Sect. 5.1) is included as a parameter in the optimization of the registration transformation. The accuracy was evaluated by the acquisition of ultrasound images of the femur and pelvis from three cadaveric specimen. Thereby, the ultrasound images were manually segmented and applied for the CT-based registration, yielding an average target registration error of 1.6 mm. A different clinical application of ultrasound to surface registration was presented by Beek et al. [9]. They developed a system to navigate the treatment of non-displaced scaphoid fractures. The trajectory of the screw to fix the fracture is planned using a CT-scan of the wrist joint and matched to the intraoperative scenario using ultrasound imaging. The accuracy of guided hole drilling was investigated in an in vitro study using 57 plastic bones and compared with conventional fluoroscopic guidance. On average, the surgical requirements were met and the accuracy of fluoroscopic fixation was exceeded. Moore et al. [44] investigated the use of ultrasound-based registration for the tracking of injection needles. The clinical goal was to treat chronic lower back pain by facet injection. In order to pinpoint the lumbar facet joint, a surgical navigation system using magnetic tracking was utilized. The registration between CT-space and the intraoperative scenario was established by paired-point matching. An experiment using a plastic lumbar spine yielded a needle placement error of 0.57 mm. Another trial using a cadaveric specimen could only be qualitatively assessed. A new concept of surface-based registration was proposed by Brounstein et al. [13]. In order to avoid the search of direct correspondences between the ultrasound-derived point cloud and the CT-segmented points, Gaussian mixture models were used. Thereby, both point sets were represented as a multidimensional Gaussian distribution and the distance between both sets was iteratively minimized using the L2 similarity metric. The accuracy of the matching was analyzed by means of ten ultrasound volumes acquired of a plastic pelvis and three volumes recorded from a patient. On average, a mean registration error of 0.49 mm was demonstrated. A different approach to solve the registration problem for the tracking of injection needles was presented by Rasoulian et al. [54]. They developed a point-based registration technique in order to align segmented point clouds from ultrasound and CT image data. In order to accomplish the registration of multiple vertebral bodies, regularization is implemented in terms of a biomechanical spring model simulating the intervertebral disks. In an experimental study with five spine phantoms and an ovine cadaveric specimen, a mean target registration error of 1.99 mm (phantom) and 2.2 mm (sheep) was yielded. An application of ultrasound-based registration for guiding a surgical robot was demonstrated by Goncalves et al. [23]. Rigid registration using iterative closest point algorithm [10] was applied to match ultrasound points extracted from the femur to a CT-segmented surface model. With ultrasound imaging, the registration could be improved from 2.32 mm for pointer-based digitization to 1.27 mm.

The generation of the virtual object from preoperative CT scans is associated with relatively high costs and a considerable burden of X-ray radiation to the patients. In order to bypass preoperative image acquisition, the virtual object could be also generated intraoperatively. A common concept is to simultaneously generate and register a statistical shape model to intraoperatively collected point clouds. These point clouds needs to be collected in a common space from the patient's anatomy. After matching the SSM to the points, the resulting patient-specific model of the anatomy can be used as a virtual object to guide the surgeon during the intervention. Therefore, several groups proposed the registration of SSMs to tracked ultrasound images. One of the first approaches was published by Stindel et al. [58] and later extended by Kilian et al. [35]. In 2004 they have launched a patented solution called Echo Morphing®, which allows registering points extracted from ultrasound images to such a generic model. The proposed method was evaluated on a dry femur model and a cadaveric specimen. Thereby, different sets of ultrasound images from the distal femur were collected and accordingly applied to the registration pipeline. The reconstructed patient-specific model was subsequently compared with the respective ground truth model segmented from CT. Even though, quantitative results were only reported for the dry bone experiment (average surface distance error of 0.9 ± 0.6 mm), it was stated that the accuracy of the cadaver trial is in the same range. Chan et al. [14] and later Barratt et al. [7] constructed a SSM from segmented CT-scans of the pelvis (ten instances) and the femur (16 instances). Following a manual alignment of the statistical mean model to the point clouds segmented from the tracked ultrasound images, the SSM was deformed and repositioned to optimally fit these points. Thereby, the statistical information inherent in such a SSM was exploited in terms of an instantiation step to achieve an optimal fit to the sparse ultrasound data. In a cadaver study, ultrasound images from two pelvises and three femurs were acquired and manually segmented. Thereby, the positions of the specimen were constantly changed to cover as many anatomical features as possible. For each bone, the respective SSM was subsequently matched to a cloud of thousands of points and non-rigidly deformed. The resulting patient-specific bone model was compared to the ground truth CT-model, yielding an average root mean square distance error of 1.52 to 1.96 mm for the femurs and 2.47 to 4.23 mm for the pelvises. In Barratt et al. [7] the registration results of six femurs and three pelvises were reported. On average a surface distance errors in the range of 2.6–4.7 mm were found. Ultrasound images from only specific pelvic regions were acquired by Foroughi et al. [18]. The bone surface is automatically segmented from the ultrasound images, though images with failed segmentations can be discarded by the user. After a random sampling step, the remaining points are used to first rigidly register the pelvic SSM and then to instantiate it. Validation experiments were performed by means of a dry bone and two cadaveric specimen. For each experiment around 500 ultrasound images were matched to the SSM constructed from 110 segmented CT-datasets. The accuracy was investigated with respect to translational and rotational difference in defining the pelvic reference system. For the dry bone experiment, a translational error of 2.63 mm and a rotational error of 0.77° were found, while errors of 3.37 mm and 0.92° were reported for the cadaver experiment. The pelvic surface could be reconstructed with an average surface distance error of 3.3 mm. Recently, we have also proposed two approaches for establishing the intraoperative pelvic reference system using SSM to ultrasound registration [56, 57]. In the former case, a registration scheme was presented for patients being operated in supine position. For the determination of the APP, three pelvic landmarks need to be identified: The bilateral anterior superior iliac spine (ASIS) and the pubis symphysis. Particularly the localization of the pubis symphysis landmark is highly sensitive to errors and can lead for obese patients to a miscalculation of the reference plane and thus to a misalignment of the cup implant during total hip replacement. The high sensitivity to localization errors is directly related to the assumption of a constant speed of sound, although the speed of sound could vary between 1475 m/s (fat) and 1580 m/s (muscle). For a constant speed of sound c, the ultrasound propagation time t solely depends on the distance d from the transducer to the reflecting bone surface and back:

$$t = \frac{2 \cdot d}{c} \tag{9}$$

Oszwald et al. [49] measured the depth of soft tissue on top of the pubis from 30 CT-scans. They found a maximum soft tissue thickness of 69 mm. For this worst case scenario a deviation in depth localization can be roughly estimated:

$$t_{est} = \frac{2 \times 0.069 \text{ m}}{1475 \frac{\text{m}}{\text{s}}} = 9.36 \times 10^{-5} \text{ s} \tag{10}$$

$$t_{exp} = \frac{2 \times 0.069 \text{ m}}{1540 \frac{\text{m}}{\text{s}}} = 8.96 \times 10^{-5} \text{ s} \tag{11}$$

Thus, the percentage of delay between the estimated and the expected propagation time can be accordingly computed:

$$\frac{\left| t_{est} - t_{exp} \right| \times 100 \text{ \%}}{t_{exp}} = 4.46 \text{ \%} \tag{12}$$

Under these circumstances, the error in depth localization can be up to 6.1 mm. A possible solution to compensate for this kind of error was developed based on the work of Barratt et al. [8]. Instead of matching the ultrasound data to a CT-scan, different levels of pelvic SSMs were used to obtain a measurement of the pelvic reference system. On the basis of a SSM of the complete pelvis, a new concept of patch SSMs was introduced. These patch SSMs represent statistical models of certain local pelvic features and were used to bridge the registration of the complete pelvic SSM. An example of progressively fitting the complete pelvic SSM is shown in Fig. 5. In order to determine the validity of our approach we acquired six in vitro

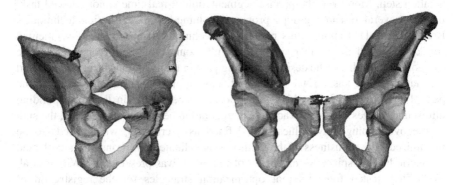

Fig. 5 Final result of fitting the global pelvic statistical shape model to sparse set of ultrasound-derived point clouds

datasets and one cadaver dataset. The in vitro experiments were conducted with a plastic pelvis model and a customized soft tissue simulation phantom. On average, an anteversion error of $2.11 \pm 1.78°$ and an inclination error of $1.66 \pm 1.14°$ were observed. For the cadaver trial, an anteversion error of $1.97 \pm 1.39°$ and an inclination error of $1.42 \pm 1.56°$ were found.

While the presented approach is only applicable for the pelvis in supine position, a different registration approach is required for the more frequently performed lateral approach. Thus, in our second contribution [57] we presented a technique to determine the APP from the pelvis operated in lateral decubitus position. The challenge of this task lies in the fact that only the hemi-pelvis is intraoperatively accessible. We solved this challenging problem by combining percutaneous pointer-based digitization and utilizing the sagittal symmetry property of the pelvic bone. Therefore, ultrasound images of the anterior and posterior region of the concerned pelvic side were collected. In addition both bilateral anterior superior iliac spine landmarks were digitized using a tracked pointer. The acquired point data was then used to initially register the pelvic SSM. Subsequently, a sagittal symmetry plane based on the bilateral landmarks was determined and the ultrasound-derived points were reflected to the contralateral side. As now evenly distributed point clouds were available for both sides, the initial matching was improved by an additional affine matching and a statistical instantiation step. A couple of experiments were conducted to estimate the error contribution for each of the described steps. The overall accuracy was determined in in vitro trials with two plastic bones and two dry bones. On average, a mean error of $3.48 \pm 1.10°$ in anteversion and of $1.26 \pm 1.62°$ in inclination was observed.

3.2.4 Volume Based Registration

All the previously presented approaches are based on point sets extracted from the tracked ultrasound images. These points need to be segmented intraoperatively and thus the segmentation adds valuable computational time to the intraoperative registration step. Moreover, the precise segmentation in real-time is not a trivial task. Instead, several research groups proposed volume-based registration techniques. Brendel et al. [12] applied this registration technique to match a CT-segmented surface model of the lumbar spine to the ultrasound volume. Preoperatively, the surgeon has to indicate the desired scanning path in terms of transducer position and orientation on the basis of the CT dataset. This information is used to estimate the part of the bone surface, which is supposed to be visible in the corresponding ultrasound images. In order to achieve a registration between both datasets, the sum of the overlapping gray values was defined as a criterion, which needs to be maximized. The robustness of the method was validated by means of an explanted cadaveric lumbar spine. An extension of this work was presented by Winter et al. [68]. They tested four different optimization strategies for the registration of CT-datasets and ultrasound images. Twelve vertebrae of five different patients were registered 1000 times to identify the best optimization method. For the multistart scenario, most reliable results were observed for Covariance Matrix Adaptation evolution strategy. This optimization strategy was able to sucessfully register the datasets in 99.97 % of the cases. An existing method for matching ultrasound and magnetic resonance images of the liver was extended to the orthopedic field by Penney et al. [51]. They proposed an intensity-based registration between ultrasound and CT images of the femur and pelvis. Thereby, both image datasets are converted to probability images using certain image features. The particular probability density functions are computed from a training set of CT images and ultrasound images, respectively. The probability image volumes are then registered using normalized cross-correlation metric. An experimental study with three cadaveric specimen was conducted to evaluate the accuracy of the proposed registration pipeline. Between 168 and 565 tracked ultrasound images were collected for each of the six femurs and three pelvises and an average root mean square error of 1.6 mm was observed. Even though the clinical motivation for the work of Wein et al. [67] was not directly related to orthopedics, the approach is of great interest and has already been adopted by other research groups. They developed a strategy to simulate ultrasound images from abdominal CT and to match it to the ultrasound data. This strategy is based on the observation that the x-ray attenuation in CT-scans is proportional to the acoustic impedance. In a preliminary clinical trial, the proposed approach was evaluated using image data of ten patients with different abdominal pathologies. More clinical cases were presented by Wein et al. [66]. They reported the study results of 25 patients having certain pathologies in liver and kidney. In 76 % of the cases the algorithm successfully converged with an average RMS target registration error of 8.1 mm. An extension of this work was proposed by Gill et al. [22] for the registration of multiple vertebrae. In order to account for the potential change in the spinal curvature of the patient between preoperative CT acquisition and surgical intervention, a group-wise registration of the individual vertebrae was implemented. In vitro tests with data collected of synthetic bone models resulted in an average target registration error of 2.08 ± 0.55 mm. This work was further improved by integrating a biomechanical model of the intervertebral disc. This model constrained the degrees of freedom of the individual vertebrae during the group-wise registration process. The overall matching accuracy was validated by means of six different synthetic bone phantoms and an ovine cadaveric specimen. On the basis of the synthetic bone models, different artificially modified spine curvatures were simulated. Across all registration trials, 98.8 % of the cases had a final target registration error of less than 3 mm (mean of 1.44 mm) and were regarded as successful. Another registration approach for the purpose of navigated needle injection was proposed by Khallaghi et al. [33]. They also adopted the ultrasound simulation method developed by Wein et al. [66]. But instead of using preoperative CT data, a volumetric statistical atlas was utilized to generate simulated ultrasound images. The statistical atlas of the L3 vertebra was constructed from 35 training instances. In a first step, simulated ultrasound images from the mean shape are rigidly registered to the intraoperatively acquired ultrasound volume. This is followed by a deformable registration in a second step. Three synthetic phantom models of the lumbar spine were available to conduct an in vitro accuracy study. For each phantom the registration was repeated 30 times with different starting positions of the mean shape, yielding an overall average target registration error of 3.4 mm. The matching of simulated ultrasound images to a volumetric statistical atlas was also investigated for identification of the pelvic reference system [19, 20]. In both papers a volumetric statistical atlas of the pelvis was constructed from 110 CT datasets. In order to initialize the registration, the mean model was manually aligned with the acquired ultrasound volume. Based on the position and orientation of the tracked ultrasound transducer relative to the tracked pelvic anatomy, 2D slices of the atlas are extracted and used to simulate ultrasound images [67]. A similarity metric is then computed between the real and the simulated ultrasound volume. First the rigid transformation is iteratively optimized, before the deformable registration is carried out. The finally matching patient-specific model is then used to set up the pelvic reference system. Experimental data was obtained from two cadaveric specimen. The pelvic reference system could be determined with a translational error of 2 and 3.45 mm, while the average axes rotation error was 3.5° and 3.9°. Additional experiments were carried out by Ghanavati et al. [20]. They reported the results of five synthetic bone models and a single dry bone. On average a translation error of 1.77 mm and a rotational error of 1.11° were found. A novel registration procedure for the pelvis was recently proposed by Hacihaliloglu et al. [25]. They proposed a rigid CT to ultrasound registration based on phase correlation. First, the bone surfaces are automatically extracted from ultrasound and CT volumes based on local phase features and projected into the Radon space. Then the rotational and translational differences are specifically solved in two consecutive steps. The proposed method was validated by means of a synthetic bone phantom and a subsequent clinical study with ten patients involved. Two ultrasound volumes were acquired for each patient from the unconcerned side of the pelvis. For the phantom study a mean surface registration error of 0.42 ± 0.17 mm was observed, while the mean error for the clinical study was 0.78 ± 0.21 mm.

A summary of the presented registration approaches is shown in Table 2. They are categorized according the used ultrasound mode and the applied imaging data. Among one category, the publications are listed chronologically. A more general survey of ultrasonic guided interventions is presented by Noble et al. [47]. For more information on registration methods for image-guided interventions, the reader is referred to the work of Markelj et al. [41].

At the end of this section a comprehensive summary of all reviewed methods is given, highlighting its clinical applicability (see Table 2).

Table 2 Summary of surveyed ultrasound-based registration approaches, published in the literature

Reference	US mode	Anatomy	Registration	Experiments
A-mode				
Maurer et al. [42]	A-mode	Skull	Point-based, US ⇔ CT	in vitro, 1× patient
Amstutz et al. [3]	A-mode	Skull	Point-based, US ⇔ CT	12× patients
Heger et al. [27]	A-mode	Femur	Point-based, US ⇔ CT	1× synthetic bone
Oszwald et al. [49]	A-mode	Pelvis	Point-based, US ⇔ CT	2× synthetic bones
Mozes et al. [45]	A-mode	Femur, tibia	Point-based, US ⇔ CT	1× synthetic bone, 3× cadaveric specimen
B-mode landmark digitization				
Parratte et al. [50]	B-mode	Pelvis	Measurements	2× cadaveric specimen
Kiefer and Othman [34]	B-mode	Pelvis	Measurements	37× patients
Wassilew et al. [65]	B-mode	Pelvis	Measurements	2× cadaveric specimen
B-mode surface-based registration: US ⇔ CT				
Ionescu et al. [30]	B-mode	Pedicle vertebra, ilio-sacral joint	Surface-based, US ⇔ CT	1× synthetic bone, 1× cadaveric specimen
Tonetti et al. [63]	B-mode	Ilio-sacral joint	Surface-based, US ⇔ CT	4× patients
Amin et al. [1]	B-mode	Pelvis	Surface-based, US ⇔ CT	1× synthetic bone, 1× patient
Barratt et al. [7]	B-mode	Femur, pelvis	Surface-based, US ⇔ CT	3× cadaveric specimen
Beek et al. [9]	B-mode	Scaphoid	Surface-based, US ⇔ CT	57× synthetic bones
Moore et al. [44]	B-mode	Lumbar spine	Point-based, US ⇔ CT	1× synthetic bone, 1× cadaveric specimen
Brounstein et al. [13]	B-mode	Pelvis	Surface-based, US ⇔ CT	1× synthetic bone, 1× patient
Rasoulian et al. [54]	B-mode	Lumbar spine	Surface-based, US ⇔ CT	5× synthetic bones, 1× sheep cadaveric specimen
Goncalves et al. [23]	B-mode	Femur	Surface-based, US ⇔ CT	1× synthetic bone
B-mode surface-based registration: US ⇔ SSM				
Kilian et al. [35], Stindel et al. [59]	B-mode	Distal femur	Surface-based, US ⇔ SSM	1× dry bone, 1× cadaveric specimen
Barratt et al.[8], Chan et al. [14]	B-mode	Femur, pelvis	Surface-based, US ⇔ SSM	3× cadaveric specimen
Foroughi et al. [18]	B-mode	Pelvis	Surface-based, US ⇔ SSM	1× dry bone, 2× cadaveric specimen

(continued)

Table 2 (continued)

Reference	US mode	Anatomy	Registration	Experiments
Schumann et al. [56]	B-mode	Pelvis	Surface-based, US ⟷ SSM	1× synthetic bone, 1× cadaveric specimen
Schumann et al. [57]	B-mode	pelvis	Surface-based, US ⟷ SSM	2× synthetic bones, 2× dry bones
B-mode intensity based registration				
Brendel et al. [12]	B-mode	Lumbar spine	Volume-based, US ⟷ CT	1× cadaveric specimen
Winter et al. [68]	B-mode	Lumbar spine	Volume-based, US ⟷ CT	5x patients/12x vertebrae
Penney et al. [51]	B-mode	Femur, pelvis	Volume-based, US ⟷ CT	3× cadaveric specimen
Wein et al. [66]	B-mode	Abdomen	Volume-based, US ⟷ CT	10× patients
Wein et al. [67]	B-mode	Abdomen	Volume-based, US ⟷ CT	25× patients
Gill et al. [21, 22]	B-mode	Lumbar spine	Volume-based, US ⟷ CT	1×/6× synthetic bone(s), 1× sheep cadaveric specimen
Khallaghi et al. [33]	B-mode	Lumbar spine	Volume-based, US ⟷ SSM	3× synthetic bones
Ghanavati et al. [19]	B-mode	Pelvis	Volume-based, US ⟷ SSM	2× cadaveric specimen
Ghanavati et al. [20]	B-mode	Pelvis	Volume-based, US ⟷ SSM	5× synthetic bones, 1× dry bone
Hacihaliloglu et al. [25]	B-mode	Pelvis	Volume-based, US ⟷ CT	1× synthetic bone, 10× patients

4   Conclusions and Outlook

In this chapter the general principles of ultrasound for computer-assisted interventions have been presented. As ultrasound imaging has many favorable advantages, its integration into surgical navigation is quite evident. Despite its benefits, the application of ultrasound imaging in orthopedic interventions also poses a certain challenge to the surgeons and the technicians. Ultrasound images only have a small field of view and are subject to various types of artifacts. A correct interpretation of ultrasound images is only feasible for skilled experts. Thus, in order to make ultrasound imaging an integral part of standard registration techniques, three major criteria need to be fulfilled:

- High precision
- Integration into surgical procedure
- Total time <10 min

Most importantly, a clinically acceptable precision and robustness of the algorithm needs to be ensured. As with any other registration technique, the application of the algorithm needs to provide a clear benefit to the surgeon and the patient. Moreover, the process of ultrasound image acquisition needs to be seamlessly integrated into the clinical workflow. Therefore, potential interactions (e.g. manual initialization) have to be reduced to a minimum or even completely eliminated. As the focus should be on the surgery and not on the ultrasound acquisition and the registration, fully automatic methods are actually favored. In addition, the computational time of the registration procedure should not delay the surgery excessively. Thus, the total time it takes to collect the ultrasound images and to perform the registration should not exceed the duration of approximately 10 min.

Compared to the published methods in the literature, only a few approaches might fulfill these criteria. Even though, most of the methods were validated not only with synthetic bone phantoms, but also with human dry bones, the clinical validity cannot be deduced. Out of the 32 reviewed methods, only seven were validated with more than one clinical patient dataset. So far, none of the methods using statistical shape models were clinically validated. Thus, it seems that CT-based methods are still more reliable and have a higher potential to be established as standard ultrasound-based registration technique in the orthopedic field. Out of the categorized registration strategies, the volume-based approaches might have the best chance to fulfill the criteria for clinical acceptance. These approaches are computationally efficient, as a non-trivial online segmentation is not required. Moreover, the required processing of the CT-scan (e.g. segmentation, simulation of ultrasound images) can be performed prior to surgery. Further improvements could be the replacement of CT by radiationless MRI, as for instance proposed by Roche et al. [55] for the application in image-guided neurosurgeries. The surface-based methods using B-mode imaging and all the remaining categories (including those using A-mode imaging) have demonstrated their powerfulness for intraoperatively establishing a reference system (primarily the anterior-pelvic plane). Only a sparse set of ultrasound images is sufficient to derive a reference system for measuring clinically relevant parameters such as the alignment of implants or drill hole trajectories. Especially for obese patients these methods are definitely superior to the common percutaneous pointer-based digitization strategies. But large-scale clinical studies further need to be conducted to prove the clinical validity. In addition to the technical challenges, also some clinical issues have to be approached to establish ultrasound imaging as a standard registration modalidy. As orthopedic surgeons are normally not used to ultrasound imaging in their everydays clinical routine, they first need to gain confidence in using this modality. Therefore, the learning curve of collecting intraoperative ultrasound images needs to be considered. Ideally, the surgeons are already involved in the preclinical trials to get familiar with this technique.

Even though, ultrasound has its major applications in obstetrics and general diagnosis, it can provide an important means for computer-assisted orthopedic surgeries. Due to its real-time capabilities, it has a high potential to diminish the application of harmful fluoroscopic imaging. Moreover, the penetration property of ultrasound waves allows locating the depth of bony surfaces through several layers of muscle and fat. More than thirty different registration methods have been analyzed with respect to their applicability and clinical validity. While most of these sophisticated approaches were only validated in in vitro and cadaver trials, a few were successfully applied to patient data, showing promising results.

References

1. Amin DV, Kanade T, DiGioia AM, Jaramaz B (2003) Ultrasound registration of the bone surface for surgical navigation. Comput Aided Surg 8(1):1–16
2. Amiot LP, Labelle H, DeGuise JA, Sati M, Rivard CH (1995) Computer-assisted pedicle screw fixation—a feasibility study. Spine 20(10):1208–1212
3. Amstutz C, Caversaccio M, Kowal J, Bächler R, Nolte LP, Häusler R, Styner M (2003) A-mode ultrasound in computer-aided surgery of the skull. Arch Otolaryngol Head Neck Surg 129(12):1310–1316
4. Anderson KC, Buehler KC, Markel DC (2005) Computer assisted navigation in total knee arthroplasty: comparison with conventional methods. J Arthroplasty 20(3):132–138
5. Audette MA, Ferrie FP, Peters TM (2000) An algorithmic overview of surface registration techniques for medical imaging. Med Image Anal 4(3):201–217
6. Azhari H (2010) Basics of biomedical ultrasound for engineers. Wiley, Hoboken, NJ
7. Barratt DC, Chan CS, Edwards PJ, Penney GP, Slomczykowsiki M, Carter TJ, Hawkes DJ (2008) Instantiation and registration of statistical shape models of the femur and pelvis using 3D ultrasound imaging. Med Image Anal 12(3):358–374
8. Barratt DC, Penney GP, Chan CS, Slomczykowski M, Carter TJ, Edwards PJ, Hawkes DJ (2006) Self-calibrating 3D-ultrasound-based bone registration for minimally invasive orthopedic surgery. IEEE Trans Med Imaging 25(3):312–323
9. Beek M, Abolmaesumi P, Luenam S, Ellis RE, Sellens RW, Pichora DR (2008) Validation of a new surgical procedure for percutaneous scaphoid fixation using intra-operative ultrasound. Med Image Anal 12(2):152–162
10. Besl PJ, McKay ND (1992) Method for registration of 3-D shapes. Int Soc Opt Photonics 1992:586–606
11. Beutel J, Fitzpatrick JM, Horii SC, Kim Y, Kundel HL, Sonka M, Van Metter RL (2002) Handbook of medical imaging, vol 3., Display and PACSSPIE Press, Washington, DC
12. Brendel B, Winter S, Rick A, Stockheim M, Ernert H (2002) Registration of 3D CT and ultrasound datasets of the spine using bone structures. Comput Aided Surg 7(3):146–155
13. Brounstein A, Hacihaliloglu I, Guy P, Hodgson A, Abugharbieh R (2011) Towards real-time 3D US to CT bone image registration using phase and curvature feature base GMM matching. Medical image computing and computer-assisted intervention (MICCAI) 2011. Springer, Berlin, pp 235–242
14. Chan CS, Barratt DC, Edwards PJ, Slomczykowski M, Carter TJ, Hawkes DJ (2004) Cadaver validation of the use of ultrasound for 3D model instantiation of bony anatomy in image guided orthopaedic surgery. Medical image computing and computer-assisted intervention (MICCAI) 2004. Springer, Berlin, pp 397–404
15. Cootes TF, Taylor CJ, Cooper DH, Graham J (1995) Active shape models—their training and application. Comput Vis Image Underst 61(1):38–59
16. Daanen V, Tonetti J, Troccaz J (2004) A fully automated method for the delineation of osseous interface in ultrasound images. Medical image computing and computer-assisted intervention (MICCAI) 2004. Springer, Berlin, pp 549–557
17. Foroughi P, Abolmaesumi P, Hashtrudi-Zaad K (2006) Intra-subject elastic registration of 3D ultrasound images. Med Image Anal 10(5):713–725

18. Foroughi P, Song G, Chintalapani G, Taylor RH, Fichtinger G (2008) Localization of pelvic anatomical coordinate system using US/atlas registration for total hip replacement. Medical image computing and computer-assisted intervention (MICCAI) 2008. Springer, Berlin, pp 871–879
19. Ghanavati S, Mousavi P, Fichtinger G, Foroughi P, Abolmaesumi P (2010) Multi-slice to volume registration of ultrasound data to a statistical atlas of human pelvis. In: Proceedings of SPIE medical imaging 2010. International Society of Optics and Photonics
20. Ghanavati S, Mousavi P, Fichtinger G, Abolmaesumi P (2011) Phantom validation for ultrasound to statistical shape model registration of human pelvis. In: Proceedings of SPIE medical imaging 2011. International society of optics and photonics
21. Gill S, Abolmaesumi P, Fichtinger G, Boisvert J, Pichora D, Borshneck D, Mousavi P (2012) Biomechanically constrained groupwise ultrasound to CT registration of the lumbar spine. Med Image Anal 16(3):662–674
22. Gill S, Mousavi P, Fichtinger G, Pichora D, Abolmaesumi P (2009) Group-wise registration of ultrasound to CT images of human vertebrae. In: Proceedings of SPIE medical imaging 2009. International society of optics and photonics
23. Goncalves PJ, Torres PM, Santos F, Antonio R, Catarino N, Martins JM (2014) A vision system for robotic ultrasound guided orthopaedic surgery. J Intell Robot Syst 1–13
24. Hacihaliloglu I, Abugharbieh R, Hodgson A, Rohling R (2008) Bone segmentation and fracture detection in ultrasound using 3D local phase features. Medical image computing and computer-assisted intervention (MICCAI) 2008. Springer, Berlin, pp 287–295
25. Hacihaliloglu I, Wilson DR, Gilbart M, Hunt MA, Abolmaesumi P (2013) Non-iterative partial view 3D ultrasound to CT registration in ultrasound-guided computer-assisted orthopedic surgery. Int Surg Comput Assist Radiol Surg 8(2):157–168
26. Hacihaliloglu I, Guy P, Hodgson AJ, Abugharbieh R (2015) Automatic extraction of bone surfaces from 3D ultrasound images in orthopaedic trauma cases. Int J Comput Assist Radiol Surg 1–9
27. Heger S, Portheine F, Ohnsorge JA, Schkommodau E, Radermacher K (2005) User-interactive registration of bone with A-mode ultrasound. IEEE Eng Med Biol Mag 24(2):85–95
28. Hendee WR, Ritenour ER (2003) Medical imaging physics. Wiley, Hoboken, NJ
29. Hoskins PR, Martin K, Thrush A (2010) Diagnostic ultrasound: physics and equipment. Cambridge University Press, Cambridge
30. Ionescu G, Lavallee S, Demongeot J (1999) Automated registration of ultrasound with CT images: application to computer assisted prostate radiotherapy and orthopedics. Medical image computing and computer-assisted intervention (MICCAI) 1999. Springer, Berlin, pp 768–777
31. Jaramaz B, DiGioia AM, Blackwell M, Nolan D (1998) Computer assisted measurement of cup placement in total hip replacement. Clin Orthop Relat Res 354:70–80
32. Jolles BM, Genoud P, Hoffmeyer P (2004) Computer-assisted cup placement techniques in total hip arthroplasty improve accuracy of placement. Clin Orthop Relat Res 426:174–179
33. Khallaghi S, Mousavi P, Gong RH, Gill S, Boisvert J, Fichtinger G, Abolmaesumi P (2010) Registration of statistical shape model of the lumbar spine to 3D ultrasound images. Medical image computing and computer-assisted intervention (MICCAI) 2010. Springer, Berlin, pp 68–75
34. Kiefer H, Othman A (2007) Ultrasound vs pointer palpation based method in THA navigation: a comparative study. Orthopedics 30(10):S153–S156
35. Kilian P, Plaskos C, Parratte S, Argenson JN, Stindel E, Tonetti J, Lavellee S (2008) New visualization tools: computer vision and ultrasound for MIS navigation. Int J Med Robot Comput Assist Surg 4(1):23–31
36. Kowal J, Amstutz C, Langlotz F, Talib H, Ballester MG (2007) Automated bone contour detection in ultrasound B-mode images for minimally invasive registration in computer-assisted surgery—an in vitro evaluation. Int J Med Robot Comput Assist Surg 3(4):341–348
37. Langlotz F, Nolte LP (2004) Technical approaches to computer-assisted orthopedic surgery. Eur J Trauma Emerg Surg 30(1):1–11

38. Lavallee S, Sautot P, Troccaz J, Cinquin P, Merloz P (1995) Computer-assisted spine surgery: a technique for accurate transpedicular screw fixation using CT data and a 3-D optical localizer. Comput Aided Surg 1(1):65–73

39. Lewinnek G, Lewis J, Tarr R, Compere C, Zimmerman J (1978) Dislocation after total hip replacement arthroplasties. J Bone Joint Surg (Am) 60:217–220

40. Maintz JB, Viergever MA (1998) A survey of medical image registration. Med Image Anal 2(1):1–36

41. Markelj P, Tomazevic D, Likar B, Pernus F (2012) A review of 2D/3D registration methods for image-guided interventions. Med Image Anal 16(3):642–661

42. Maurer CR, Gaston RP, Hill DL, Gleeson MJ, Taylor G, Fenlon MR, Edwards PJ, Hawkes DJ (1999) AcouStick: a tracked A-mode ultrasonography system for registration in image-guided surgery. Medical image computing and computer-assisted intervention (MICCAI) 1999. Springer, Berlin, pp 953–963

43. Mercier L, Langø T, Lindseth F, Collins LD (2005) A review of calibration techniques for freehand 3-D ultrasound systems. Ultrasound Med Biol 31(2):143–165

44. Moore J, Clarke C, Bainbridge D, Wedlake C (2009) Image guidance for spinal facet injections using tracked ultrasound. Medical image computing and computer-assisted intervention (MICCAI) 2009. Springer, Berlin, pp 516–523

45. Mozes A, Chang TC, Arata L, Zhao W (2010) Three-dimensional A-mode ultrasound calibration and registration for robotic orthopaedic knee surgery. Int J Med Robot Comput Assist Surg 6(1):91–101

46. Noble JA, Boukerroui D (2006) Ultrasound image segmentation: a survey. IEEE Trans Med Imaging 25(8):987–1010

47. Noble JA, Navab N, Becher H (2011) Ultrasonic image analysis and image-guided interventions. Interface Focus 1(4):673–685

48. Nolte LP, Visarius H, Arm E, Langlotz F, Schwarzenbach O, Zamorano L (1995) Computer-aided fixation of spinal implants. Comput Aided Surg 1(2):88–93

49. Oszwald M, Citak M, Kendoff D, Kowal J, Amstutz C, Kirchhoff T, Hüfner T (2008) Accuracy of navigated surgery of the pelvis after surface matching with an A-mode ultrasound probe. J Orthop Res 26(6):860–864

50. Parratte S, Kilian P, Pauly V, Champsaur P, Argenson JN (2008) The use of ultrasound in acquisition of the anterior pelvic plane in computer-assisted total hip replacement. J Bone Joint Surg (Br) 90(2):258–263

51. Penney GP, Barratt DC, Chan CS, Slomczykowski M, Carter TJ, Edwards PJ, Hawkes DJ (2006) Cadaver validation of intensity-based ultrasound to CT registration. Med Image Anal 10(3):385–395

52. Poon TC, Rohling RN (2006) Three-dimensional extended field-of-view ultrasound. Ultrasound Med Biol 32(3):357–369

53. Rajamani KT, Styner MA, Tablib H, Zheng G, Nolte LP, Gonzales Ballester MA (2007) Statistical deformable bone models for robust 3D surface extrapolation from sparse data. Med Image Anal 11(2):99–109

54. Rasoulian A, Abolmaesumi P, Mousavi P (2012) Feature-based multibody rigid registration of CT and ultrasound images of lumbar spine. Med Phys 39(6):3154–3166

55. Roche A, Pennec X, Malandain G, Ayache N (2001) Rigid registration of 3-D ultrasound with MR images: a new approach combining intensity and gradient information. IEEE Trans Med Imaging 20(10):1038–1049

56. Schumann S, Nolte LP, Zheng G (2012) Compensation of sound speed deviations in 3-D B-mode ultrasound for intraoperative determination of the anterior pelvic plane. IEEE Trans Inf Technol Biomed 16(1):88–97

57. Schumann S, Nolte LP, Zheng G (2012) Determination of pelvic orientation from sparse ultrasound data for THA operated in the lateral position. Int J Med Robot Comput Assist Surg 8(1):107–113

58. Stindel E, Briard J, Lavallee S, Dubrana F, Plaweski S, Merloz P, Troccaz J (2004) Navigation and robotics in total joint and spine surgery. Springer, Berlin, pp 39–45

59. Stindel E, Briard JL, Merloz P, Plaweski S, Dubrana F, Lefevre C, Troccaz J (2002) Bone morphing: 3D morphological data for total knee arthroplasty. Comput Aided Surg 7(3):156–168
60. Stöckl B, Nogler M, Rosiek R, Fischer M, Krismer M, Kessler O (2004) Navigation improves accuracy of rotational alignment in total knee arthroplasty. Clin Orthop Relat Res 426:180–186
61. Suri J, Kathuria C, Chang RF, Molinari F, Fenster A (2008) Advances in diagnostic and therapeutic ultrasound imaging. Artech House
62. Thomas JG, Peters RA, Jeanty P (1991) Automatic segmentation of ultrasound images using morphological operators. IEEE Trans Med Imaging 10(2):180–186
63. Tonetti J, Carrat L, Blendea S, Merloz P, Troccaz J, Lavallee S, Chirossel JP (2001) Clinical results of percutaneous pelvic surgery. Computer assisted surgery using ultrasound compared to standard fluoroscopy. Comput Aided Surg 6(4):204–211
64. Wachinger C, Wein W, Navab N (2007) Three-dimensional ultrasound mosaicing. Medical image computing and computer-assisted intervention (MICCAI) 2007. Springer, Berlin, pp 327–335
65. Wassilew GI, Heller MO, Hasart O, Perka C, Südhoff I, Janz V, König C (2012) Ultrasound-based computer navigation of the acetabular component: a feasibility study. Arch Orthop Trauma Surg 132(4):517–525
66. Wein W, Brunke S, Khamene A, Callstrom MR, Navab N (2008) Automatic CT-ultrasound registration for diagnostic imaging and image-guided intervention. Med Image Anal 12 (5):577–585
67. Wein W, Khamene A, Clevert DA, Kutter O, Navab N (2007) Simulation and fully automatic multimodal registration of medical ultrasound. Medical image computing and computer-assisted intervention (MICCAI) 2007. Springer, Berlin, pp 136–143
68. Winter S, Brendel B, Pechlivanis I, Schmieder K, Igel C (2008) Registration of CT and intraoperative 3-D ultrasound images of the spine using evolutionary and gradient-based methods. IEEE Trans Evol Comput 12(3):284–296
69. Zheng G, Kowal J, Gonzales Ballester MA, Caversaccio M, Nolte LP (2007) Registration techniques for computer navigation. Curr Orthop 21(3):170–179

Medical Robotics for Musculoskeletal Surgery

Sanghyun Joung and Ilhyung Park

Abstract Bony structure has low shape deformity comparing to soft tissue. This fact has been made many trials of developing a robotic system for musculoskeletal surgery. ROBODOC was firstly used on human for total hip replacement in 1992 and was commercialized at 1994. It provides fully-automated surgery and has showed improved surgical precision. However, its usage was declined due to safety concerns. Trends have been changed to semi-automatic, a small size, and a bone-mountable robotic system. Nowadays, surgeons have some options on robotic surgery for total hip replacement, total knee replacement, unicompartmental knee replacement, and spine surgery. On the other hand there is not a commercialized robotic system for fracture surgery despite surgeon's strong request. They want to increase precision in fracture-reduction and reduce a radiation exposure and fatigue with a robotic system. Several research groups including our group have developed robotic systems for this purpose. This chapter will introduce clinical facts and opinions about commercialized robotic systems, such as ROBODOC, RIO, and MAZOR. Robotic systems for fracture surgery under developing will be also introduced and some highlight data will be shared.

1 Introduction

General advantages of a robotic application are quality and safety. The robots provide more precise motion, high power and high speed to do something. Accuracy of work and productivity will be increased. The robots give safety to users by

S. Joung (✉)
Medical Device and Robot Institute of Park, Kyungpook National University,
1006 GlobalPlaza, 80, Daehak-Rd, Daegu 41566, South Korea
e-mail: shjoung@mdrip.knu.ac.kr

I. Park
Department of Orthopaedic Surgery, School of Medicine, Kyungpook National University,
1006 GlobalPlaza, 80, Daehak-Rd, Daegu 41566, South Korea

© Springer International Publishing Switzerland 2016
G. Zheng and S. Li (eds.), *Computational Radiology*
for Orthopaedic Interventions, Lecture Notes in Computational
Vision and Biomechanics 23, DOI 10.1007/978-3-319-23482-3_15

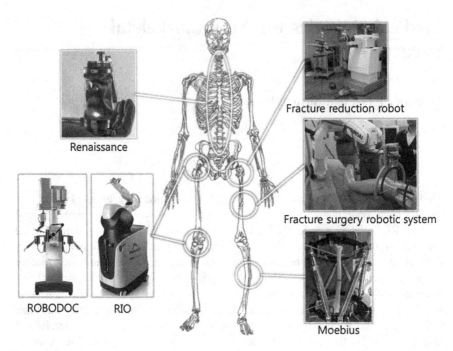

Fig. 1 Medical robots for musculoskeletal surgery

replacing human who is enduring heavy load or works hazardous condition. These benefits of the robots are suited for orthopedic surgery. Surgeons repair a functional or anatomical structure of musculoskeletal system by cutting, drilling, milling, repositioning and other actions, the motion of which are similar to machine works. The robots could improve surgical output of musculoskeletal system as they could be improved quality and safety in machining. Many robotic systems have been applied to orthopedic surgery and spine surgery since ROBODOC is firstly used to human in 1992. Figure 1 shows the robotic systems for the orthopedic and spine surgery.

Major clinical application of robotic systems is arthroplasty. Now, there are two commercially available robotic systems; one is ROBODOC (Curexo Thechnology Corp., Fremont, California) and the other is RIO (MAKO Surgical Corop., FortLauderdale, Florida). Other application is a guidance system for screw insertion to spine. Renaissance (MAZOR Robotics Inc., Orlando, Florida) is the only commercial product in this field. Several robotic systems have been developed for trauma surgery though they are not commercially available until now. Next section, we briefly review three commercialized robotic systems. And we will treat safety issues related to robot-assisted fracture reduction with the fracture reduction robot. The developing process and strategy will be introduced with the fracture surgery robotic system that provides total solution for fracture surgery. Finally, Moebius^{TM} robotic system that is multi-purpose bone surgical robot will be introduced.

1.1 ROBODOC

The development of the ROBODOC was started from two doctors' idea, which was to make a precise cavity in a femur for an accurate positioning of artificial hip implant and restore a proper joint biomechanics. In 1986, they cooperated with IBM research center, and IMB funded a startup company, ISS (Integrated Surgical Systems) to commercialize the idea. ISS had developed ROBODOC by customizing an industrial robot of NIDEC SANKYO Inc. and firstly tried to use ROBODOC to human surgery under FDA approval of feasibility study in 1992. ROBODOC was firstly sold in Europe at 1996 while the FDA approval was pending. ISS was taken over by CUREXO Inc. at 2006, and then clinical trial in USA was completed with a subsidiary of CUREXO. ROBODOC was finally approved for commercial sale under a 510(k) notification by FDA at 2008.

A surgical planning procedure using 3D CT data is required before surgery. If ROBODOC is connected to patient's bone and a registration procedure was finished, it automatically milling the bone for exact contact with implants.

Bach et al. reported that ROBODOC surgery did not impair hip abductor function in spite of a wider exposure of the proximal femur, a rigid fixation of leg from comparing gait of patients after ROBODOC and conventional total hip arthroplasty [3]. Honl et al. [15] reported that the robotic assisted technology had advantage in accuracy from a prospective study, but there are disadvantage such as high revision rate, the amount of muscle damage, and longer surgical time.

Nishihara et al. evaluated the clinical accuracy of femoral canal preparation using 75 consecutive total hip arthroplasties performed with ROBODOC system. They compared the preoperative planning with the postoperative CT data at one month after surgery. Results show a high degree of accuracy with less than 5 % in canal fill, less than 1 mm in gap, and less than 1 degree in alignment [35]. They also reported that in the robotic milling group does not occur intraoperative femoral fracture and this group shows a radiographically superior implant fit, and shows significant superior Merle D' Aubigne hip score at two year after surgery [34]. As the comparison study of Nakamura et al., robotic-milling shows slightly better clinical scores until 3 years after surgery. This difference was no longer present at 5 years after surgery, but robotic milling groups showed less variance in limb-length inequality and less stress shielding of the proximal femur [33].

Schulz et al. reported that the results of total hip arthroplasty of 97 hips with ROBODOC were equaled compared to a manual technique. However, they found technical complications directly related to the robotic device in nine cases, such as fine halted milling process, two femoral shaft fissures, one damage to the rim of the acetabulum, and one defect at the greater trochanter [39].

The benefits of ROBODOC surgery for total hip arthroplasty are still controversial. The robotic assisted milling of a bone certainly improved accuracy, and surgical outcome were equaled compared to a conventional technique. However, the robot-assisted surgery has shown clinical and technical complications.

1.2 RIO robotic arm

Another approach from an automatic robotic system like ROBODOC for robotic arthroplasty is semi-active, and this method is adopted by Arcrobot Company Ltd., and MAKO Surgical Corp. MAKO was relatively recently founded in 2004 than Acrobot that was founded in 1999. Though their robotic systems use active constraint, which gives surgeon's hand resistance force for keeping the safety zone during knee arthroplasty, MAKO's robotic system does need rigid fixation to bony structure. MAKO recently acquired Acrobot as a settlement in an intellectual property litigation [37]. MAKO got the FDA clearance of RIO Robotic Arm Interactive Orthopedic System in 2008, and is merged with Stryker Medical in 2013.

Lonner et al. [25] reported that the robotic arm-assisted unicompartmental knee arthroplasty (UKA) shows more accurate and less variable in initial results from comparing 37 UKA using robotic arm-assisted bone preparation with 27 UKA using conventional technique. Pearle et al. [36] also agreed that haptic guidance in combination with a navigation module allows for precise planning and execution in 10 patients with UKA.

Citak et al. [7] evaluated whether the robotic system with dynamic bone tracking that was newly updated function would provide more accurate implant from six fresh-frozen cadaver studies (six knees with robotic UKK and six knees with conventional method). Robotic UKA showed the decreased RMS error in both position and orientation for the tibial and femoral components.

1.3 Renaissance

Mazor robotics Inc. was founded at 2001 based on research of Israel Institute of Technology. Their early product, SpineAssist that was a mechanical guidance system for spine surgery, got an approval from FDA at 2004, and the company commercially released new product, Renaissance guidance System in 2011.

Renaissance can improve the insertion accuracy of pedicle screws using a hexapod-type robot that guides mechanically an entry point of the screw. Surgical procedure is quite simple; this is plan, mount, registration, and operation. Surgeon plans for the ideal surgery with 3D images before surgery. In operation, surgeon rigidly attaches a guide rail to patient, and then takes two fluoroscopic images to registration between patient and CT data. If Renaissance is positioned at indicated location on the rail, it starts guiding tools and implants to the planned position.

From a prospective randomized comparison between a robot-assisted placement of lumbar and sacral pedicle screws and a conventional freehand screw implantation, Ringel et al. [38] reported the conventional method was more accurate than the robot-assisted method, and radiation exposure was equaled. But, they thought the modification of the robotic system would increase accuracy of screw position.

In recent study, Dravel et al. [8] reported the use of the robotic system enables minimally invasive, percutaneous transpedicular interventions with safety and a high accuracy of screw placement. The adequacy of quality control of robot-assisted pedicle screw fixation accuracy was also reported from a cumulative summation test [20].

2 Robot Assisted Fracture Surgery

2.1 Minimally Invasive Fracture Surgery

The principles of fracture surgery are accurate alignment of bone fragments, rigid fixation, avoidance of soft tissue injury, and early recovery. Generally, it has priority to restore a bony structure between a proximal bone fragment and a distal bone fragment than to fit all pieces of bone fragment of diaphysis. A repositioning to normal anatomical structure is important in upper limbs fracture, whereas a restoration of mechanical axis is emphasized in lower limbs fracture. Bone fragments should be rigidly fixed not to make movement at the fracture site.

There are two main methods for fixation of bone fragments; a closed intramedullary nailing and a plate osteosynthesis. The closed intramedullary nailing is now considered as a standard treatment for a long bone shaft fracture. The plate osteosynthesis has showed particularly advantageous when an intramedullary nail may be technically not feasible.

Lee et al. reported that MIPO (Minimally Invasive Plate Osteosynthesis) had the shorter bony union time and the longer operation time than interlocking intramedullary nailing. Two groups did not show statistical significance in clinical results [24]. Apivatthakakul et al. [2] reported that although the biomechanics of the plate fixation are less stable compared to the intramedullary nail, the mechanical stability is stable enough for bone healing. Guo et al. [13] conclude that both an intramedullary nailing and a percutaneous locked compression plate can be used safely to treat distal metaphyseal fractures of the tibia from a prospective comparison study. They prefer the intramedullary nailing, because of shorter operating and radiation time, and easy removal of the implant.

If surgeon makes incision at fracture site for accurate bone alignment and rigid fixation, this incision may be cause of a soft tissue injury, a failure or delay of a bone healing, and an infection. Over the past decades surgeons have tried to fix bone fragments by inserting implant away from the fracture site through minimally invasive incision. This surgical technique, called minimally invasive fracture surgery, can keep blood flow to the injured tissues and reduce the risk of infection. Consequentially, it shows good surgical results and early recovery than the conventional surgical methods.

In early state of MIPO (minimally invasive plate osteosynthesis), David et al. were confident that the MIPO technique for the treatment of distal tibial fractures would be a feasible and worthwhile while avoiding the severe complications [14].

Mahmood et al. [29] reports that minimally invasive technique shows less blood loss, minimal soft tissue destruction, shorter hospital stay, and early mobilization comparing to conventional methods in dynamic hip screws for fixation of inter-trochanteric fractures of femur. Wong et al. [46] concluded that MIDHS (Minimally invasive Dynamic Hip Screw) fixation for intertrochanteric femoral fractures is superior to the conventional technique from their double-blind, prospective, ran-domized, and controlled clinical trial. The MIDHS produces less blood loss, less pain and a shorter rehabilitation period, while still achieving good radiological outcome.

MIPO for mid-distal humeral shaft fractures could effectively treat with advantages of shorter fracture union time and lower incidence of iatrogenic radial nerve palsies but with similar functional outcomes to the conventional open plating technique [1].

2.1.1 Related Issues

Though the minimally invasive fracture surgery has many advantages as describe above, it has also disadvantages, such as technical demanding, malreduction, malalignment, and radiation exposure.

Malreduciton

Krettek et al. [23] discussed that MIPPO (Minimally Invasive Percutaneous Plate Osteosynthesis) technique is technically demanding and the intraoperative deter-mination of limb alignment must be improved, though the technique yields clinical results comparable to those achieved with the traditional plating techniques. Khoury et al. [19] pointed out that for the MIPO technique, reduction should be performed cautiously due to the tendency of sagittal plane malreduction. Apivatthakakul et al. [2] also agreed that major complications of MIPO were malalignment and screw breakage. Buckley et al. [6] reported that the incidence of malrotation was 38.5 and 50 % respectively, following fixation of distal femoral and proximal tibial fractures with the minimally invasive percutaneous osteosynthesis technique.

Radiation Exposure

Muller et al. reported that the recommended dose limit of 500 mSv to the dominant hand of the primary surgeon and first assistant would be exceeded if more than 407 intramedullary nailing procedures. But the radiation dose to thyroid by wearing lead protection were very lower compare to 300 mSv per year. Despite the relatively low dosage, they recommend to reduce the radiation dose to a minimum, considering a risk of incidental radiation damage [32].

Madan and Blakeway [26] showed that it was within acceptable limits that the overall radiation to patient's gonads and surgeon's hands in intramedullary nailing of the lower limb. On the other hand, they warned a surgeon to avoid live fluoroscopy, because they do not yet know the long-term effects of radiation.

Thomas et al. found that the hands are at higher risk and additional hazard is created for the less experienced surgeons [4]. Michael et al. concluded that the emission of radiation depends on the fracture type and the experience; higher emission rates are surveyed at the type C fractures and inexperienced team members [22].

Kim and Kim [21] reported that surgery must be more cautious about radiation exposure during fracture management from their experimental result, which the estimated annual equivalent dose outside the lead apron was close to or higher than the maximum limit of radiation exposure. Particularly, radiation exposure times of the minimally invasive intramedullary nailing and MIPO were higher than other orthopaedic surgeries.

Physical Burden

Weight of limbs and stiffening of muscles are other problems. Medical team endures to pull out a lower limb for fracture reduction. The weight itself makes difficult for traction of limbs, and the stiffening makes it more difficult.

Maeda et al. measured forces and torques that were applied to the lower limbs of 62 healthy and young volunteers with the robotic system, the end effector of which connected to the patient's lower limbs using a boot. The average of maximum traction force was 232.9 N, range from 114.0 to 311.0 N, and the maximum torque was 6.31 Nm in external rotation and 7.69 Nm in internal rotation [28]. Maximum traction forces were 267.7 N in male group and 201.6 N in female group. Their following works, including subjects of seven female patients with intertrochanteric fractures, reported that the average traction force and rotation torque needed for reduction were 215.9 N, ranged from 146.3 to 294.9 N, and 3.2 Nm, respectively [27].

On the other hand, Gosling et al. measured forces and torque during fracture reduction in seven patients with eight fractures of the femoral shaft using a load cell that is connected to distal bone fragment with two Schanz screws. Results showed that the maximum resulting force was 411 N and the maximum resulting torque 74 Nm using the load cell [10].

In short, the minimally invasive fracture surgeries have shown similar outcomes to the conventional surgical techniques at many kinds of fracture surgeries. And it has many advantages such as less blood loss, minimal soft tissue damage, shorter rehabilitation, shorter hospital stay, and lower incidence of complication. On the other hand, the surgeons should be experienced in an accurate bone alignment and screw insertion. They give attention to malalignment and radiation exposure. And they must also put up with the large force required to pull out the lower limbs during fracture reduction.

Table 1 Minimally invasive fracture surgery and benefits of robotic system

Minimally invasive fracture surgery		Benefits of robot
Advantage	Disadvantage	
Good or similar outcome to the conventional technique Avoiding severe complication Less blood loss Minimal soft tissue destruction Early mobilization Shorter rehabilitation Shorter fracture union time Lower revision incidence	Technically demanding Malalignment/Malrotation Difficult Screw insertion Additional radiation exposure	Standardization of surgeon's skill Precise motion Work at hazardous condition Heavy load

2.1.2 Suggestions

Advantages and disadvantages of the minimally invasive fracture surgery are listed up in Table 1. It would be a good suggestion to use a robotic system which may utilize its advantage while overcome its disadvantages in the minimally invasive fracture surgery. As mentioned before, the robot system can provide the precise motions for the accurate bone positioning and the exact determination of insertion point of fixation screws. The robot has sufficient power to pull out a lower limb and to endure its position during a fixation procedure of bone fragments. Of course the robot can work in hazardous condition like high radiation exposure area. In this case, the robot can be controlled in remote, or it will automatically work. With these points, several research groups have developed the robotic system for fracture surgery.

2.2 Previous Studies for Robot Assisted Fracture Srugery

The robot system has two main roles in fracture surgery; one is a guidance or insertion of a needle or a screw, and the other is a positioning the bone fragment for a fracture reduction.

Though there are already some commercialized surgical navigations for determining the insertion positon, only a few laboratory-level's robotized systems are reported. K. Bouazza-Marouf et al. first introduced robot assisted orthopedic surgery. Their robot allowed a drill-bit guide to be automatically aligned with an intra-operatively planed drilling trajectory [5]. Shoham et al. [41] have developed a bone-mounted miniature robot to precisely position and orient a drill or a needle for the same use.

The robotic system for assisting the fracture reduction is also not commercially available until now. A reduction robot system called "RepoRobo" was firstly introduced by Fuchtmeier et al. [9]. They converted a commercial industrial robot for medical use by appropriate modification. The requirements for using the industrial robot for the reduction of femoral shaft fractures are well described.

Westphal et al. also tried to use an industrial robot for medical use. They showed that robot assisted fracture reduction for the femur provides a high precision in alignment, while reducing the amount of intra-operative imaging from the statistical analysis [11]. Then, they have developed a surgical telemanipulator system to support long bone fracture reduction procedures [44, 45]. A joystick with force feedback was installed in their system as the control device for the telemanipulator, and the interaction for 3D fraction reduction with a 2D joystick input device was described. They showed the very accurate reduction results in their experiments with the telemanipulator.

Gramham et al. introduced a parallel robot for long bone fracture reduction [12, 31]. They used a foot holster to fix the leg to the parallel type robot. The geometric model of the bone fragments were visualized in a monitor. They also introduced force modeling during fracture reduction to determine the requirements for a robotic device.

Ye et al. [47] suggested a hybrid robotic system to balance the accuracy, payload and workspace. And Tang et al. [42] showed a hexapod computer-assisted fracture reduction system could reduce long-bone diaphyseal fractures effectively.

Mitsuish et al. had developed a fracture reduction assisting robotic system, which consists of a newly designed robot for only medical user, and a reduction path navigation system based on 3D CT imaging [30, 43]. The foot is fixed using a boot like the conventional fracture table. Clinical data of reduction forces/torques are reported with this robotic system by Maeda et al. [28] as mentioned before. An automatic reduction for fracture surgery was also developed by modifying this system [16]. The robot was fixed to a distal bone fragment through two fixation pins and the goal position and the reduction path were generated from a navigation system based on surgeon's opinion. The phantom experiment results were very promising.

3 Safety Issues of Robot Assisted Fracture Reduction

In case of an industrial robot, the major strategy for ensuring human safety is to physically separate the robot from vulnerable human by creating a safety zone from robot's workspace. However, this strategy is obviously inappropriate to surgical robots which interact with surgeons. Moreover, the surgical robots need to touch an affected part or to be coupled to patients for the treatment.

3.1 Related Studies

Fuchtmeier et al. [9] suggested some safety features of "Reporobo". First, the robotic system has the velocity limits using the large reduction gear ratios, though the general operating speed of the robot may be controlled via software. At the same time, "back-driven" should be possible so that the surgeon can physically move the robot arm away from the patient in an emergency. "Harmonic drives" were applied to their system in this end. Secondly, they used a robotic gripper with integrated compliance to prevent the robot from pulling the fixator out. Lastly, they designed the control software for the force sensor to enable a free definition of the load limits to avoid over-correction and over-extension of the extremity.

Westpal et al. [44] took safety precautions by dividing the project into two layers, a software layer and a hardware layer, while developing the surgical te-lemanipulator. The software layer checks all the variable sensor information against predefined thresholds. If one safety threshold is exceeded, a stop command is sent to the robot control unit as soon as possible. For instance, if forces/torques exceeds a defined threshold, the motion of the robot is immediately stopped, and subsequently only those motions that lead to a reduction of the applied forces are allowed. Position information is used to limit the translational or rotational speeds. The hardware layer also limits the forces or torques with a load limiter (ULS 100, IPR-Intelligente Peripherie fur Roboter GmBh, Schwaiger, Germany). As soon as the forces or torques applied to this device the thresholds, the robot controller performs an emergency stop.

Gramham et al. [12] explained that parallel robots have inherent safety such as an increased stiffness, a high gear ratio, and high accuracy. Thus, they used the parallel robot mechanism for fracture reduction. And it was said that they plan to incorporate safety features such as a watchdog timer, a dead-man switch, force monitoring, encoder redundancy, and software motion limits.

3.2 Hazard Analysis

Here, safety and system design methods will be described based on developing experience at the University of Tokyo [17]. First of all, inherent hazards of the surgery and the robot-induced hazard should be listed up. Figure 2 shows the hazards related to a robot-assisted fracture reduction. An indirect reduction means that connection part does not directly connect to the bone fragment. In this way, limbs are just grabbed by hands or seized with a connection part like a boot. The conventional surgical technique belongs to the indirect reduction. A direct reduction method was suggested to improve reduction accuracy by fixing an end effector to a bone fragment using fixation screws. And a robotic system was introduced for more accurate and safe fracture reduction. Though the accuracy of the reduction can be increased by applying the new technique to the conventional methods, the extended

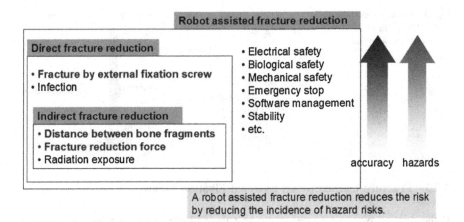

Fig. 2 Hazards related to robot assisted fracture reduction

system also results in an increase of number of hazards. However, if the robotic system is able to control the hazards satisfactorily, the application of this system could increase reduction accuracy with lower risk. Some hazards that must be controlled and its preventive measures will be suggested.

3.2.1 Distance Between the Bone Fragments

Excessive traction may injure the sciatic nerve, which starts in the lower back and runs through the buttock and down the lower limb and serves nearly the whole of the skin of the leg, the muscles of the back of the thigh, and those of the leg and foot. Though traction of the limbs is required, it must be a safe range. Surgeons recommend restricting the traction distance between a distal and a proximal bone within 10 mm. A navigation system can measure the traction distant. The navigation system sounds an alarm for over-traction and sends a stop signal the robot for stopping. The motions of the robot are spatially constrained to minimize a reduction motion.

3.2.2 Fracture Reduction Force

Excessive reduction force causes the injury of the soft tissues. The robot needs to have functions that can limit the reduction force to a safe range. The safe range can be controlled in reasonable limit because it is not easy to define the safe range using the previously reported data. In literature, traction force and torque is approximately 300 N and 5 Nm. Mechanical failsafe units and a software force limiter is designed for this end.

Fig. 3 Bending load and
pull-out force in condition of
finite element analysis

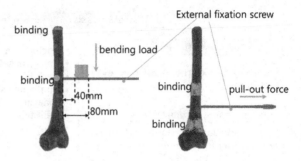

3.2.3 Fracture Caused by an External Fixation Screw

If the fixation pins were used to connect the end effector of a robot with a bone
fragment, these pins have possibility to crack around its insertion point. A finite
element analysis was used to estimate this problem.

Finite analysis models of two bones were prepared based on the CT-data of a
healthy bone and a fracture bone of a patient. The fixation pin model was generated
from the CAD model and its material was set as titanium, which is generally used in
clinical application. The pins were inserted into five sites into the healthy bone and
into seven sites on the fracture bone. Bending load and pull-out force were applied
to the pins as shown in Fig. 3.

While the bending load and pull-out force was varied, the number of destroyed
elements was estimated. The bending load was varied from 5 to 40 kg, and the
pull-out force was varied from 20 to 200 kg.

The destroyed elements were fist found at the bending load of 25–30 kg and at
the pull-out force of 120–200 kg in the case of healthy bone, while they were found
at the bending load of 20–25 kg and the pull-out force of 40–120 kg in the fracture
bone. Though the fracture bone shows lower strength, there are recommended
points of fixation pin insertion on the bone.

3.3 System Design

Configuration of the fracture-reduction system is shown in Fig. 4. The fracture
reduction system consists of a fracture reduction robot and a navigation system. The
surgical bed and the fracture reduction robot are arranged in a line. One side of
the surgical bed is used for the navigation system, and the other side is open for the
surgeons.

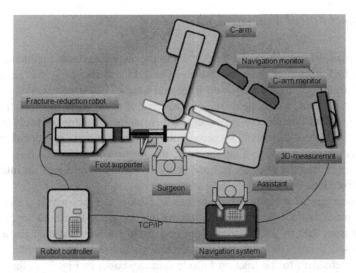

Fig. 4 Configuration of the fracture-reduction system in an operation room

3.4 Fracture Reduction Robot

The fracture robot and its kinematic model are drawn in Fig. 5. The fracture reduction robot has six DOFs (i.e., three translation DOFs and three rotation DOFs). Three rotational axes intersect each other at one point for easy robot control. Two mechanical failsafe units and a force sensor are installed at y-axis. A customized jig is used to fix the bone fragment to the robot. The user controls the robot with a tough panel. A four-color LED bar shows the robot status: Power On, Ready, Operating, and Emergency stop. The size of robot that is 640 mm (width) * 1084 mm (length) * 1317 mm (Height) is suitable to transport using a normal passenger elevator.

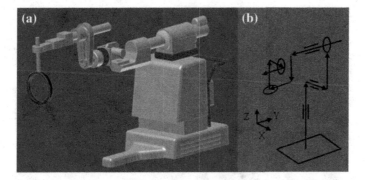

Fig. 5 Fracture reduction robot; **a** outline, two blue parts show installation location of two mechanical failsafe units and **b** kinematic model and coordinates of the robot

3.5 Mechanical Failsafe Unit

Two mechanical failsafe units are designed to prevent an excessive reduction force. The installed positon of them are illustrated with blue parts in Fig. 5a. These units maintain rigidity within the allowed force and torque. But if an excessive force is applied to the unit, it decouples the end effector of robot from the actuation unit. The longitudinal direction of the bone fragment coincides approximately with the traction direction of the robot. Surgeon should ensure that these positional relations are correct before surgery. Consequently, the traction failsafe unit can limit the traction force and the rotation failsafe unit can limit the torque of an internal or external rotation.

The structure of a traction failsafe unit is a plunger type, which a steel roller pushed into a hollow by a spring as shown in Fig. 6. A threshold force can be adjusted from 200 to 400 N by tighten or loosen a screw that changes a spring compression force. The rotational failsafe unit mounted on the end effector have similar mechanism to the traction failsafe unit as shown in Fig. 7. A threshold is adjustable from 20 to 40 Nm. When the unit is decoupled, the rotational angle of end effector is constrained by mechanical stoppers, the position of which can be varied from 30° to 120°.

A threshold can be calculated by considering equilibrium of force and moment with two parameters, the spring force and the contact angle between a roller and a hollow. Figure 8 shows the forces acting on the roller. F_s is the spring force, and F_{ex} is an external force. N is the vertical component of force acting on a contact point between the roller and the hollow and is equal to F_s. W denotes the horizontal component of force acting on the contact point and is equal to F_{ex}. The following equation explains this equilibrium at the center of the roller.

$$F_s r \sin\theta = F_{ex} r \cos\theta \tag{1}$$

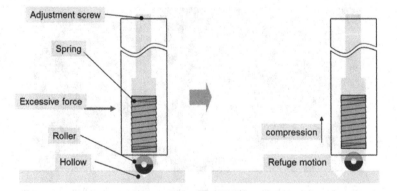

Fig. 6 Structure and mechanism of a translational failsafe unit

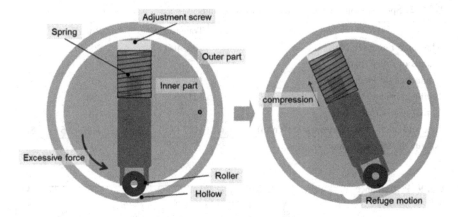

Fig. 7 Structure and mechanism of a rotational failsafe unit

Fig. 8 Forces acting on a roller and a hollow

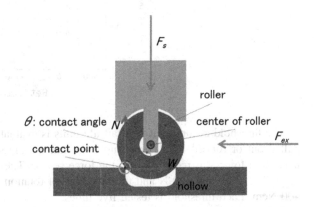

Here r is the radius of the roller and θ is the contact angel. From (1), the spring force for constraining the given F_{ex} is

$$F_s = kx = \frac{f_{ex}}{\tan\theta} \tag{2}$$

Here, k denotes the spring constant and x is the displacement of spring. A similar idea can be applied to explain the rotational failsafe units, which gives

$$F_s = \frac{f_{ex}}{R\tan\theta} \tag{3}$$

where, R is the radius of the inner part of the rotational failsafe unit. Specifications for the mechanical failsafe units list up in Table 2.

Table 2 Specification for the mechanical failsafe units

	Translation	Rotation
Threshold range (N, Nm)	200–400	20–40
Radius of roller (mm)	9.5	8
Contact angle (degree)	34	45
Spring constant (N/m)	121	204
Spring displacement (mm)	14.7	6.8
Size and pitch of adjustment screw (mm)	M27 (3.0)	M22 (2.5)
Radius of inner part (mm)		50

Fig. 9 Performance of failsafe units, n = 5

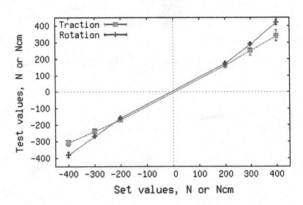

The threshold accuracy of two failsafe units is evaluated. The end effector was pulled out or rotated by a human until the failsafe units were activated, and the maximum force was recorded with the force sensor. The threshold of traction was set at 200, 300, and 400 N and the threshold of rotation was set at 200, 300, and 400 Ncm. Each threshold is tested five times.

Figure 9 shows the evaluation results. The error bars means the variation in each of the five trials. The activated forces of the traction failsafe unit were smaller than the set threshold values. This differences results from the weakened offset tension for the spring, which occurs when the robot is reassembled for correction of movement. The differences can be reduced by adjusting the offset tension.

3.6 Software Force Limiter

A software force limiter is designed to control a velocity of a bone movement against a reduction force as shown in (4)

$$G(t) = \begin{cases} 1 & F(t) < Th1 \\ \frac{(F(t) - Th1)^2}{(Th2 - Th1)^2}, & Th1 \le F(t) < Th2 F(t) \\ 0 & F(t) \ge Th2 \end{cases} \tag{4}$$

Here, $F(t)$ is the reduction force, $Th1$ and $Th2$ are the first and second thresholds of the software force limiter, and $G(t)$ is the velocity gain. Two thresholds are used to avoid the sudden stopping of the robot and to forewarn an operator of the increased reduction force by decelerating the movement. The control speed given by (5) is slowed down according to a quadratic curve between the two thresholds.

$$V(t) = \alpha G(t) F(t) \tag{5}$$

Here, a is a weighting factor and $V(t)$ is the control velocity of the robot. Two thresholds are set under the limitation of the mechanical failsafe units so that it works only the software has treble. Two thresholds are set under the limitations of the mechanical failsafe units so that these units work only when the software has error.

The software force limiter was evaluated using a static obstacle placed beside the end effector. The robot moved in the direction of the x-axis with speed of 10 mm/s. Two thresholds were set at 100 N and 150 N, respectively. The movement and the reaction force are recorded with a frequency of 50 Hz.

Results, variations of force and velocity against time, are shown in Fig. 10. The end effector contacted the obstacle approximately 4 s, and then the force is slowly increased. The robot start to reduce its velocity when the force reached first threshold and it stop at second threshold.

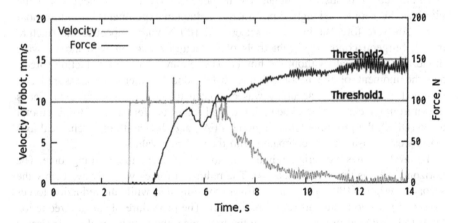

Fig. 10 Evaluation result for the software force limiter

3.7 Spatial Constraint

In femoral shaft fractures, a distal bone fragment is pulled up to the hip area and externally rotated due to the influence of soft tissues. Surgeons need to pull out the distal bone fragment and internally rotate it. The robot should generate similar movements of the bone. The coordinates of a bone is set an origin locates on a fracture section and its one axis matches a longitudinal direction of the bone. The spatial constraint algorism generates the robot's movements along the bone coordinates. Surgeons can control the robot using the handle that is attached the end effector through a force sensor of six DOFs. The measured data by the force sensor are transformed to the bone coordinates. The directions of a resulting force and a resulting moment is intended movements, and their magnitudes are proportional to the velocity of bone movements. The motion of the robot can be calculated using the relationship of the bone coordinates and the robot coordinates.

The spatial constraint algorism was evaluated. The coordinates of the bone, the force sensor, and the robot were set using an optical tracking system (Polaris; NDI, Waterloo, Ontario, Canada). The movements of origin of the bone coordinates were measured while an operator controls the robot. The ideal movement of it should be zero. As the results, we reduce the movement of the origin within 2 mm. Main reason for this error is caused by play of the robot's axes.

3.8 Fracture Reduction Experiments

The safety features of the fracture reduction robot were evaluated overall using the hip fracture models that are made by cutting a neck of femur model, and then attaching rubber bands to simulate the influence of the gluteus medius and the iliopsoas. Surgeons tried eight times fracture reduction experiments. Two threshold of the software force limiter were set 60 and 100 N with respect to the fracture model. Surgeon conducted eight trials of a fracture reduction with the software limiter and eight trials without the limiter. The required time, the reduction forces, and the moment of the robot and the distal bone fragment are measured and recorded. The reduction accuracy was evaluated using a mechanical axis that is drawn from the center of the knee joint to the head of the femur. The distal femoral angle (DFA), the proximal femoral angle (PFA), and the length of mechanical axis (MA) were measured and compared with their normal values.

Figure 11 shows the variation in the traction distance and the resulting force. The horizontal axis is time (milliseconds). The reduction forces were always below the second threshold (100 N) with the software force limiter, while the reduction forces exceeded it without the software force limiter. The procedure of fracture reduction tends to be divided in three stages. In the first stage, the translational displacement was increased as the distal bone fragment is pulled out. As a result, the resulting

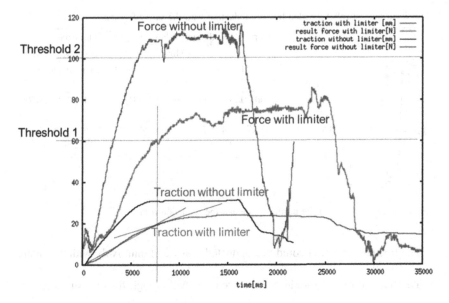

Fig. 11 Variation in traction distance and reduction force during the robot assisted fracture reduction

force is increased. The bone fragment is rotated during the second stage. Fine alignment of the bone is conducted in the third stage.

Figure 12 shows the fracture model after reduction experiment with the safety features. The average time for trial was 82.5 s, and the reduction accuracy shown in Table 3. The average defenses in PFA, DFA, and MA were 0.79°, 0.34°, and 1.06 mm, respectively (n = 8). Surgeon is comments are

- The reduction accuracy is enough comparing to the conventional technique.
- The restriction of the reduction force within the intended limit was good.

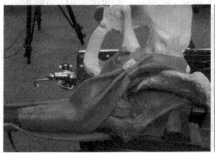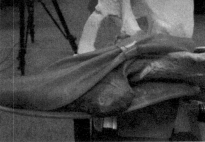

Fig. 12 Robot assisted fracture reduction with the safety features; *left* shows before reduction and *right* shows after reduction

Table 3 Results of robot assisted fracture reduction with the safety features

		Normal	Reduction value			Difference		
			Ave	Min	Max	Ave	Min	Max
	PFA (degree)	88.14	87.81	86.84	89.28	0.70	0.32	1.30
	DFA (degree)	90.60	90.89	90.45	91.34	0.34	0.04	0.74
	MA (mm)	426.78	427.76	426.44	428.88	1.06	0.26	2.10

() is the difference between the normal value and the reduction results

- The fracture reduction could be conducted easily and intuitively with the spatial constraint
- The fracture model should be enhanced to have a rough fracture surface.

4 Fracture Surgery Robotic System: Total Solution for Fracture Surgery

We have developed the fracture surgery robotic system for repairing damaged musculoskeletal region in limbs with improved accuracy and safety, which consists of a bone positioning robot and a bone tunneling robot system having 2 mm level precision. The developing strategy is diagramed in Fig. 13. First of all, we modeled a procedure of the conventional surgery and discussed what should be improved using the robotic system. The role of the robot was defined. And then, the robotic system was designed and manufactured considering safety features, and basic functions of the system was evaluated. Tentative clinical trials were conducted by medical teams and its results give feedback to engineers. We have also developed a fracture model for a proper validation. Now, we are preparing clinical trial through the repetitive developing processes.

4.1 Surgical Procedure Modeling

As the first step of develop, we analyzed the surgical procedure and define the role of the robotic systems at each procedure and defined the additional procedure. We evaluated four cases of fracture surgeries on-site; they are three femur fractures and

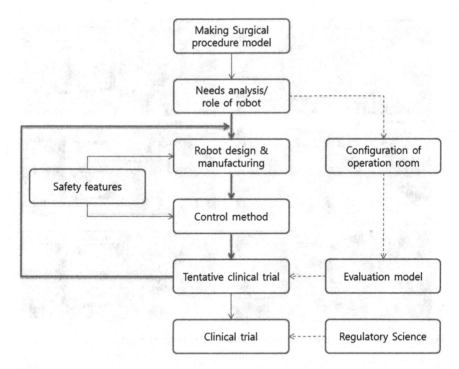

Fig. 13 Developing strategy of the fracture surgery robotic system

one tibia fracture. And we made a surgical procedure model of those and validated it by reviewing with orthopedic surgeons. In addition, the roles of robotic system were discussed.

The simplified surgical model is shown in Fig. 14; left side shows real surgery scenes, and right side shows fluoroscopic images in each step. A fluoroscope was mainly used in step 2 and 5 in this technique. We found two roles associated with the robotic system. The bone positioning robot should assist the fracture reduction with high power and precision in step 2. The tunneling robot has to bore guide holes for insertion of interlocking fixation screws. A navigation system could reduce the usage of the fluoroscope.

4.2 Surgical Procedure and General Setting

The surgical procedure with the robotic system is redefined as block diagram in Fig. 15. Blue boxes are additional procedures for the navigation system, and red boxes show the robot assisting procedures.

A general setting of the robotic system in a surgical room is illustrated as Fig. 16. A patient is laid supine upon the operation table so that the surgeon can easily insert a nail or a plate after fracture reduction. The bone positioning robot is located at the

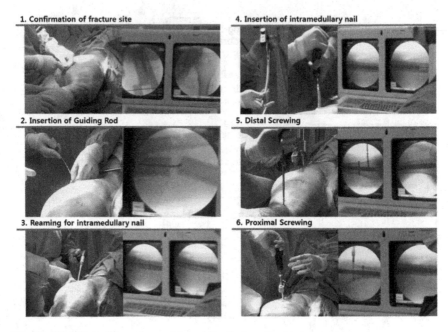

Fig. 14 Modelling of procedures of the minimally invasive fracture surgery

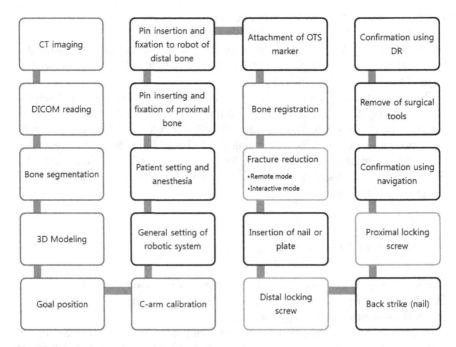

Fig. 15 Surgical procedures with the robotic system

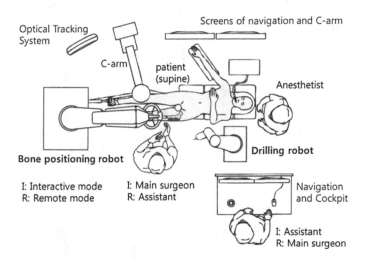

Fig. 16 General setting of the fracture surgery robot

foot side and the tunneling robot is located beside of trunk of the patient. The positions of surgical team and surgical equipment are not different to the conventional setting.

4.3   Fracture Surgery Robotic System

The surgical robotic system consists of a bone positioning robot, a tunneling robot, a navigation system, a motorized fluoroscope, and some customized devices for bone fixation. Each robot has six degrees of freedom and its power is designed suitably for traction of a bone fragment possessing a femur and for guidance of a drilling position. Two robots can be controlled by joystick-like devices from the side or remote position. The navigation system shows the surgical procedure, and relative position of bone fragments and surgical instruments. The motorized fluoroscope is also designed for having six degrees of freedom and can be remotely controlled. The fixation pins to bone fragment are specially designed to insert an intramedullary nail without removing it after fracture reduction.

4.4   Reduction Strategy with the Robot

Femoral fractures show the three types due to the anatomical characteristics of the femur as illustrated in Fig. 17 [18]. First, if fracture is occurred 1/3 part of proximal, the proximal bone fragment shows external rotation, abduction, and flexion. Second,

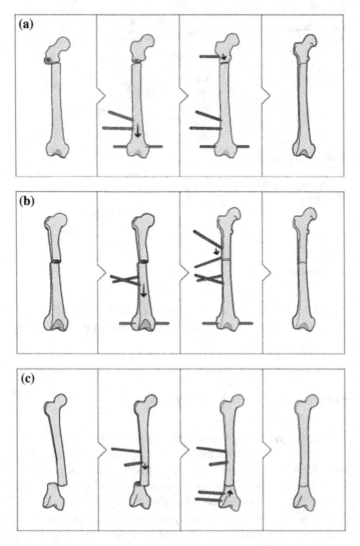

Fig. 17 Strategy of fracture reduction with a bone positioning robot with respect to fracture types; *Left* show fracture patterns, *Two middle* show pin insertion position and reduction methods, *Right* show final reduction position. **a** Fracture on 1/3 part of proximal. **b** Fracture on 1/3 part of diaphyseal. **c** Fracture on 1/3 of distal fracture

if fracture is shown at 1/3 part of diaphysis, bone fragments are shortened by overlapping due to the strong muscles around fracture site. Third, 1/3 part of distal, shows extreme posterior angulation of a distal bone fragment.

For the fracture reduction with the robotic system, the distal bone fragment is connected to the bone positioning robot and the proximal bone fragment is fixed to the operation table using a passive arm having six degree of freedoms. Two middle illustrations of Fig. 17a–c show the methods of pin insertion and reduction.

In case of 1/3 part of proximal, the proximal bone fragment showing extreme deformation is manually relocated to its approximate original position after pulling the distal bone fragment using the bone positioning robot. And then, the proximal bone fragment is fixed with the passive arm. The bone positioning robot moves the distal bone fragment by motions of internal rotation, adduction, and extension.

The main role of the bone position robot is longitudinal traction of the distal bone fragment for reduction of overlapping in case of the 1/3 part of diaphysis. For this purpose, strong traction using a transfixing pin/screw through the distal femur or the proximal tibia is required. If 1/3 part of distal case, reduction would be easy by flextion of knee with tibia, of which motion can correct extreme posterior angulation of a distal bone fragment

4.5 Tentative Clinical Trial

Tentative clinical trial was performed to verify the surgical procedure using the fractured femur phantom by orthopedic surgeons and engineers. Orthopedic surgeons verified all procedure with engineers' assistance. Procedures of concern are "pin insertion and connection of distal bone to robot" and "locking screw insertion". We should check interferences between devices, and fluency of the procedures.

Figure 18a shows an overall view of the tentative clinical trial and Fig. 18b shows close-up view of bone fragments connected to the positioning robot. Procedures are good enough to conduct robot-assisted fracture reduction, and a surgeon can reduce the fracture. We confirm that the intramedullary nail could be inserted without removal of the fixation pins. However, we find out some problems related to the devices; stiffness of the fixation pin, interference between connection devices and the tunneling robot, and artifact to the fluoroscope.

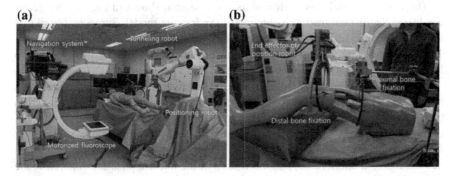

Fig. 18 Tentative clinical trial of the fracture surgery with robotic system; **a** overview and **b** close-up view of fixation of bone fragments

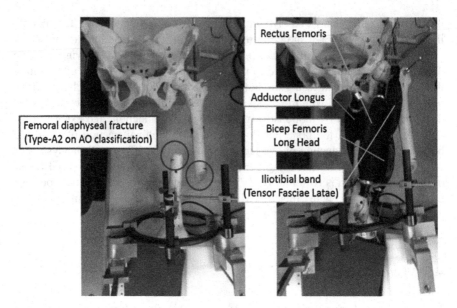

Fig. 19 The fracture model with four pneumatic actuators

4.6 Evaluation Model

A model of femoral fracture simulating muscular contraction force has been developed [40]. It is difficult to evaluate the robotic system using cadavers or animals; the mechanical characteristics of human muscular tissues are significantly changed after death and the animals are anatomically different from human. Four pneumatic actuators were used to simulate action of rectus femoris, long head of biceps femoris, adductor longus, and tensor fasciae late. The pneumatic pressure was controlled by the force-length properties of Hill's muscle model as shown in Fig. 19.

The fracture model was evaluated as configuration shown in Fig. 20. Surgeons conducted the manual fracture reduction of the fracture model. Surgeons found that the fracture model had similar elastic property to human but different initial traction. The fracture model was upgraded to simulate an initial position with respect to fracture type, and a time response of muscle.

The surgical robotic system to assist the fracture surgery, and its application procedure is introduced. The robot system shows possibility of clinical trial from the tentative clinical trial although it needs to modify some devices. Above all, orthopedic surgeons satisfied that they could insert the intramedullary nail after fracture reduction with the robot system. Future works are new designs of fixation devices to avoid interference, and fabrication of the fracture phantom generating the reduction force, and the second tentative clinical trial.

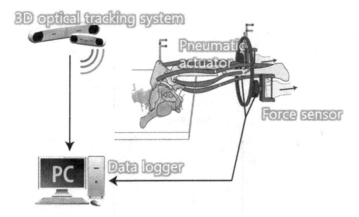

Fig. 20 Experimental setup of the fracture model

5 MOEBIUS Robot: Multi-purpose Robotic System for Bone Surgery

We have introduced the fracture surgery robotic system capable of total solutions. Although the robotic system gives many functions and advantages, it is very big to be installed into an operation room. The other approach is a small system with restricted function. Here, "small" means small size, easy to use, and simple and convenient installation. We designed a multipurpose robotic system for musculo-skeletal surgery based on Stewart platform to give the shape to this "small" concept. A parallel type robot generates big power comparing to a serial type robot; this makes the physically small size, while it gives enough power for positioning bone fragments. The robotic system has master-slave control structure, and the slave robot is assembled with its modular units. The master device is designed as similar structure to the slave robot for intuitive control. The motions of the master device to position a bone are similar to "infinity character", and the salve robot follows these motions. The robotic system was named "Moebius robot" after these motions. Modular units help to give easy installation to a patient and various clinical applications, such as bone deformity correction, bone lengthening, and fracture reduction as shown in Figs. 21 and 22.

5.1 Design

An operation concept of the Moebius robot is illustrated in Fig. 23. The application target in this concept is a fracture reduction of tibia. One ring frame is fixed to a proximal bone fragment and the other ring frame is fixed to distal bone fragment.

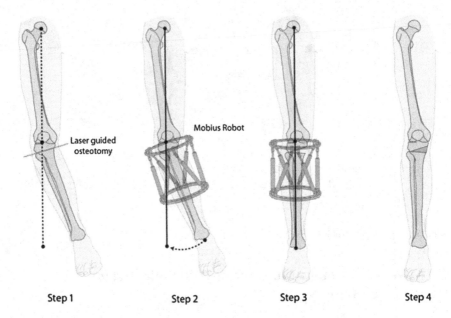

Fig. 21 Bone deformity correction with Moebius robot

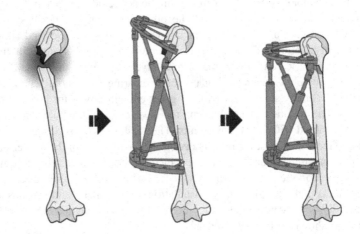

Fig. 22 Fracture reduction with Moebius robot

A fluoroscope is used to monitor a bone fragments. A surgeon operates the robot behind a lead-glass to avoid radiation exposure.

The modular units of the salve robot consist of three types of an actuator axis and three types of a ring frame, the size of which corresponds to the affected part. The modular units of the salve robot consist of three types of an actuator axis and three types of a ring frame with respect to the affected parts such as a femur, a tibia, and a

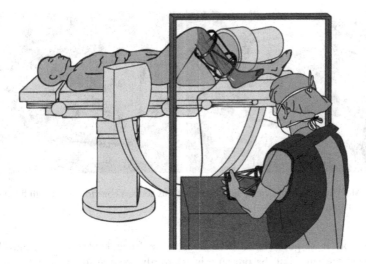

Fig. 23 Operation concept of Moebius robot

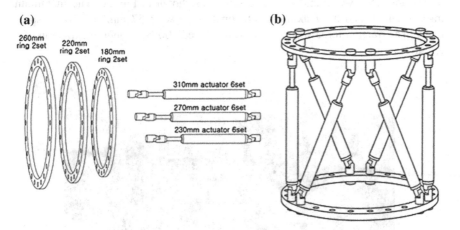

Fig. 24 Modular design of the slave robot; **a** modular units, and **b** the assembled slave robot

humerus. Figure 24a shows the designed modular units and Fig. 24b shows the slave robot that is basically assembled with two ring frames and six actuator axes. The ring frame can be separated into two half-ring frames so that it provides a convenient installation to a long bone. The actuator axis is packed in a ball-screw, a brushless motor, an encoder and a motor driver. The maximum speed and force is 4.8 mm/s and 154 N, respectively. The upper and lower ring frames are connected to bone fragments using two Shanz pins as shown in Fig. 25a.

The master device has similar structure to the slave robot as shown in Fig. 25b, but it does not have an actuator in its axes. Each axis has an encoder and a pneumatic friction shaft. The combination of six encoders generates motions of the

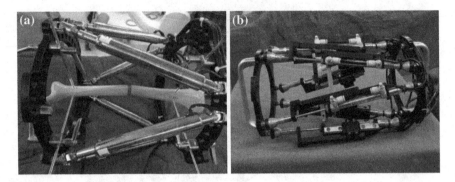

Fig. 25 Moebius robotic system; **a** the slave robot was connect bone fragments with fixation pins, and **b** the master device

slave robot. It is possible to give a user the force feedback by controlling a pneumatic pressure, but the pressure is arbitrarily fixed at this moment.

The workspace and the traction force of the slave robot were evaluated with the optical tracking system and the force sensor as shown in Fig. 26. The maximum traction length and the maximum rotational angle were 57 mm and 30°, respectively. The traction force was confirmed to exceed 800 N, enough for traction of a femur.

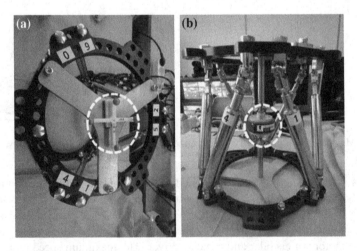

Fig. 26 Evaluation of the slave robot; **a** optical tracking maker to measure workspace, and **b** force sensor to measure the traction force

5.2 Tentative Clinical Trials

In a real operation room, tentative clinical trials were conducted with the Moebius robotic system as shown in Fig. 27. We prepared a femur shaft fracture model of polyurethane. The model did not have any soft tissue. A C-armed fluoroscope was used to confirm the fracture site.

Figure 28 shows the femur before and after the reduction procedure with Moebius robotic system. And Table 4 shows the reduction error that was smaller

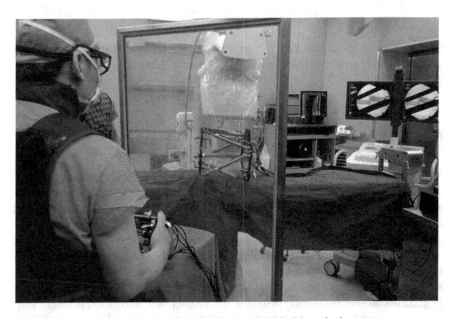

Fig. 27 Tentative clinical trial for fracture reduction with Moebius robotic system

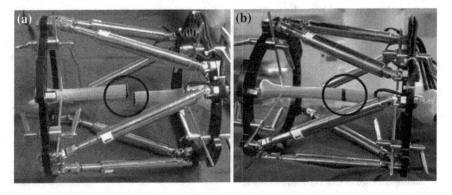

Fig. 28 Reduction results with Moebius robotic system; **a** before reduction, and **b** after reduction

Table 4 Reduction error

Axis		Error (Ave/SD)
	Translation of X axis (mm)	0.6/0.2
	Translation of Y axis (mm)	0.7/0.3
	Translation of Z axis (mm)	0.7/0.2
	Rotation around X axis (degree)	1.2/0.4
	Rotation around Y axis (degree)	0.9/0.4
	Rotation around Z axis (degree)	1.1/0.6

than 1.0 mm in translation and 1.2° in rotation. We found that surgeons with this system for first time would complete the fracture reduction only with the fluoroscopic imaging.

6 Conclusion

In this chapter, we introduced various robotic systems for musculoskeletal tissues. The brief history, the features, and the clinical outputs of three commercially available robotic systems were introduced. Safety issues and its management methods were mentioned with the fracture reduction robot. With the newly developing fracture surgery robotic system, we introduced the developing method and strategy for musculoskeletal robotic system. Finally, "small robot" concept was introduced from Moebius robotic system. We are still developing these robotic systems for musculoskeletal tissues and preparing clinical trials. Though we did not treat cost-benefit of the robotic system in this chapter, it would be important to keep in mind to develop the robotic system considering consumer. In musculoskeletal tissues, surgeons are more significant consumers than the patients. In this point, surgeon initiative tentative clinical trials would be the essential developing process, and we believe it could accelerate the developing processing for commercialization.

References

1. An Z, Zeng B et al (2010) Plating osteosynthesis of mid-distal humeral shaft fractures: minimally invasive versus conventional open reduction technique. Int Orthop 34(1):131–135
2. Apivatthakakul T, Chiewcharntanakit S (2009) Minimally invasive plate osteosynthesis (MIPO) in the treatment of the femoral shaft fracture where intramedullary nailing is not indicated. Int Orthop 33(4):1119–1126
3. Bach CM, Winter P et al (2002) No functional impairment after Robodoc total hip arthroplasty. Acta Orthop 73(4):386–391

4. Blattert TR, Fill UA et al (2004) Skill dependence of radiation exposure for the orthopaedic surgeon during interlocking nailing of long-bone shaft fractures: a clinical study. Arch Orthop Trauma Surg 124(10):659–664
5. Bouazza-Marouf K, Browbank I et al (1996) Robot-assisted invasive orthopaedic surgery. Mechatronics 6(4):381–397
6. Buckley R, Mohanty K et al (2011) Lower limb malrotation following MIPO technique of distal femoral and proximal tibial fractures. Injury 42(2):194–199
7. Citak M, Suero EM et al (2013) Unicompartmental knee arthroplasty: is robotic technology more accurate than conventional technique? Knee 20(4):268–271
8. Dreval O, Rynkov I et al (2014) Results of using spine assist mazor in surgical treatment of spine disorders. Interv Transpedicular Fixations 5(6):9–22
9. Füchtmeier B, Egersdoerfer S et al (2004) Reduction of femoral shaft fractures in vitro by a new developed reduction robot system 'RepoRobo'. Injury 35(1):113–119
10. Gosling T, Westphal R et al (2006) Forces and torques during fracture reduction: Intraoperative measurements in the femur. J Orthop Res 24(3):333–338
11. Gosling T, Westphal R et al (2005) Robot-assisted fracture reduction: a preliminary study in the femur shaft. Med Biol Eng Compu 43(1):115–120
12. Graham AE, Xie SQ et al (2006) Design of a parallel long bone fracture reduction robot with planning treatment tool. In: International conference on intelligent robots and systems, IEEE/RSJ, IEEE
13. Guo JJ, Tang N et al (2010) A prospective, randomised trial comparing closed intramedullary nailing with percutaneous plating in the treatment of distal metaphyseal fractures of the tibia. J Bone Joint Surg Br 92B(12):1717
14. Helfet DL, Shonnard PY et al (1997) Minimally invasive plate osteosynthesis of distal fractures of the tibia. Inj Int J Care Injured 28:Sa42–Sa48
15. Honl M, Dierk O et al (2003) Comparison of robotic-assisted and manual implantation of a primary total hip replacement: a prospective study. J Bone Joint Surg 85(8):1470–1478
16. Joung S, Kamon H et al (2008) A robot assisted hip fracture reduction with a navigation system. In: Medical image computing and computer-assisted intervention–MICCAI 2008. Springer, Berlin. pp 501–508
17. Joung S, Liao H et al (2010) Hazard analysis of fracture-reduction robot and its application to safety design of fracture-reduction assisting robotic system. In: IEEE International conference on robotics and automation (ICRA), IEEE
18. Joung S, Park CW et al (2013) Strategy for robotization of lower limb fracture reduction. In: 44th International symposium on robotics (ISR), IEEE
19. Khoury A, Liebergall M et al (2002) Percutaneous plating of distal tibial fractures. Foot Ankle Int 23(9):818–824
20. Kim H-J, Lee SH et al (2015) Monitoring the quality of robot-assisted pedicle screw fixation in the lumbar spine by using a cumulative summation test. Spine 40(2):87–94
21. Kim JW, Kim JJ (2010) Radiation exposure to the orthopaedic surgeon during fracture surgery. J Korean Orthop Assoc 45(2):107–113
22. Kraus M, Röderer G et al (2013) Influence of fracture type and surgeon experience on the emission of radiation in distal radius fractures. Arch Orthop Trauma Surg 133(7):941–946
23. Krettek C, Schandelmaier P et al (1997) Minimally invasive percutaneous plate osteosynthesis (MIPPO) using the DCS in proximal and distal femoral fractures. Injury Int J Care Injured 28: Sa20–Sa30
24. Lee K, SY S et al (2008) A comparison between minimally invasive plate osteosynthesis and interlocking intramedullary nailing in distal tibia fractures. J Korean Fract Soc 21(4):286–291
25. Lonner JH, John TK et al (2010) Robotic arm-assisted UKA improves tibial component alignment: a pilot study. Clin Orthop Relat Res 468(1):141–146
26. Madan S, Blakeway C (2002) Radiation exposure to surgeon and patient in intramedullary nailing of the lower limb. Injury 33(8):723–727
27. Maeda Y, Sugano N et al (2008) Robot-assisted femoral fracture reduction: preliminary study in patients and healthy volunteers. Computer Aided Surgery 13(3):148–156

28. Maeda Y, Tamura Y et al (2005) Measurement of traction load and torque transferred to the lower extremity during simulated fracture reduction. International Congress Series, Elsevier
29. Mahmood A, Kalra M et al (2013) Comparison between Conventional and Minimally Invasive Dynamic Hip Screws for Fixation of Intertrochanteric Fractures of the Femur. ISRN orthopedics
30. Mitsuishi M, Sugita N et al (2005) Development of a computer-integrated femoral head fracture reduction system. In: IEEE international conference on mechatronics, ICM'05, IEEE
31. Mukherjee S, Rendsburg M et al (2005) Surgeon-instructed, image-guided and robot-assisted long bone fractures reduction. In: 1st international conference on sensing technology
32. Müller L, Suffner J et al (1998) Radiation exposure to the hands and the thyroid of the surgeon during intramedullary nailing. Injury 29(6):461–468
33. Nakamura N, Sugano N et al (2010) A comparison between robotic-assisted and manual implantation of cementless total hip arthroplasty. Clin Orthop Relat Res 468(4):1072–1081
34. Nishihara S, Sugano N et al (2006) Comparison between hand rasping and robotic milling for stem implantation in cementless total hip arthroplasty. J Arthroplasty 21(7):957–966
35. Nishihara S, Sugano N et al (2004) Clinical accuracy evaluation of femoral canal preparation using the ROBODOC system. J Orthop Sci 9(5):452–461
36. Pearle AD, O'Loughlin PF et al (2010) Robot-assisted unicompartmental knee arthroplasty. J Arthroplasty 25(2):230–237
37. Ponnusamy KE, Golish SR (2013) Robotic surgery in arthroplasty
38. Ringel F, Stüer C et al (2012) Accuracy of robot-assisted placement of lumbar and sacral pedicle screws: a prospective randomized comparison to conventional freehand screw implantation. Spine 37(8):E496–E501
39. Schulz AP, Seide K et al (2007) Results of total hip replacement using the Robodoc surgical assistant system: clinical outcome and evaluation of complications for 97 procedures. Int J Med Rob Comput Assist Surg 3(4):301–306
40. Sen S, Ando T et al (2014) Development of femoral bone fracture model simulating muscular contraction force by pneumatic rubber actuator. In: 36th annual international conference of the IEEE engineering in medicine and biology society (EMBC), IEEE
41. Shoham M, Burman M et al (2003) Bone-mounted miniature robot for surgical procedures: concept and clinical applications. Rob Autom IEEE Trans 19(5):893–901
42. Tang P, Hu L et al (2012) Novel 3D hexapod computer-assisted orthopaedic surgery system for closed diaphyseal fracture reduction. Int J Med Rob Comput Assist Surg 8(1):17–24
43. Warisawa S, Ishizuka T et al (2004) Development of a femur fracture reduction robot. In: Proceedings of IEEE international conference on robotics and automation, ICRA'04, IEEE
44. Westphal R, Winkelbach S et al (2006) A surgical telemanipulator for femur shaft fracture reduction. Int J Med Rob Comput Assist Surg 2(3):238–250
45. Westphal R, Winkelbach S et al (2009) Robot-assisted long bone fracture reduction. Int J Robot Res 28(10):1259–1278
46. Wong T-C, Chiu Y et al (2009) A double-blind, prospective, randomised, controlled clinical trial of minimally invasive dynamic hip screw fixation of intertrochanteric fractures. Injury 40(4):422–427
47. Ye R, Chen Y et al (2012) A simple and novel hybrid robotic system for robot-assisted femur fracture reduction. Adv Robot 26(1–2):83–104

A Cost-Effective Surgical Navigation Solution for Periacetabular Osteotomy (PAO) Surgery

Silvio Pflugi, Li Liu, Timo M. Ecker, Jennifer Larissa Cullmann, Klaus Siebenrock and Guoyan Zheng

Abstract In this chapter a low-cost surgical navigation solution for periacetabular osteotomy (PAO) surgery is described. Two commercial inertial measurement units (IMU, Xsens Technologies, The Netherlands), are attached to a patient's pelvis and to the acetabular fragment, respectively. Registration of the patient with a pre-operatively acquired computer model is done by recording the orientation of the patient's anterior pelvic plane (APP) using one IMU. A custom-designed device is used to record the orientation of the APP in the reference coordinate system of the IMU. After registration, the two sensors are mounted to the patient's pelvis and acetabular fragment, respectively. Once the initial position is recorded, the orientation is measured and displayed on a computer screen. A patient-specific computer model generated from a pre-operatively acquired computed tomography (CT) scan is used to visualize the updated orientation of the acetabular fragment. Experiments with plastic bones (7 hip joints) performed in an operating room comparing a previously developed optical navigation system with our inertial-based navigation system showed no statistical difference on the measurement of acetabular component reorientation (anteversion and inclination). In six out of seven hip joints the mean absolute difference was below five degrees for both anteversion and inclination.

S. Pflugi (✉) · L. Liu · G. Zheng
Institute for Surgical Technology and Biomechanics, University of Bern, Bern, Switzerland
e-mail: silvio.pflugi@istb.unibe.ch

T.M. Ecker · K. Siebenrock
Department of Orthopedic Surgery, Inselspital, University of Bern, Bern, Switzerland

J.L. Cullmann
Department of Radiology, Inselspital, University of Bern, Bern, Switzerland

© Springer International Publishing Switzerland 2016
G. Zheng and S. Li (eds.), *Computational Radiology for Orthopaedic Interventions*, Lecture Notes in Computational Vision and Biomechanics 23, DOI 10.1007/978-3-319-23482-3_16

1 Introduction

1.1 Periacetabular Osteotomy Surgery

Chronic abnormal hip mechanics often lead to osteoarthrosis and are associated with instability or impingement stemming from a surgically treatable anatomic abnormality (e.g. hip dysplasia). The success of joint-preserving interventions depends on whether the mechanical environment could be normalized and the degree of irreversible articular damage [1]. Hip preservation surgery (e.g. peri-acetabular osteotomy (PAO)) is performed in early stages of hip diseases when an active lifestyle is demanded. PAO is a demanding surgical procedure for the treatment of adult hip dysplasia [2]. Several cuts separate the acetabular fragment from the rest of the pelvis so that it can be reoriented to improve femoral coverage. The view of the surgeon is strongly limited and some cuts have to be made without overseeing the whole area [3].

1.2 Surgical Navigation

Recent advancements in computation power, better understanding of the anatomy and new imaging modalities made it possible to merge different innovative technologies like 3D modelling, image registration and instrument tracking to support a physician pre- and intra-operatively in the diagnosis and treatment of pathologies [4]. Real-time tracking and visualization of surgical instruments and anatomy intra-operatively allows to obtain a precise digital representation of the procedure and improves accuracy and patient safety [5]. These new techniques are summarized as "computer assisted surgery (CAS)". DiGioia et al. [6] were the first to introduce a surgical navigation system for the accurate positioning of the acetabular fragment during PAO in 1998. After their success a continued enthusiasm was born in the field of CAS [7]. Langlotz et al. [8] and Jaeger et al. proposed a CT-based PAO navigation system. Liu et al. [9] proposed a computer-assisted planning and navigation system for PAO surgery including a range of motion optimization. Investigations showed that the use of CAS systems improve accuracy compared to traditional procedures for PAO [10–12]. Nevertheless, modern CAS systems are not yet widely used. Surgical navigation is mainly based on optical tracking which has an inherent disadvantage that two cameras need a constant visual connection to the instruments and the patient, limiting the working area of involved surgeons in the OR [13]. Additionally, the optical stereo camera takes up a lot of space in the already cluttered working area of a surgeon. Different approaches were presented to overcome this disadvantage. One such approach is electromagnetic tracking [14, 15] which is comparably expensive but provides lower accuracy than optical tracking due to magnetic field distortions.

1.3 Inertial Measurement Units

Recent advancements in microelectromechanical systems (MEMS) made it possible to use inertial sensor-based navigation systems [16–19]. Such inertial measurement units (IMU) are small and comparably cheap. IMUs usually consist of several internal sensors and estimate a full 3D orientation and translation information with respect to an Earth-fixed coordinate system. IMUs are already widely used for navigational tasks, for example car navigation systems based on GPS: in cases when the navigation system lost its connection to the GPS satellite, inertial guidance fills the gap in coverage with what is called dead reckoning [20]. An IMU usually consists of several internal sensors and the orientation and translation is estimated by fusing the information from all internal sensors. Accelerometers which are sensitive to the Earth's gravity are combined with gyroscopes which measure the rate of rotation around a specific axis to estimate the tilt of the device. If full 3D orientation is measured, the sensor setup is usually completed by magnetometers which sense the magnetic field around the device to estimate the device's heading information. More advanced setups exist where the IMU is extended with other sensors such as barometers or GPS receivers.

The orientation and translation is then estimated by fusing the information from all internal sensors which are usually three orthogonally aligned accelerometers, three gyroscopes and three magnetometers to sense the magnetic field in three directions. The output is the orientation with respect to an Earth-fixed coordinate system defined by the magnetic North (x-axis) and the gravity vector (z-axis). The orientation is usually represented as three rotation matrices, the rotation around the x-axis (roll), the rotation around the y-axis (pitch) and the rotation around the z-axis (yaw, see Fig. 1). Estimating the translation is a bit trickier since a double integration of the accelerometers output is necessary to get translational information. The error might seem small when performing this process a few times but rapidly increases over time and generally doesn't give the accuracy many are looking for.

The fusion of the data from all internal sensors is usually performed using a Kalman filter [21]. A Kalman filter is a set of equations which try to estimate the

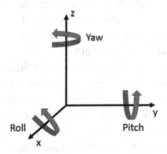

Fig. 1 The three rotations necessary to describe full 3D orientation. The IMU outputs the rotation around the Earth-fixed reference coordinate system's x-, y- and z-axis

state of a discrete-time controlled process. In a recursive matter, it estimates the states in the past, present and in the future, even when the precise nature of the modeled system is unknown [22]. The estimated state at some point in time is adjusted by feedback in the form of potentially noisy measurements from the sensors. This way, a continuous cycle is achieved by first projecting into the future and then improving the value based on actual measurements. The forward projection is based on past values which means that in order for the IMU to produce reliable output, it is necessary for the Kalman filter to build up a history first [23]. This initial warm-up time depends on the environment where the sensor works in.

Behrens et al. [16] proposed an inertial sensor-based guided navigation tool for tumor re-identification. They make use of the fact that in cytoscopic interventions the anatomy of the bladder reduces the degrees of freedom (the bladder opening with the urethra forms a fixed pivotal point). The endoscope movement they try to estimate is therefore limited to a vertical, horizontal and axial rotation as well as translation. Integrating one sensor into endoscopes with different view angles and applying an extended endoscope model they are able to accurately measure orientation angles but experience inaccuracies trying to measure the translation of the endoscope due to temporal drift effects that occur due to double integration of the accelerometer values.

Ren et al. [19] developed a prototype IMU including an extended Kalman filter for tracking hand-held surgical instruments and compared it to optical tracking. In their work, they considered the problem of estimating the gravity and magnetic field under small perturbations and their relationship to the measurements of the gyroscope sensors. They were able to achieve very accurate results, however, they experienced problems with environmental interference and their gyro-assisted compensation was not able to fully compensate for it.

O'Donovan et al. [17] used an inertial and magnetic sensor-based technique for joint angle measurements. The technique makes use of a combination of IMUs attached to the lower extremities to compute joint angles and the ankle joint. Since the IMUs are not depending on a fixed reference coordinate system, their technique may be suitable for use in a dynamic system such as a moving vehicle.

All these systems report problems estimating the heading information from the IMU due to magnetic field distortions. The rotation around the z-axis (heading information) is heavily relying on accurate magnetic field measurements. If the surrounding magnetic field is distorted, the employed Kalman filter relies more on the measurements from the gyroscope and accelerometer, introducing drift errors due to the above mentioned double integration.

Hybrid systems were proposed to overcome the limitations of a single tracking technology [13, 24–26]. Haid et al. [13] presented novel methods combining image processing routines with inertial sensors for surgical navigation and Ren et al. [27] proposed an integrated tracking system for endoscopic surgery combining inertial sensors with electromagnetic tracking to reduce the effect of environmental distortions. However, these systems have an increased complexity and are significantly more expensive.

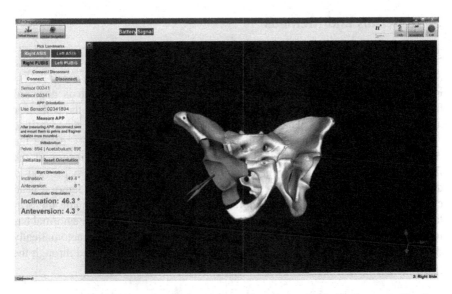

Fig. 2 Our application to visualize the re-orientation of the acetabular fragment. The four landmarks can easily be picked on the computer model. Battery level and signal strength of the sensors is indicated using green (or *red*) marks at the *top*. The updated inclination and anteversion values are shown on the side below the starting values. The acetabular fragment is rotated on the screen to give the surgeon a 3D view of the current state (Color figure online)

In this book chapter, we propose a low-cost system to measure the acetabular fragment orientation during PAO surgery which is solely based on inertial measurements units Fig. 2. We make use of a newly designed device to measure the orientation of the patient's anterior pelvic plane (APP) and to register the orientation of the patient's pelvis and acetabular fragment, respectively, we are able to compute the inclination and anteversion of the acetabular fragment intra-operatively in real-time. Compared to work reported in [16], we only measure rotations and combine readings from two IMUs. Walti et al. [28] use an inertial measurement sensor to track the surgical instrument during pedicle screw placement. Their application makes use of just one sensor and the whole procedure to place the pedicle screw only takes a short amount of time reducing the influence of drift errors. Our work can be compared to [17] with the difference that we use the system in a clinical setting inside the operating room (OR) for a demanding open surgery. To our knowledge, we are the first to use the IMUs to measure the acetabular orientation during PAO surgery, successfully removing the line-of-sight limitation of optical tracking systems.

2   Materials and Methods

2.1   Pre-operatively

For visualization purposes a patient-specific computer model which is acquired from segmented CT data is used. Next, four landmarks are picked on the computer model before the surgery. These landmarks—left and right anterior superior iliac spine (ASIS) as well as the left and right pubic tubercles (PUBIS)—define the APP. Reorientation of the acetabular fragment is modeled as a rotation of the fragment around the center of the femoral head. With the help of a previously developed comprehensive PAO planning system [9] we estimate the femoral head center by fitting a sphere to the femoral head. Additionally, to compute the ace-tabular orientation (anteversion and inclination) the acetabular cup plane normal has to be known. To compute the cup plane, the acetabular rim points are automatically detected using the PAO planning software and a plane is linearly fitted through the points to get the plane normal [9].

2.2   Inertial Measurement Units

Two commercially available MEMS inertial sensors (Xsens Technologies, The Netherlands) fuse data from three internal accelerometers, three gyroscopes and three magnetometers as well as a barometer to output full 3D orientation data in an Earth-fixed coordinate system. The sensor fusion is performed using a variation of the Kalman filter [21]. The orientation data is read into our application and pro-cessed using the software development kit provided with the sensors. As the sensor system used in this work is a commercially available system and therefore not open, we refer the reader to other works using Kalman filter equations and IMUs for more on this topic [29, 30]. It is possible to read the orientation from the sensor in three different representations (Euler angles, rotation matrix and quaternion) which are all equal and represent the same orientation in the Earth-fixed coordinate system. We did internal processing using the quaternion representation, however, in this book chapter, for the sake of simplicity, we use rotation matrices to explain all the computations performed.

2.3   Registration

After acquiring all pre-operative information, the computer model's orientation is registered to the patient's pelvis' orientation. For that task, we designed a new device which aligns a single IMU with the patient's APP (Figs. 3 and 4). Aligned with the device's top plate, the sensor's orientation (local z-axis) represents the APP

Fig. 3 APP measuring device. Three pillars are placed on the *right* and *left* anterior superior iliac spine (ASIS) and one of the two pubic tubercles. The sensor is placed on the top plate which is aligned with the APP

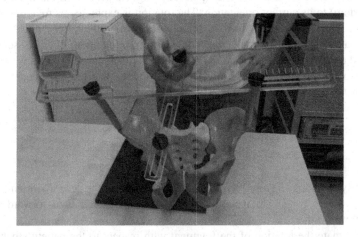

Fig. 4 The newly designed APP measuring device is placed on the patient's pelvis so that the three levers are placed on the two ASIS and one pubic tubercle. The *top* plate will then be aligned with the APP and the sensor can measure its orientation

normal since all three pillars are of equal length and parallel to each other. The computer model is then transformed in a way that its APP (known from the picked landmarks), and therefore the whole model, has the same orientation as the patient's APP/pelvis. One hundred data packets from the wireless sensor (quaternion orientations) are recorded and the orientation of the APP is estimated by taking the mean of all quaternion orientations to filter out noisy data. The algorithm proposed by Markely et al. [31] is used to compute the average quaternion by performing an eigenvalue/eigenvector decomposition of a matrix composed of the given quaternions.

2.4 Sensor Setup

After registration, the two sensors are mounted to the patient: one sensor is attached to the patient's pelvis (Sensor A) and the other is mounted to the acetabular fragment (Sensor B, see Fig. 5). The output from Sensor A and B at time t, represented using a rotation matrix, specifying the orientation of a sensor with respect to an Earth-fixed reference orientation O is then:

$$R_{OA_t} = R_{A_{yaw}} * R_{A_{pitch}} * R_{A_{roll}} \tag{1}$$

$$R_{OB_t} = R_{B_{yaw}} * R_{B_{pitch}} * R_{B_{roll}} \tag{2}$$

With R_{OA_t} and R_{OB_t} being the rotation matrices of sensor A (pelvis) and B (fragment) at time t representing the rotations necessary to bring sensor A and B from the reference orientation O (perfectly aligned with the Earth-fixed coordinate system) to its current orientation. The rotation is expressed as rotations around the three axes of the Earth fixed coordinate system (roll—x-axis, pitch—y-axis, yaw—z-axis). Before reorientation, the starting orientation of each sensor is recorded ($t = 0$). At every time point t after the initialization step ($t = 0$), we take the current rotation matrices from sensor A and B to compute the incremental update between $t - 1$ and t for each sensor:

$$R_{dt_A} = R_{OA_t} * R_{OA_{t-1}}^T \tag{3}$$

$$R_{dt_B} = R_{OB_t} * R_{OB_{t-1}}^T \tag{4}$$

Intuitively speaking, R_{dt_A} and R_{dt_B} define a rotation from the sensor's orientation at time point $t - 1$ back to the reference orientation O and then forward to the orientation at time t.

To estimate the rotation of the fragment with respect to the patient's pelvis we must account for movements of the pelvis during surgery. The rotation of the pelvis (sensor A) is the same as an inverse rotation of the fragment (sensor B) and vice versa. This way, movements of the patient intra-operatively can be compensated by treating the incremental update of the pelvis R_{dt_A} inversely, with the assumption that the rotation center is the same. Therefore, to compute the incremental update of the fragment, we first rotate the fragment by the incremental update from sensor B and then apply an inverse rotation using the incremental update from sensor A:

$$R_{F_{dt}} = R_{dt_A}^T * R_{dt_B} \tag{5}$$

Using $R_{F_{dt}}$, we rotate the acetabular cup plane normal at every time point t and recompute the acetabular orientation as described below.

2.5 Acetabular Orientation

In our application we compute radiographic inclination and anteversion angles [32] which are defined as:

$$inclination = r2d\left(acos\left(dot\left(\overrightarrow{proj}, -\vec{n}_{axial}\right)\right)\right) \tag{6}$$

$$anteversion = r2d\left(acos\left(dot\left(\overrightarrow{proj}, \vec{n}_{cup}\right)\right)\right) \tag{7}$$

where $r2d$ is the function to convert from radians to degree, $\vec{n}_{cup}$ is the normal of the acetabular cup plane defined by the rim points, $\vec{n}_{axial}$ represents the axial plane which is defined as the cross product between the APP normal and the vector connecting the right and left anterior superior iliac spine and $\overrightarrow{proj}$ is defined as follows:

$$\overrightarrow{proj} = \vec{n}_{cup} - dot\left(\vec{n}_{cup}, \vec{n}_{APP}\right) * \vec{n}_{APP} \tag{8}$$

Using $R_{F_{dt}}$ (Eq. 5) we are able to update the orientation of the acetabular cup plane at every time point and then re-compute the necessary plane normal and vectors to compute the inclination and anteversion values. The overall procedure is outlined in below.

- Pre-operative

  - Acquire CT and reconstruct a patient specific 3D computer model of the hip and proximal femur
  - Extract the acetabular rim points from the computer model and fit a plane through these points to get the acetabular cup plane normal [9]
  - Fit a sphere to the femoral head and extract the center of this sphere as the rotation center
  - Pick landmarks on the 3D computer model

- Intra-operative

  - Connect to both sensors and move them slowly for several minutes to warm-up the Kalman filter
  - Measure the orientation of the APP using the new measurement device
  - Mount the two sensors to the patient's pelvis and the acetabular fragment and record initial orientation relative to each other
  - Start re-orientation

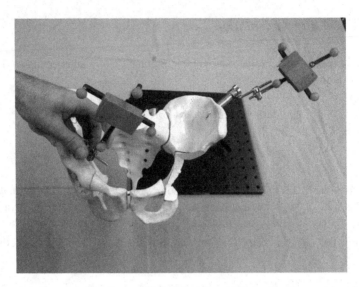

Fig. 5 The sensor setup with one sensor attached to the acetabular fragment and the other to the pelvis. In our setup, we attached the sensors to the optical marker shields to directly compare the optical tracking system with our inertial-based system

2.6 Experiments

In order to validate our system we performed a plastic bone study using 7 hip joints (4R, 3L). The experiments were performed inside an OR to simulate an as-real-as-possible magnetic field environment for the sensors. The osteotomies were directly drawn onto the plastic bones by an orthopedic surgeon and then cut using a coping saw. For comparison and validation purposes, we simultaneously ran a previously developed navigation system using an optical tracking camera as ground truth. After registering the pelvis' APP with the model and recording the starting orientation (Fig. 5), the acetabular fragment was slowly rotated to new positions and every couple of seconds a second person recorded the inclination and anteversion values of both systems simultaneously.

2.7 Statistical Evaluation

To evaluate the feasibility of our inertial sensor-based PAO navigation system for measuring acetabular orientation we defined the following hypotheses:

1. The measurements performed during surgery (anteversion, inclination) of our inertial sensor-based application are **not** significantly different than the measurements performed using the optical tracking-based system during the same procedure.

2. The mean absolute difference between our system and the optical tracking-based system is **less** than five degrees.

For hypothesis number one we treat anteversion and inclination values separately and compare them to the optical tracking-based system using a paired t-test. The hypothesis will be rejected if the p-value is smaller than 0.05. The second hypothesis will be rejected if the mean absolute difference between the two systems is higher than five degrees for anteversion of inclination. Five degrees as a threshold was chosen based on input from our clinical partners to be a reasonable cutoff to evaluate accuracy.

3 Results

Tables 1 and 2 show the results for all seven hip joints (inclination and anteversion). The results from both systems are never statistical significantly different (p-value > 0.05) and therefore we can accept the first hypothesis for all seven hip

Table 1 Experiment results for inclination

Subject	Side	Mean Diff (°)	Min Diff (°)	Max Diff (°)	p-Value	Duration (mm:ss)
P1	R	3.1	0.0	10.7	0.920	09:00
P1	L	3.1	0.1	9.7	0.444	08:48
P2	R	3.3	0.1	9.1	0.365	10:44
P2	L	4.1	0.1	9.7	0.608	08:23
P3	R	3.1	0.0	9.2	0.066	11:48
P3	L	5.2	0.1	14.3	0.065	07:44
P4	R	2.2	0.0	6.3	0.658	16:00

The mean absolute, minimum and absolute difference is shown compared to the optical tracking system. The duration is without warm-up time and simply refers to the time spent on re-orienting the acetabular fragment. There was no significant difference to the optical tracking system observed ($p < 0.05$)

Table 2 Experiment results for anteversion

Subject	Side	Mean Diff (°)	Min Diff (°)	Max Diff (°)	p-Value	Duration (mm:ss)
P1	R	1.6	0.0	6.0	0.644	09:00
P1	L	2.6	0.1	13.5	0.697	08:48
P2	R	3.3	0.1	13.7	0.604	10:44
P2	L	2.3	0.2	7.3	0.755	08:23
P3	R	1.8	0.0	5.4	0.867	11:48
P3	L	2.4	0.0	9.5	0.276	07:44
P4	R	2.3	0.0	6.1	0.406	16:00

The mean absolute, minimum and absolute difference is shown compared to the optical tracking system. The duration is without warm-up time and simply refers to the time spent on re-orienting the acetabular fragment. There was no significant difference to the optical tracking system observed ($p < 0.05$)

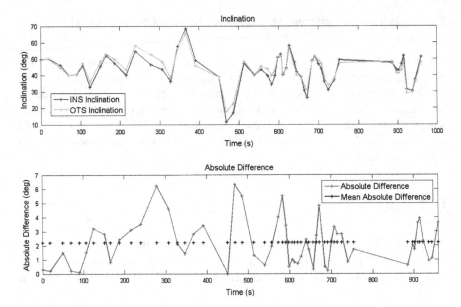

Fig. 6 *Top* Comparison of inclination values between our inertial sensor-based system (*blue*) and the current gold-standard, optical tracking (*magenta*). *Bottom* The absolute difference between the inertial-based and the optical tracking-based system (*red*). The mean absolute difference is shown in *black* (Color figure online)

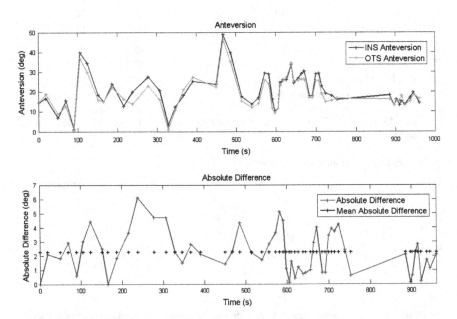

Fig. 7 *Top* Anteversion comparison between our system (*blue*) and the optical tracking-based system (*magenta*). *Bottom* The absolute difference between the inertial-based and the optical tracking-based system (*red*). The mean absolute difference is shown in *black* (Color figure online)

joints. As for hypothesis number two, we can observe that in one case the mean absolute difference was slightly higher than five degrees. Figures 6 and 7 show the comparison with the optical tracking system for every measurement point for one case (Pelvis 4, right) as well as the absolute difference between the two systems over time. The error does not increase over time which would be the case if a systematic drift error would be present from the sensors.

Additionally we calculated correlation for inclination and anteversion separately. The mean correlation was 0.92 for inclination and 0.87 for anteversion, respectively.

4 Discussion

In this work, we demonstrated the feasibility of a low-cost system using IMUs to measure the orientation of the acetabular fragment during PAO surgery, which we compared to the current gold-standard, optical tracking. We defined two hypotheses which were accepted in all but one case (inclination). For that case, the mean absolute difference was higher for inclination ($5.21°$, p-value $= 0.065$) than our goal of less than $5°$. The reason can be found in the acquisition scheme. As can be seen in Figs. 6 and 7, the maximum difference between the two systems is sometimes quite high, increasing the mean absolute difference. For the person holding the acetabular fragment it is difficult to keep a steady position for several seconds. Additionally, the measurement rate of the two systems is quite different and including the visualization update, the update rate of our system is by far lower than the update rate of the optical tracking system. This results in a delay and can lead to big differences for single measurements when recording the values for both systems as the inertial-based system lags behind with the update. This explains why a higher absolute difference in a single measurement can decrease in the next couple of measurements. This update lag will not have an influence in a real surgery: the acetabular fragment can be better kept in a certain orientation for a couple of seconds to check the current rotation since it is stabilized by the femoral head and the joint capsule even though it is completely separated from the pelvis.

We register the patient's APP with the computer model using a newly designed device (Fig. 1). It has three levers which are place on the patient's left and right ASIS and one pubic tubercle (Fig. 3). This registration method is very convenient but might not always be that accurate in a real surgery due to an unknown amount of soft tissue between the bone and the skin surface. In our plastic bone study, this problem is not apparent since we can directly place the device onto the necessary bony structures. However, due to the way we fixated the plastic bones, it could happen that while measuring the APP the whole construct was unstable, which could lead to measurement errors due to shaky movements during registration.

The main concern using IMUs is their systematic error due to temporal drift effects [14]. We experienced issues with magnetic field distortions as was experienced by others. O'Donovan et al. [17] report differences in accuracy depending on if the measured joint angle is mainly based on the sensor's heading information or not and therefore relying on a stable magnetometer measurement. Ren and Kazanzides [19] also report that the magnetometer is vulnerable to environmental interferences. The first couple of seconds, the sensors estimate the quality of the magnetic environment and the Kalman filter adjusts to small magnetic field distortions by trusting less on the magnetometer output and relying more on the output of the gyroscopes to compute the heading information. The sensor needs some time to adjust internally (building up a history for the Kalman filter) and output a stable heading value. According to the manufacturer, the necessary time for the Kalman filter to warm-up is less than two minutes. In our experience, this is not enough to get stable measurements for the IMUs heading value. We set the warm-up time to 12 min which resulted in a stable yaw angle even in an OR. Twelve minutes seem like a lot of time. A possible approach would be to start the connection to the sensors shortly before the osteotomies are performed (OR assistant can establish the connection and slowly move them around for a short time).

Our system successfully removes the line of sight impediment of optical tracking-based navigation systems. However, using IMUs it is not possible to achieve the high accuracy that camera-based systems offer and the warm-up time must be reduced which is part of future work. Nevertheless, the increased convenience and low-cost of our proposed system may help to further push navigation systems into clinical routine.

5 Conclusion

In this chapter we presented a low-cost system to measure the orientation of the acetabular fragment during PAO surgery. We showed the feasibility of using IMUs in a surgical environment using a plastic bone study performed in an OR. Future work will include a cadaver study to validate the clinical usefulness of the proposed system and reducing the warm-up time to improve convenience in the OR.

References

1. Millis M, Kim Y (2002) Rationale of osteotomy and related procedures for hip preservation: a review. Clin Orthop Relat Res 405:108–121
2. Hsieh PH, Chang YH, Shih CH (2006) Image-guided periacetabular osteotomy: computer-assisted navigation compared with the conventional technique: a randomized study of 36 patients followed for 2 years. Acta Orthopaedica 77(4):591–597. doi:10.1080/17453670610012656
3. Langlotz F, Stucki M, Baechler R, Scheer C, Ganz R, Berlemann U, Nolte L-P, Mueller M (1997) The first twelve cases of computer assisted periacetabular osteotomy. Comput Aided Sur 2:317–326
4. Khanduja V, Villar R (2006) Arthroscopic surgery of the hip—current concepts and recent advances. J Bone Joint Surg Br 88-B(12):1557–1566. doi:10.1302/0301-620x.88b12

5. Hazan E (2003) Computer-assisted orthopaedic surgery—a new paradigm. Tech Orthop 18(2):221–229
6. DiGioia A, Jaramaz B, Blackwell M, Simon D, Morgan F, Moody J, Nikou C, Colgan B, Aston C, LaBarca R, Kischell E, Kanade T (1998) An image guided navigation system for accurate alignment in total hip replacement surgery. The Robotics Institute, Carnegie Mellon University
7. Dorr L, Hishiki Y, Wan Z, Newton D, Yun AC (2005) Development of imageless computer navigation for acetabular component position in total hip replacement. The Iowa orthopaedic journal 25:1
8. Langlotz F, Bachler R, Berlemann U, Nolte LP, Ganz R (1998) Computer assistance for pelvic osteotomies. Clin Orthop 354:92–102
9. Liu L, Ecker T, Schumann S, Siebenrock KA, Nolte LP, Zheng G (2014) Computer assisted planning and navigation of periacetabular osteotomy with range of motion optimization. MICCAI 2:643–650
10. Jolles D, Genoud P, Hoffmeyer P (2004) Computer assisted cup placement techniques in total hip arthroplasty improve accuracy of placement. Clin Orthop 426(1):174–179
11. Nogler M, Kessler O, Prassl A (2004) Reduced variability of acetabular cup positioning with use of an imageless navigation system. Clin Orthop 426(1):159–163
12. Ryan JA, Jamali AA, Bargar WL (2010) Accuracy of computer navigation for acetabular component placement in THA. Clin Orthop Related Res 468(1):169–177. doi:10.1007/s11999-009-1003-7 (PubMed PMID: 19629609; PubMed Central PMCID: PMC2795805)
13. Haid M, Kamil M, Chobtrong T, Guenes E (2013) Machine-vision-based and inertial-sensor-supported navigation system for the minimal invasive surgery. AMA conferences 2013—SENSOR 2013. doi:10.5162/sensor2013/P5.3
14. von Jako R, Carrino J, Yonemura K, Noda G, Zhue W, Blaskiewicz D, Rajue M, Groszmann D, Weber G (2009) Electromagnetic navigation for percutaneous guide-wire insertion: accuracy and efficiency compared to conventional guidance. NeuroImage 47(2):127–132
15. Zhang H, Banovac F, Lin R, Glossop N, Wood B, Lindisch D, Levy E, Cleary K (2006) Electromagnetic tracking for abdominal interventions in computer aided surgery. Comput Aided Surg 11(3):127–136
16. Behrens A, Grimm J, Gross S, Aach T (2011) Inertial navigation system for bladder endoscopy. In: Engineering in medicine and biology society, Engineering in Medicine and Biology Society, EMBC, Annual International Conference of the IEEE
17. O'Donovan KJ, Kamnik R, O'Keeffe DT, Lyons GM (2007) An inertial and magnetic sensor based technique for joint angle measurement. J Biomech 40(12):2604–2611. doi:10.1016/j.jbiomech.2006.12.010
18. Rebello K (2004) Application of MEMS in surgery. Proc IEEE 92(1):43–55
19. Ren H, Kazanzides P (2012) Investigation of attitude tracking using an integrated inertial and magnetic navigation system for hand-held surgical instruments. IEEE/ASME Trans Mechatron 17(2):210–217
20. Scannell B (2012) MEMS Enable Medical Innovation. Analog Devices, Technical Article MS-2393
21. Kalman RE (1960) A new approach to linear filtering and prediction problems. Journal of Fluids Engineering 82(1):35-45
22. Welch G, Bishop G (2006) An introduction to the Kalman filter 2006. Available from: http://www.cs.unc.edu/~welch/media/pdf/kalman_intro.pdf
23. Welch G, Bishop G (2006) An introduction to the kalman filter. University of North Carolina: Chapel Hill, North Carolina, US
24. Beller S, Eulenstein S, Lange T, Hunerbein M, Schlag PM (2009) Upgrade of an optical navigation system with a permanent electromagnetic position control: a first step towards "navigated control" for liver surgery. J Hepato-Biliary-Pancreatic Surg 16(2):165–170. doi:10.1007/s00534-008-0040-z

25. Claasen G, Martin P, Picard F (2011) High-bandwidth low-latency tracking using optical and inertial sensors. In: Proceedings of the 5th international conference on automation, robotics and applications, pp 366–371

26. Mahfouz M, Kuhn M, To G, Fathy A (2009) Integration of UWB and wireless pressure mapping in surgical navigation. IEEE Trans Microw Theory Tech 57(10):2550–2564

27. Ren H, Rank D, Merdes M, Stallkamp J, Kazanzides P (2012) Multisensor data fusion in an integrated tracking system for endoscopic surgery. IEEE Trans Inf Technol Biomed 16(1):106–111

28. Walti J, Jost G, Cattin P (2014) A new cost-effective approach to pedicular screw placement. MICCAI—Lect Notes Comput Sci 8678:90–97

29. Roetenberg D, Luinge HJ, Veltink PH (2005) Compensation of magnetic disturbances improves inertial and magnetic sensing of human body segment orientation. Neural Systems and Rehabilitation Engineering, IEEE Transactions on, 13(3):395–405

30. Sabatini A (2006) Quaternion-based extended Kalman filter for determining orientation by inertial and magnetic sensing. IEEE Trans Biomed Eng 53(7):1346–1356. doi:10.1109/TBME.2006.875664

31. Markley F, Cheng Y, Crassidis J, Oshman Y (2007) Averaging Quaternions. J Guid Control Dyn 30(4):1193–1196

32. Murray D (1993) The definition and measurement of acetabular orientation. J Bone Joint Surg Br 75-B(2):228–232

Computer Assisted Hip Resurfacing Using Patient-Specific Instrument Guides

Manuela Kunz and John F. Rudan

Abstract Hip resurfacing is considered to be a viable alternative to total hip replacement in the treatment of osteoarthritis, especially for younger and more active patients. There are, however, several disadvantages reported in the literature, due to difficult surgical exposure and the technical challenges of the intraoperative procedure. Surgical errors, such as notching of the femoral neck, tilting of the femoral component in excess varus, or improper prosthesis seating, can result in early failure of the procedure. In this chapter we discuss the use of patient-specific instrument guides as an accurate and reliable image-guided method for the placement of the femoral and acetabulum components during hip resurfacing. The outcome of patient-specific guided procedures depends on many factors, starting with the accurate depiction of the anatomy in a preoperative image modality, the careful selection of registration surfaces for the guide, the accuracy of the guide creation, as well as the reliability of the guide registration intraoperatively. We will discuss in detail how current research is addressing these points in patient-specific instrument guided hip resurfacing applications.

1 Introduction

Total Hip Arthroplasty (THR) is considered the number one treatment choice for patients with advanced hip disease. However, when younger and more active patients started to seek hip replacement in an attempt to restore an active life style, disadvantages of THR—such as limited revision possibilities and partial range of hip motion—became a restrictive factor and accelerated research into more bone-preserved hip replacement options. In 1977, Amstutz et al. [1] first introduced

M. Kunz (✉)
School of Computing, Queen's University, Kingston, ON, Canada
e-mail: kunz@queensu.ca

J.F. Rudan
Department of Surgery, Queen's University, Kingston, ON, Canada

© Springer International Publishing Switzerland 2016
G. Zheng and S. Li (eds.), *Computational Radiology
for Orthopaedic Interventions*, Lecture Notes in Computational
Vision and Biomechanics 23, DOI 10.1007/978-3-319-23482-3_17

a method for total hip articular replacement, in which a cemented cobold (Co) and chromium (Cr) femoral component articulated with an all-polyethylene acetabulum component. However, with a 10-year survival rate of only 61.5 % [2], the prosthesis was far inferior to THR. In the early 90s it was established that increased poly-ethylene wear debris, generated by large femoral heads articulating with ultra-high-molecular-weight polyethylene (UHMWPE), was the main reason for failure of the early designs, and low-wear bearing metal-on-metal became the standard material for Hip Resurfacing Arthroplasty (HRA). Since then, 10-year survival rates of between 92.9 % [2] and 99.7 % [3] have been reported, making hip resurfacing arthroplasties a valuable alternative to total hip arthroplasies, especially for younger and active patients.

The advantages of HRA compared to THR include: preservation of femoral bone stock (ease of revision) [4], lower dislocation risk [5], as well as improved Range of Motion [6]. Studies measuring patient satisfaction found that the outcome of the procedure met patients' expectations in terms of pain relief and improvement of hip function [7, 8]. On the other hand, the literature reports a series of postoperative problems unique to HRA such as femoral neck fractures, adverse local tissue reaction, and increased blood metal-ion levels.

1.1 Femoral Component Positioning

Many studies have reported femoral neck fractures as the main reason for early failures of hip resurfacing procedures [9–11]. Notching of the femoral neck during preparation of the proximal femur, it was found, increases the risk for postoperative fractures [12]. Finite element analysis [13] has demonstrated that the stress riser created by the femoral component is due to the disparity of the high-stiffness CoCr prosthesis versus the low-stiffness femoral neck. The same study showed that superior femoral neck notching will further concentrate stress at the rim of the femoral component, increasing the risk of neck fracture even further.

Further, mal-alignment of the femoral component was also identified as a stress riser for the femoral neck, leading to greater risk for femoral neck fractures. Various studies have shown the importance of aligning the femoral component in a valgus orientation (with respect to the anatomical neck-shaft angle) for optimal postop-erative outcome [3, 9, 11, 14].

Besides the alignment, the seating of the femoral component at the prepared proximal femur influences the postoperative fracture risk. For optimal stabilization of the trabecular bone, the femoral component needs to completely cover the reamed cancellous proximal femur bone [11, 15].

During the preparation of the femur, a minor amount of bone of the proximal head is resected (Fig. 1) to allow for the thickness of the femoral component. The under- or over-resection of the proximal femoral head can result in changes in the femoral

Fig. 1 Depth of the resection
of the proximal head to allow
for the femoral component
thickness

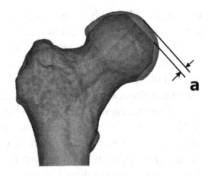

offset[1] and leg-length postoperatively. The literature demonstrates that for an optimal functional outcome, leg-length restoration [16, 17] and femoral offset [18–20]
play an important role. Recently, various studies have shown offset and leg-length
changes after HRA. Silva et al. [21] found that compared to preoperative values, the
postoperative offset significantly decreased by a mean of 0.8 mm, while the postoperative length of the treated leg significantly increased by a mean of 4.6 mm. In an
in vivo study, Loughead et al. [22] found a significant reduction of offset from a
preoperative average of 49.4 to 44.9 mm postoperatively. This reduction in postoperative femoral offset was also observed by Girard et al. [23]: they measured an
average decrease of 3.3 mm. The authors of this study also observed an average
shortening of leg length of 1.9 mm. In contrast, a study with 28 patients found an
average postoperative leg length increase of 4.9 mm [24]. The authors of this study
also observed a decrease of femoral offset by 1.3 mm on average.

Although the published results for offset and leg-length changes following HRA
are often significantly smaller compared to THR studies, they still indicate
non-optimal postoperative biomechanical reconstruction of the hip. In particular the
decrease in femoral offset can negatively influence the hip's function.

1.2 Acetabular Component Alignment

Although the role of the femoral component alignment with respect to postoperative
outcome is well known, the role of acetabulum component alignment is less certain.
Recent studies suggest that improper alignment of the acetabulum component might
be connected to increased metal ion levels, as well as unexplained postoperative
pain. Abnormal periprosthetic soft-tissue lesions (also know as pseudotumor, or
adverse reaction to metal debris) have been reported after metal-on-metal HRA with
an incidence rate of up to 4 % [25] for asymptomatic hips and were linked to an
increased blood, serum and hip aspirate level of cobold (Co) and chromium

---

[1]The femoral offset is defined as the distance between the center of rotation of the femoral head
and the long shaft axis of the femur, measured perpendicular to the femoral long axis.

(Cr) [26, 27]. This correlation indicated that these abnormal periprosthetic soft-tissue lesions could have been the result of increased component wear, a hypothesis confirmed in retrieval studies [28, 29]. These studies also indicated that edge-loading[2] with the loss of fluid-film lubrication may be a likely reason for the wear. Follow-up studies have investigated this connection successfully [30, 31]. Various authors have found connections between acetabulum component alignment and the occurrence of edge-loading and high metal-ion levels [27, 31–34]. Recent publications investigate optimal acetabulum component alignment. Langton et al. concluded [27] as a result of an in vivo study that there is a correlation between acetabulum inclination angle and high levels metal ions, but that the maximum limits for inclination depended on the prosthesis type. In 2013 Liu and Gross [35] suggested that the "safe zone" for the acetabulum inclination angle varies depending on component size. In 2015 Mellon et al. [12] suggested that optimal acetabulum component alignment is patient-specific.

1.3   Image-Guided Hip Resurfacing

Most of the limitations and problems discussed above are controlled by the surgeon's actions intraoperatively, with only small margins of error. Therefore, it is not surprising that HRA is deemed a technically challenging procedure with a significant learning curve [11, 36]. Various authors have discussed the advantages of using image-guided, or computer-assisted systems to help surgeon improve the accuracy of HRA.

In 2004, the first clinical results using a fluoroscopy-based navigation system for positioning of the femoral component during hip resurfacing was published [37]. In this study, four intraoperatively registered fluoroscopy images of the hip were used to navigate the insertion of the femoral central pin. Deviation between the intraoperative targeted and the postoperatively achieved central pin alignment in 31 cases were measured using standard postoperative x-rays. The reported average was 2.6°, with no postoperative complications reported. However, the authors did identify an increase in operating time of 10–15 min in the last 10 cases of the series.

El Hachmi and Penasse [38] compared mid-term postoperative outcomes between patients where the conventional method for femoral guide pin placement was used and patients, in which the femoral guide pin was placed using the BrainLab Hip Essential navigation system (BrainLAB AG, Feldkirchen, Germany). In this imageless navigation system a point/surface acquisition is used to generate a three-dimensional (3D) model of the femoral head and neck and to measure the anatomical 3D neck-shaft angle. Although patients from both groups showed significant mid-term clinical improvement, the authors found a significant reduction of

---

[2]Contact between the femoral and acetabulum component at the edge of the acetabulum component.

outliers in femoral pin placement for the navigated group in both anterio-posterior (AP) and lateral postoperative x-rays.

In a similar study by Bailey et al. [39], 37 patients who underwent HRA, in which the femoral central pin was navigated using the imageless BrainLab Ci ASR navigation system (BrainLAB AG, Feldkirchen, Germany), were followed up with a 4-month postoperative CT scan. The measured average deviation between the targeted and postoperatively measured femoral component angle in the frontal plane was 0.5°.

In 2007, Hodgson et al. [40] reported improved repeatability in varus/valgus placement of the femoral central-pin when using a CT-based, opto-electronic navigation system versus manual technique. Further, there was no significant dependence on surgeon skill level (in contrast to the manual technique), and surgical time was significantly reduced in the navigated group. However, the authors of this in vitro study also noted a reduced reproducibility in version alignment of the central-pin of the navigated group, compared to the manual group.

The influence of using a navigation system on the learning curve for femoral central-pin placement during HRA was investigated in an in vitro study by Cobb et al. [41]. Students inexperienced in HRA performed central-pin placement, either with conventional methods or by using the CT-based, mechanically tracked Wayfinder system (Acrobot Co Ltd, London, UK). After each procedure the femur model was reregistered and a CT-based navigation system (Acrobot Co Ltd, London, UK) was used for 3D evaluation of the pin placement. The authors found that students using the navigation method performed the placement three times more accurately compared to the non-navigated group, and suggested that navigation may play a major role in reducing the length of the learning curve for HRA.

Similar observations about the reduction of learning curves using the imageless BrainLab navigation system were published by Romanowski and Swank [42], as well as Seyler et al. [43].

These published results suggest that computer assistance can help achieve higher accuracy and/or precision during the intraoperative process, and could reduce the learning curve for inexperienced surgeons in hip resurfacing. However, disadvantages with computer assistance include additional technical equipment in the operating theater; matching the intraoperative action to the imaging modality; and a time-consuming intraoperative registration process.

In 1994, Rademacher et al. [44] first described individual (or patient-specific) templates as an easy-to-use and cost-effective alternative for computer-assisted orthopaedic surgeries. The principle of the individualized templates was to customize surgical templates based on 3D reconstruction of patient-specific bone structures. Small reference areas of these bone structures were integrated into the template and an instrument guidance component (e.g., drill sleeve, etc.) was attached. By this means, the planned position and orientation of the instrument guide in spatial relation to the bone was stored in a structural way, which could then be reproduced intraoperatively by fitting the references areas of the template to the bone. Other authors subsequently published the results of various applications of individualized templates including pedicle screw fixation [45], computer-guided reduction of acetabulum fractures [46], and cervical pedicle screw placement [47].

Although this novel method was well-received in the research community, it did not transfer to clinical use, mostly due to the inaccessibility of the prototype technology used to create such templates. However, with the development of inexpensive and more accessible additive manufacturing technologies in the last 10 years, patient-specific templates have made a strong comeback in orthopaedic research and clinical use [48–52].

In the following sections, we describe how individualized templates, also known as patient-specific instrument guides, are currently used to navigate hip resurfacing procedures.

2   Patient-Specific Instrument Guides for Femoral Central-Pin Placement

Most published applications for patient-specific instrument guides in hip resurfacing are in the placement of the femoral central-pin. In the majority of hip resurfacing systems, the placement of the femoral central-pin (also know as the guide-wire) is a crucial step for the accuracy of femoral component alignment since it identifies the final femoral component orientation, as well as 2 of the 3 degrees of freedom for femoral component positioning. Only the position of the component along the central-pin axis is not directly determined during the central-pin placement.

Conventional instrument sets provide mechanical guides designed to help the surgeon to connect intraoperative actions to preoperatively planned pin alignment. However, the preoperative planning is often performed on a 2D x-ray, which may result in possible projection errors. Furthermore, the mechanical guidance tools are used in a very limited surgical exposure and are therefore challenging to use. Last but not least, mechanical guidance tools are designed to provide navigation for an average patient population, and are therefore not necessarily optimal for a specific patient.

Various research groups have published methods and results for patient-specific femoral central-pin guidance tools to replace general mechanical guidance tools. In a review of the literature it is apparent that the proposed methods can be split into two different groups of patient-specific femoral pin placement guides. They mainly differ in the anatomical area in which the guide is registered. While some authors choose the femoral head as a registration surface, other authors prefer the femoral neck area for registration of the guide. So far, published accuracy studies do not show a substantial difference between the two types of guides.

In 2013 Du et al. [53] published the result of a randomized study in which the accuracy of femoral central-pin placement using a patient-specific guide compared to conventional pin placement techniques was investigated. In this study, the guide was registered to the femoral head surface. In an postoperative radiographic analysis of 34 patients, the authors found that using the guide significantly improved the accuracy of the femoral component neck shaft angle with an average of 137° (targeted angle was 140°), compared to an average angle of 121° for the conventional surgical group. Using a similar design for the patient-specific instrument guide, Zhang et al. [54] also found a significant increase of accuracy in femoral component placement. In their randomized study of 20 patients, the conventional method for femoral pin placement was compared to the placement using a patient-specific guide, which was registered to the medial aspect of the femoral head. Measurements of postoperative x-rays showed, that the use of the guide in this study significantly reduced the average deviation between the neck-shaft angle and the actual implanted short neck-shaft angle from 10.2° for the conventional group to 1.3° in the navigated group.

In an in vitro study, Olsen et al. [55] compared the outcome of conventional, imageless optoelectronic-guided and patient-specific instrument-guided techniques. The guide in this application registered mainly to the femoral head, but it also contained a small registration area around the femoral head-neck junction. Pin placement accuracy was assessed by anteroposterior and lateral radiographs and was defined as the absolute mean deviation from the planned alignment values. Results of this study did not show an improvement in the femoral pin placement in the coronal plane using the patient-specific guide (average error of 6.4°) compared to the conventional method (average error of 5.5°). However, the authors found the average error of 1.3° for the imageless optoelectronic guided group was significantly smaller compared to both the conventional and patient-specific guided group. In contrast, when comparing the deviations in version angle, the patient-specific guided alignment was, with an average error of 1.0°, significantly better compared to the conventional group (5.6°). The same authors also compared the femoral component alignment error in the coronal plane using patient-specific guided and imageless optoelectronic guided procedure in a clinical study, and found no significant difference between methods.

All three groups—Du et al., Zhang et al., and Olsen et al.—suggested the use of a patient-specific femoral alignment guide, which was mainly registered on the femoral head. In comparison, Sakai et al. and Kitada et al. proposed the use of a patient-specific guide for femoral central-pin alignment, which was designed to fit on a small portion of the femoral head, to the posterior aspect of the femoral neck, as well as to the intertrochanteric region. Kitada et al. [56] tested the accuracy of femoral central-pin placement using the patient-specific guide in a laboratory study with synthetic femoral bone models, and compared the outcomes to errors in pin placement using both a CT-based optoelectronic navigation system and a conventional method. The analysis of postoperative CT obtained from the femoral bone models showed that the patient-specific guides had a significant influence in reducing the error of the stem-shaft angle: 2.4° compared to the stem-shaft angle error of −5.3° for the conventional method. No significant changes in the measurements of femoral pin placement were found between the patient-specific and optoelectronic guided methods. Using the same guide design, Sakai et al. [57]

investigated in an in vitro study the influence of the size of the surface for guide registration on the accuracy of femoral central-pin placement. The authors found that guides in which the contact area was approximately 25 % larger significantly reduced the absolute error of the insertion point for the central-pin compared to guides registering at the same anatomical surfaces, but with smaller contact areas. However, the authors did not see any significant changes in the orientation errors of the central-pin between the two guided groups, with absolute errors for neck-shaft angle of 2.6° in the smaller contact guides and 0.9° in the guides with larger contact areas.

Over a period of more than 6 years of research with patient-specific guides for femoral central-pin alignment, we changed our design from a guide that had an equal amount of femoral head and neck registration area to a design that relies mainly on the registration on the femoral neck (Fig. 2). In an early accuracy study [58] with 45 patients in which guides of the earlier design were used, we measured using a CT-based optoelectronic tracking system that in average, the intraoperatively achieved central-pin alignment deviated 1.1° in the frontal plane and 4.3° in the transverse plane, compared to the planned pin alignment. For the entrance point, we measured an average error of 0.1 mm in the frontal plane and 3.5 mm in the transverse plane. Analyzing our results, we suspected that a mis-registration of the guide on the femoral head was the reason for the substantially higher deviations in the version alignment and the anterior-posterior placement of the entrance point.

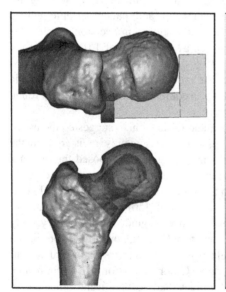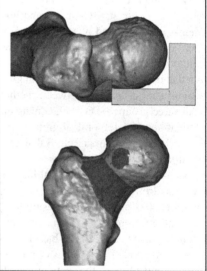

Fig. 2 Changes in patient-specific guide design for femoral central-pin placement. On the *left side* an older design, in which the guide registered to equal parts of the head and neck. On the *right side* the newer design, in which only a minor part of head is part of the registration surface. The *green areas* in the lower images represent the registration surfaces for each design (Color figure online)

Based on this hypothesis, we changed the design to minimize the involvement of the femoral head anatomy in the registration of the guide, and instead increased the contact area in the lateral aspect of the femoral neck. In 2011 [59], we reported on a consecutive study with 80 patients in which the alignment error for the central-pin was reduced to 0.05° in the frontal plane and 2.8° in the transverse plane. We found errors in the entrance point for the central-pin of 0.47 mm in the frontal plane and 2.6 mm in the transverse plane. Both alignment and positioning of the central-pin in the transverse plane seemed to have improved using the new guide design. However, we did not perform a direct comparison study in which we statistically evaluated the differences between the two guide designs.

From the 80 patients, 72 were operated with an anterolateral approach and 8 patients with a posterior approach. When analyzing our data with respect to approach, we found a tendency for the final pin alignment to be more retroverted and in valgus for the posterior approach and anteverted for the anterolateral approach. Also the direction for the deviation in the entrance point seem to be depending on the approach; as we found a more superior pin entrance for the posterior approach and more anterior for the anterolateral approach. A further investigation of the anterolateral approach cases revealed, that there was a significant correlation between increased anteversion deviation and anterior misplacement of the entrance point. This correlation, together with results from changes in error directions between both approaches, suggests that the errors in the anterolateral approach were the result of inaccurate registration of the patient-specific guide in the medial part of the anterior femoral neck and/or parts of the femoral head. Although we eliminated most of the articular surface from the registration, our data implies that there are still segmentation uncertainties in the femoral neck area. However, on the anterior femoral neck, especially on the junction between head and neck, there are often osteophytes (small bony protrusions), which might be the reason for the increased segmentation errors in this area. To investigate this relationship we performed further studies. The results of our investigation will be discussed in a later section of this chapter.

The following case report uses a patient-specific guide for femoral pin replacement, registered to the anterior aspect of the femoral neck.

2.1 Case Report

The patient, a 42-year-old male, presented in the orthopaedic clinic with a two-year history of right hip pain and a moderate degree of arthritic change in his hip. Once treatment options were discussed and the risks and benefits of hip arthroplasties reviewed, the patient decided on an HRA (Birmingham Hip Resurfacing System, Smith and Nephew, Memphis, USA) for the right hip, and consented to use of a patient-specific guide for femoral central-pin placement.

2.1.1 Preoperative Planning

A Computed Tomography (CT) scan of the patient's pelvis was obtained prior to the scheduled procedure. The scans were obtained in helical mode, with a slice thickness of 2 mm at 120 Kvp. During the scan, the patient was positioned as Feet-First-Supine. Using a commercially available software package Mimics (Materialise, Leuven, Belgium), three-dimensional surface models of the proximal femur and the acetabulum of the affected side were created (Fig. 3) and saved in steriolithography format (stl). For planning of size, position and orientation of femoral component, virtual 3D models were loaded into Mimics software and their position and orientation was manually manipulated until the surgeon was satisfied with the positioning.

The planned component size of 54 mm in diameter and the position of the femoral component were verified by virtual reaming of the femoral head (subtraction of the component model from the femur model) to identify any potential risk for notching. Furthermore, outlines of the planned component were superimposed onto orthogonal views of the CT dataset to identify any potential postoperative risk due to bone cysts in the femoral head (Fig. 3).

Based on the final planning for the femoral component, the central-pin orientation was determined with 134° varus and 4.05° anteversion.

For creation of the patient-specific guide, the virtual femur model, as well as the central-pin planning data, were loaded into custom-made software and displayed to

Fig. 3 Preoperative planning. *Top left* A model of the femoral component is superimposed onto the femur model. *Top right* Femur model is virtual reamed. *Bottom* Outlines of the femoral component are superimposed onto the CT dataset of the hip

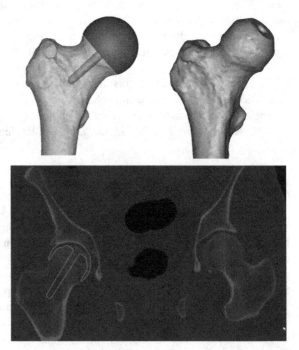

the user. The user selected the size and position of the registration component of the guide. Bone surface for guide registration was chosen with respect to the following criteria:

- Surgical approach: The registration surface for the guide should be completely accessible. The chosen approach for this patient was anterolateral.
- Registration stability: A sufficient number of significant anatomical landmarks had to be covered to allow for unique positioning of the guide intraoperatively.
- Segmentation uncertainty: Anatomical surface areas with a higher potential for segmentation uncertainties (due to a missing joint gap between femur and acetabulum, or known region of osteophytes such as the femoral head-neck junction) must be avoided.

Based on these criteria, the registration part of the guide was constructed from two subcomponents. The first component was oriented along the anterior femoral neck, which ensured stable position and orientation of the guide along the neck axis. Because of the segmentation uncertainty in the articular surface of the head and head-neck junction, these regions were, as best as possible, eliminated from the registration surface, as shown in Fig. 2. The second registration subcomponent was a region oriented perpendicular to the neck axis, and positioned on the lateral aspect of the femoral neck. This ensured rotational stability around the neck (Fig. 4). To increase the overall stability of guide registration, this second registration component enveloped roughly 120° of the lateral femoral neck, measured with respect to the femoral neck axis.

Both registration subcomponents were united and a drill-guidance component was attached to the medial side of the guide. A drill guide channel was inserted into this guidance component, which was oriented along the planned central-pin trajectory as shown in Fig. 4.

A physical model of the patient-specific guide was created using a rapid prototype machine (dimension SST, Statasys, Inc., Eden Prairie, MN, USA). The material used for this 3D printing process was a thermo-plastic acrylonitrile

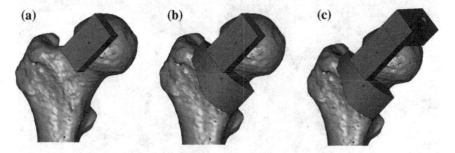

Fig. 4 Patient-specific instrument guide for femoral central-pin placement. **a** First registration subcomponent along the neck axis; **b** Second registration subcomponent on the lateral aspect of the anterior neck; **c** Attachment of the drill-guide component

butadiene styrene (ABS). Finally, the patient-specific drill guide and a plastic model of the patient's femur were gas plasma sterilized (STERRAD Sterilization System, Advanced Sterilization Products, a division of Ethicon US, LLC., Irvine, CA, USA), and labeled before being sent to the operating theater.

2.1.2 Intraoperative Procedure

After the patient was brought into the operating theater, a spinal anesthetic was administered and he was placed in a peg board in the left lateral decubitus position with the right side up. His right hip was prepped and draped in the usual fashion. An antero-lateral approach was performed and the femoral head was dislocated (Fig. 5a). Inspection of the femoral head and neck showed no significant osteophytes.

The surgeon fitted the patient-specific drill guide to the corresponding bone surface at the proximal femur (Fig. 5b) and a conventional metal drill sleeve was inserted into the drill guide channel (Fig. 5c). The central-pin was then drilled using a conventional power drill (Fig. 5d), after which the drill sleeve and the patient-specific guide was removed. In accordance with the conventional procedure,

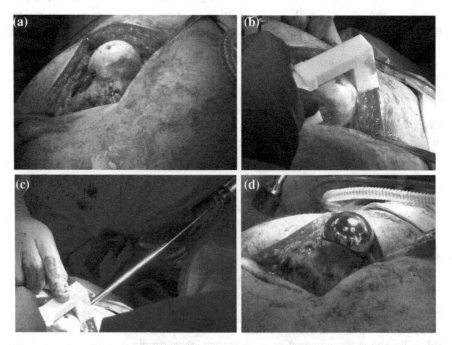

Fig. 5 Intraoperative procedure. **a** Surgical exposure of the proximal femur. **b** Registration of the patient-specific guide. **c** Drilling of the femoral central-pin, navigated by the patient-specific guide. **d** Final femoral component in place

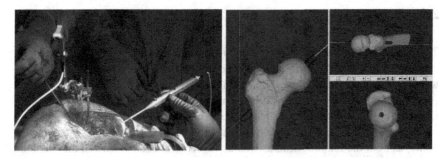

Fig. 6 Accuracy measurement for achieved central-pin alignment. *Left side* After the attachment of an optoelectronic marker and successful registration a tracked pointing device was aligned inside the central-pin hole. *Right* Position and orientation of the tracked pointer were stored with respect to the CT coordinate system and deviation between planned and achieved central-pin alignment were determined

the central-pin was now overdrilled and the guide pin was removed. To measure the achieved central-pin alignment for the purpose of accuracy measurement, we utilized an optoelectronic tracking system (Certus Optotrak, Northern Digital Inc., Waterloo, Canada), with a company reported 3D accuracy of 0.1 mm and a resolution of 0.01 mm. After an optoelectronic marker was attached to the proximal femur, a registration between anatomy and CT model was performed, using a robust method for surface-based registration [60] with reported submillimeter root-mean-square error in the presence of spurious data. After successful registration a tracked pointing device was used to capture the 3D position and orientation of the femoral guide hole (Fig. 6) in the CT coordinate system. After storing this information, the optoelectronic marker was detached from the femur and the remaining steps for the preparation of the proximal femur were performed in the conventional manner. After final preparation and verification, the femoral component was filled with bone cement and impacted into position (Fig. 5, bottom right).

2.1.3 Results

The intraoperatively achieved central-pin alignment was 134.1° varus and 7.05° anteversion, only slightly more varus (0.12°) and 3° more anteverted compared to the planned alignment. When comparing the entrance point on the proximal head we found that the final central-pin position was 0.8 mm more distal and 0.2 mm more anterior compared to the planned positioning.

During the final steps of the proximal femur preparation, the surgeon chose to change the planned femoral component size from 54 mm diameter to a 52 mm diameter femoral head.

2.2 Discussion

The application of patient-specific instrument guides for the navigation of femoral central-pin placement during HRA procedures shows promising results in laboratory, in vitro and in vivo studies. So far, all studies that compared outcome between patient-specific guided to conventional methods have shown an increase in accuracy for pin placement using patient-specific instrument guides.

There seems to be so far no agreement about optimal registration areas for the guide, as both the femoral head as well as the femoral neck is used in the literature and similar results are measured with both guide designs. In any case, we believe that patient-specific guide for the navigation of femoral central-pin placement has the strong potential for routine clinical use, since it provides for a complex and challenging surgical action an easy-to-use and accurate solution. However, for the introduction into a clinical routine use, more research in the complete system, containing the preoperative preparation of the guide, as well as the intraoperative procedure, needs to be done. So far studies have only investigated the intraoperative procedure and have not taken time and complexity of the preoperative guide design into account. It might be from interest to review and develop procedure and algorithms for improved and efficient preoperative patient-specific guide design methods.

3 Patient-Specific Instrument Guides for Femoral Resection Depth Navigation

Although the use of patient-specific instrument guides for central-pin placement is well discussed in the literature, so far, the use of image-guided methods for navigation of proximal bone resection, and with this the position of the femoral component along the central-pin, is not investigated widely. In a randomized pilot clinical trial, we investigated whether adding simple navigation features to our femoral central-pin patient-specific guide could improve the accuracy of proximal femur head resection.

We performed a randomized trial with 10 patients scheduled for a patient-specific instrument guided HRA. The same surgeon operated on all patients and the prosthesis used in all procedures was the Birmingham Hip Resurfacing Implant (Smith and Nephew, Memphis, USA). In 6 of these 10 patients, the patient-specific guide used to navigate the femoral central-pin placement also had the resection depth navigation marks integrated as described below; while in the remaining 4 patients, these marks were missing. In these patients, the surgeon used the conventional methods to identify bone resection levels. Selection was randomized and the surgeon was not aware at the time of the planning whether the patient was in the conventional or navigated depth group.

During the preoperative planning, the surgeon identified the final position of the femoral component in addition to the location and orientation of the femoral central-pin and the femoral component size. To aid the surgeon in this planning step, a virtual 3D model of the femoral component was superimposed onto the femur model (Fig. 3). Changes in positioning along the central-pin axis were supported by the user interface. If required, horizontal and vertical offset could be measured in a frontal view.

During the virtual creation of the patient-specific instrument guide, notches were incorporated into the anterior head-neck part of the guide, which were aligned with the lower edge of the planned femoral component (Fig. 7). Intraoperatively, after the guide was registered to the proximal femur, an electrosurgical knife and/or surgical marking pen was used to transfer the position of the integrated notches onto the femoral head/neck junction (Fig. 8a, b).

During the use of the sleeve cutter, used to ream the femoral head, these marks were used to identify the depth of the reaming (Fig. 8). Once the teeth of the sleeve cutter were aligned with the marks on the femoral head, reaming was stopped. In the Birmingham Hip Resurfacing system, the sleeve cutters, which are unique for each femoral component size, have lines that mark the resection height for the proximal femur cut. With the sleeve cutter advanced to the navigated reamer stop, a surgical marking pen was used to mark the resection line on the bone surface through the 'window' in the sleeve cutter (Fig. 8d). For resection of the proximal bone, the plane cutter is advanced over the guide rod until it is aligned with the resection line on the bone surface. Final steps of preparing the proximal femur were performed in the conventional manner.

In all 10 cases, an optoelectronic navigation system was utilized for 3D evaluation of proximal femoral resection height. A Certus Optotrak (Northern Digital

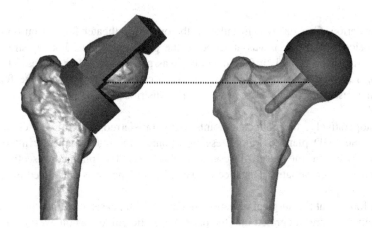

Fig. 7 Navigation of proximal femoral resection depth. Notches in the patient-specific instrument guide were aligned with the lower edge of the planned femoral component, to mark the stop of the required reaming of proximal bone

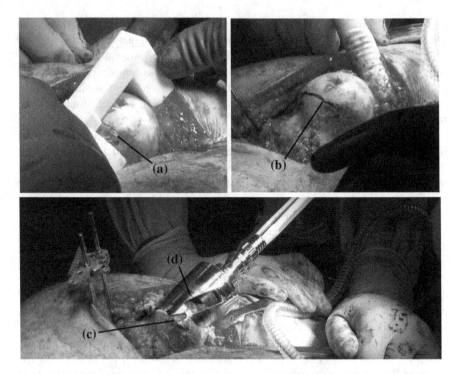

Fig. 8 Navigated reaming. *Top left* With the patient-specific instrument guide in place, the level of the integrated marks were transferred onto the femoral head using a electrosurgical knife (**a**). *Top right* A surgical pen was used to make the marks more apparent on the head (**b**). *Bottom* The sleeve cutter was advanced until the teeth aligned with the femoral head marks (**c**). In this position, the *marks* in the sleeve cutter (**d**) were used to mark the resection line for the plane cutter on the proximal head

Inc., Waterloo, Canada) was installed in the operating theater for this purpose and an optoelectronic marker was attached to the proximal femur. For registration, a combined pair-point and surface matching was performed. After successful registration, a tracked pointing device was used to collect 4–7 points on the femoral proximal resection plane. Care was taken to distribute these points as widely as possible.

Postoperatively, the collected points were transferred into the CT coordinate system and a 3D plane (surgical resection plane) was mathematically fitted. The distance between the surgical resection plane and the planned resection was determined and evaluated. Distance between both planes was calculated along the central-pin axis.

We found that the surgical resection height in the six cases in which the resection height marks were integrated into the patient-specific guide was on average 0.4 mm larger (range, 3.2 mm smaller to 1.9 mm larger) versus the planned resection height. In comparison, in the group in which resection height was navigated using conventional methods, the resection height was on average 3.6 mm larger (range, 0.7 mm larger to 6.5 mm larger) compared to the planned resection height. The difference in resection height errors in these two groups was significant ($p < 0.02^3$).

3.1 Discussion

Proximal femoral resection height influences the postoperative femoral offset and leg length changes, and further, this has a direct link to the biomechanical reconstruction of the joint [17, 18, 20]. Like previous studies [21–24], we observed a tendency to over-resect proximal femoral bone during conventional resection of the femoral head, consequently decreasing the femoral offset. A postoperatively decreased femoral offset is linked to various functional problems of the hip, such as soft-tissue tension, impingement, and loss of abductor muscle strength. However, we found that simple features integrated into the patient-specific instrument guides which link the preoperatively planned component position to the intraoperative procedure of preparing the proximal femoral head can significantly decrease the error in proximal resection height. Our hypothesis is that such increased accuracy in femoral head resection can ensure a better postoperative biomechanical outcome for the hip resurfacing procedure.

4 Patient-Specific Instrument Guides for Acetabulum Cup Orientation

The importance of the acetabulum component orientation for the long-term success of hip resurfacing has been discussed more often recently and has triggered interest in easy-to-use and accurate intraoperative navigation possibilities for acetabulum cup orientation. However, to date most published research for patient-specific acetabulum component alignment guides is focused on total hip arthroplasty applications [57, 61–65]. Although the procedure for preparation and impacting of the acetabular component during a THR is similar to the procedure performed during a HRA, a different surgical exposure of the acetabulum during HRA might not allow an easy translation of patient-specific acetabulum component alignment guides from THR to HRA. For example, we believe that it might not be possible to register a guide at the acetabulum rim during a hip resurfacing procedure, due to the limited access to this feature. To our knowledge, so far only one group has published an acetabulum guide solution designed for HRA procedure. In 2011 Zhang et al. [54] introduced a concept for a patient-specific acetabulum component alignment guide for hip resurfacing procedures. As a registration area for the guide,

---

[3]Unpaired, one-tailed Student's t-Test.

the authors proposed the articular surface of the acetabulum cup. With the guide in place, the insertion of a 3.2 mm Kirschner wire in the center of the acetabulum cup was navigated using an integrated guidance cylinder. After removing the guide, the wire was used to guide a custom-designed cannulated reamer. The accuracy of the proposed method was compared to the conventional method for cup alignment in a study with 20 HRA patients, using postoperative CT scans of the hip. The measured average error in the abduction angle of 1.2° and the average error of anteversion angle of 2.1° were both significantly smaller compared to the errors in the conventional group.

5 Osteophytes: Limitations for Patient-Specific Instrument Guides Procedures?

As discussed in previous sections, a majority of authors report a significant overall improvement of intraoperative accuracy using patient-specific instrument guides. However, there are repeated reports of outliers, in which the intraoperatively achieved instrument trajectory was notably deviated from its planned position and/or orientation [66–68]. The most likely reason for such problems is poor registration of the instrument guide intraoperatively. An exact registration between the guide and the anatomy requires that the registration surface of the guide be chosen with sufficient registration features. Furthermore, the registration surface of the guide has to be a mirror image of the anatomy. To achieve such geometrical fit, the preoperative image modality (such as CT or MRI) from which the guide is designed needs to depict the real anatomy accurately. Particularly in patients with osteoarthritis (which is the main reason for joint replacement surgeries such as hip resurfacing), this consideration might be vital, as the disease is characterized by the breakdown of articular cartilage accompanied by the changing of local bone anatomy [69]. One example of such bone alterations is the development of osteophytes—abnormal osteo-cartilagenous tissue that grows along joints [70], which are considered a major limiting factor for accuracy in image-guided surgeries. Research into osteophyte development has shown that osteophytes grow in four distinct stages [70, 71]. The osteophyte tissue in stages I and II has cartilage and fibrocartilage characteristics; in stage III an ossification process (impregnation with calcium) starts in the deepest cell layers of the osteophyte (closest to the bone). Stage IV is characterized by extended ossification within the central core. The resulting high variability in density and composition can interfere with an accurate depiction in medical image modalities, such as CT scans. The following clinical case will demonstrate the significance of this issue.

The patient was a 46-year-old man, scheduled for a resurfacing procedure of the right hip. Six weeks prior to the surgery, a CT scan of the hip was obtained. The scan was taken with a LightSpeed VCT scanner (GE Healthcare, Waukesha, USA) in helical mode, with a slice thickness of 2.5 mm, a peak kilovoltage of 120 kVp, and a pitch factor of 0.984375. During the scan, the patient was positioned Feet First-Supine and a bone imaging convolution kernel was used. The pixel size for the scan was 0.7 × 0.7 mm.

A 3D virtual surface model of the femur (Fig. 9) was created using the commercially available software package Mimics (Materialize, Leuven, Belgium, version 15). The CT images were loaded and a semi-automatic segmentation was performed, using thresholding algorithms as well as manual editing functions. An experienced user with the goal of representing the anatomical surface with the highest possible accuracy, including osteophyte depiction, performed manual editing, consisting of adding and/or deleting voxels to the threshold segmentation.

At this stage the position and orientation of the central-pin was determined and the femoral component size chosen. A patient-specific instrument guide for navigation of femoral central-pin placement was designed and a physical model of the guide was produced using a Rapid Prototyping machine (dimension SST, Stratasys, Inc., USA). Using the same thermo-plastic production process, a plastic model of the patient's proximal femur was created (Fig. 10). Both the plastic femur model, as well as the patient-specific instrument guide, were sterilized and sent to the operating theater.

During the surgery, an antero-lateral approach was performed and the proximal femur was exposed. The surgeon inspected the exposed anterior aspect of the femoral neck and compared it to the corresponding surfaces on the plastic femur model (Fig. 10). In this patient, the surgeon detected a visible difference in osteophyte depiction between the real anatomical surface and the segmented femur

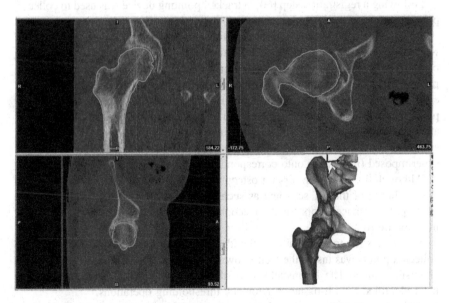

Fig. 9 Segmentation of preoperative CT dataset. The *blue outlines* show the final segmentation of the femur. In the *bottom right* window the final virtual models for proximal femur and acetabulum are displayed to the user for verification (Color figure online)

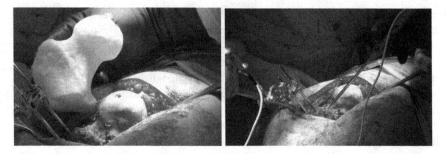

Fig. 10 Intraoperative collection of osteophytes. *Left side* A plastic model of the proximal femur was used to compare the segmentation to the real anatomy and identify undetected osteophytes. *Right side* A tracked pointing device was used to collect osteophyte border points intraoperatively

Table 1 Segmentation errors and Hounsfield units for osteophyte points collected intraoperatively

	1	2	3	4	5
Distance in (mm)	2.0	3.7	3.2	5.1	0.7
Hounsfield unit	290	65	125	125	160

model. To quantitatively measure this difference, an optoelectronical tracking system was used (Optotrak Certus Motion Capture System, Northern Digital Inc., Waterloo, Canada).

Following a registration step [60], a tracked pointing device was used to collect 5 points on the surface of undetected osteophytes (Fig. 10).

By using the registration information from the optoelectronical tracking system, we were able to transfer the collected positions into the CT coordinate system of the patient and evaluate the distance between the segmented femur surface and the points (which represent the real outlines of the osteophytes). Table 1 contains the error value for each collected point in mm. For a better understanding of the problem, we also identified the Hounsfield units (also known as CT numbers or CT attenuation numbers) of the collected positions in the preoperative CT scan, which are given in Table 1. In Fig. 11 the position of the five collected points are superimposed (red circle) onto corresponding axial slices of the CT dataset.

Although the distance errors for osteophyte 1 (2 mm) and 5 (0.7 mm) are within the resolution of the CT scan and as such might be the result of a partial volume effect [72], distance errors for the osteophytes 2–4 indicate depiction and/or segmentation errors.

It is interesting to note that only the identified Hounsfield unit of one of the five collected points was inside the well-known predefined bone threshold of 226–3070 Hounsfield units. The Hounsfield units of the remaining four points would have been identified as soft-tissue in automatic thresholding operations.

If these segmentation errors for the osteophytes had gone unnoticed, the deviation between the real anatomical registration surface and the integrated registration

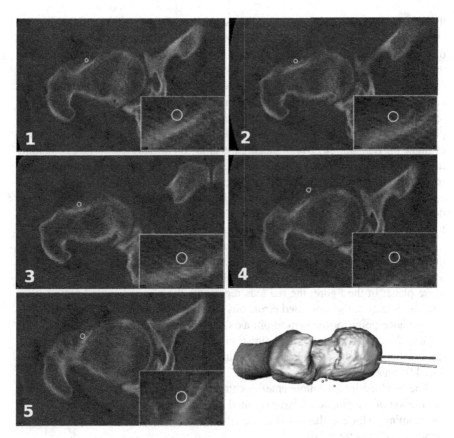

Fig. 11 Positions of the digitized osteophytes superimposed onto the CT dataset. The *numbers* in the *left lower corner* correspond to the number of osteophytes in Table 1. The *inlays* in the *right lower corner* are zoomed regions in the osteophyte area. The *lower right corner* of the figure illustrates the simulation of error to the central-pin placement. The *red axis* demonstrates the planned central-pin alignment, the *green line* the simulated central-pin alignment. The *green spheres* represent the digitized osteophyte points (Color figure online)

surface of the patient-specific guide could potentially result in errors in registration of the guide—which consequently would result in an error for the guided central-pin position and alignment.

We simulated this error for the patient[4] by virtually modifying the registration surface of the femur model based on the intraoperatively collected osteophyte information. To determine the deviation between the two registration surfaces (segmented and surgical), a point cloud for each surface was created. Initially, the collected osteophyte positions were added to the surgical surface point cloud and

---

[4]To guarantee the best possible surgical treatment for the patient, the surgeon removed the osteophyte completely or partially to correct the fit of the guide during the surgery.

the corresponding closest points from the segmented femur model to the segmented surface point cloud. In addition, about 20 points were collected on the lateral aspect of the femoral neck registration surface, which were added to both the surgical and the segmented surface point clouds. The reason for adding these points to both surface clouds was that the lateral aspect was unaffected by osteophytes and so was considered to be a stable registration area. Therefore, segmented and surgical registration surfaces in this area are represented identically. Taking these two sorted point clouds, a pair-point registration using Horn's method [73] was performed to estimate the registration error of placing the guide on the surgical surface containing the osteophytes compared to the planned registration on the segmented surface. The resulting transformation between the segmented and surgical surfaces was then applied to the position and orientation of the planned central-pin alignment and the deviation of the resulting pin position and orientation were evaluated in the transverse and frontal anatomical planes of the femur.

As Fig. 11 (bottom right) demonstrates, the registration of the patient-specific guide without the removal of the undetected osteophyte could have resulted in a substantial alignment error for the central-pin placement, especially in the transverse plane. In the figure, the red axis represents the planned central-pin position, and the blue axis, the simulated erroneous central-pin axis. In this case, the error in the entrance point for the central-pin axis on the femoral head was simulated with 1.2 mm deviation in the proximal direction and 4.9 mm in the anterior direction. For the orientation of the central-pin, we simulated the resulting error with 0.3° valgus and 4.9° anteversion.

The combination of a more anterior entrance point, as well as a more anteversion orientation of the pin, could have resulted in a notching of the femoral head during the reaming, which is discussed in the literature as a risk factor for postoperative femoral neck fractures [10, 11].

5.1 Discussion

The presented case demonstrated that undetected osteophytes in registration surfaces could present a substantial source of error for the postoperative outcome in patient-specific instrument guided procedures. On the other hand, the Hounsfield units of the undetected osteophytes ranged between 65 and 290, which suggests that the osteophyte borders were imaged with a wide range, indicating changing tissue properties along the osteophyte surface. This noteworthy high standard deviation of Hounsfield units on osteophyte borders was confirmed in a study we performed on 35 patients, collecting osteophyte data intraoperatively [74]. Here we found that the osteophyte points were depicted in a clinical CT dataset with an average of 156 Hounsfield units and a standard deviation of 112. In this study, about 80 % of the collected osteophyte border points had a Hounsfield unit outside of the defined threshold for bone and would have presented as soft-tissue in the scan. Considering the development of osteophytes from an abnormal cartilaginous mass [70] in stage I

to hypertrophy and endochondral bone formation in stage IV, soft-tissue characteristics in HU representation is not unexpected. However, it does demonstrate the
difficulties in segmenting osteophytes in preoperative CT datasets and supports the
careful selection of registration surfaces for patient-specific instrument guides for
femur and acetabulum excluding known osteophyte regions, such as femoral
head-neck junctions or acetabulum rim.

6 Conclusions

Hip resurfacing procedures have many advantages. However, they are surgically
challenging techniques with a small margin of error. This, plus the need for careful
patient selection, leaves room for surgical errors that can result in minor as well as
major postoperative complications. In this chapter we have discussed how
patient-specific instrument guides can be utilized to create easy-to-use navigation
systems for hip resurfacing arthroplasty. Although to date most applications are in
the guidance of the femoral central-pin drilling, recent research in applying
patient-specific guides to navigate acetabulum component alignment and proximal
femoral resection depth are discussed as well. Many of the studies discussed in this
contribution found that the application of patient-specific instrument guides significantly increased the accuracy of the intraoperative procedure compared to
conventional surgical methods. Such improvement of intraoperative accuracy might
lead to fewer postoperative complications and improved stability and functionality
of HRA. However, further studies are needed to investigate the relationship
between the use of image-guided technologies, such as patient-specific instrument
guides, and the mid- or long-term outcome for HRA.

In addition to improved accuracy during the intraoperative procedure, the guides
also have the potential to reduce surgical time for hip resurfacing procedures. For
example, Zhang et al. [54] found that the average surgical time was significantly
shortened, by 22 min, due to the use of patient-specific guides for femoral and
acetabular component alignment. Olsen et al. [55] also found in an in vitro study
that the time required for insertion of the femoral central-pin was significantly
reduced using a patient-specific guide (1.3 min), compared to both the conventional
placement technique (5.4 min) and the imageless optoelectronic guidance method
(7.3 min). The elongation of surgery time using conventional image-guided
methods (such as optoelectronic) is often discussed as a drawback for using this
technology in the surgical room [75]. Patient-specific instrument guides in HRA not
only avoid increased surgical time, but reduce it–advantages likely to make these
guides more acceptable for routine clinical use.

A requirement for the application of patient-specific instrument guides is the
preoperative detailed reconstruction of the patient anatomy, which allows generation of the contact surfaces of the guides. In all of the studies and applications for
patient-specific instrument guides discussed in this chapter, a CT scan of the patient
hip was the preoperative image modality. CT scans are relatively fast and easy to

obtain and depict cortical and dense bone with a high resolution, which makes them an optimal image modality for preoperative planning prosthesis component positioning and the creation of patient-specific instrument guides. On the other hand, the use of CT also raises concerns, such as uncertainties in depiction of osteophytes and soft tissues (cartilage, etc.), as well as radiation exposure for the patient. For application of patient-specific instrument guides in other areas of orthopaedic surgery, such as Total Knee Arthroplasty, the use of Magnetic Resonance Imaging (MRI) as the preoperative image modality for surgical planning and guide creation is employed. Further investigations have to show whether MRI might also be a valuable alternative as an image modality for patient-specific instrument guides in hip resurfacing.

Metal-on-Metal Hip Resurfacing is a relatively young treatment option and research is still ongoing with respect to optimal component alignment. An accurate and widely available image-guided technology (such as patient-specific guides) to reliable implant components in predefined positions and orientations could be used in support of clinical studies as a valuable research tool.

Research into the application of patient-specific instrument guides for HRA combines expertise from many different fields, involving orthopaedic surgeons, computer scientists, radiologists, mechanical engineers and materials engineers, and demonstrates how interdisciplinary research can accelerate new developments in the medical fields.

Acknowledgments The authors are grateful to Paul St. John and Joan Willison for their always-enthusiastically technical support, and would like to thank the surgical and perioperative teams of Kingston General Hospital, as well as the Human Mobility Research Centre at Queen's University and Kingston General Hospital for their support.

References

1. Amstutz HC, Clarke IC, Christie J, Graff-Radford A (1977) Total hip articular replacement by internal eccentric shells: the Tharies approach to total surface replacement arthroplasty. Clin Orthop 128:261–284
2. Amstutz HC, Le Duff MJ (2012) Hip resurfacing; a 40-year perspective. HSS J 8(3):275–282
3. Amstutz HC, Takamura K, Le Duff M (2011) The effect of patient selection and surgical technique on the results of conserve plus hip resurfacing—3.5 to 14 year follow-up. Orthop Clin North Am 142(2):133–142
4. Bradley GW, Freeman MA (1983) Revision of failed hip resurfacing. Clin Orthop Relat Res 178:236–240
5. Klotz MCM, Breusch SJ, Hassenpflug M, Bitsch RG (2012) Results of 5 to 10-year follow-up after hip resurfacing. A systematic analysis of the literature on long-term results. Orthopade 41(6):442–451
6. Vail TP, Mina CA, Yergler JD, Pietrobon R (2006) Metal-on-metal hip resurfacing compares favorably with THA at 2 years followup. Clin Orthop Relat Res 453:123–131
7. Sandiford NA, Ahmed S, Doctor C, East DJ, Miles K, Apthrop HD (2014) Patient satisfaction and clinical results at a mean eight years following BHR arthroplasty: results from a district general hospital. Hip Int 24(3):249–255

8. Bow JK, Rudan JF, Grant HJ, Mann SM, Kunz M (2012) Are hip resurfacing arthroplasties meeting the needs of our patients? A 2-year follow-up study. J Arthroplasty 27(6):684–689

9. Mont MA, Ragland PS, Etienne G, Seyler TM, Schmalzried TP (2006) Hip resurfacing arthroplasty. J Am Acad Orthop Surg 14(8):454–463

10. Amstutz HC, Campbell PA, Le Duff MJ (2004) Fracture of the neck of the femur after surface arthroplasty of the hip. J Bone Joint Surg 86-A(9):1874–1877

11. Siebel T, Maubach S, Morlock MM (2006) Lessons learned from early clinical experience and results of 300 ASR hip resurfacing implantations. Proc Inst Mech Eng [H] 220(2):345–353

12. Mellon SJ, Grammatopoulos G, Andersen MS, Pandit HG, Gill HS, Murray DW (2015) Optimal acetabular component orientation estimated using edge-loading and impingement risk in patients with metal-on-metal hip resurfacing arthroplasty. Biomechanics 48(2):318–323

13. Davies ET, Olsen M, Zdero R, Papini M, Waddell JP, Schemitsch EH (2009) A biomechanical and finite element analysis of femoral neck notching during hip resurfacing. J Biomech Eng 131(4):041002-1–041002-8

14. Beaule PE, Lee JLL, Le Duff MJ, Amstutz HC, Ebramzadeh E (2004) Orientation of the femoral component in surface arthroplasty of the hip. J Bone Joint Surg 86-A(9):2015–2021

15. Long JP, Barel DL (2006) Surgical variables affect the mechanics of a hip resurfacing system. Clin Orthop Relat Res 453:115–122

16. Konyves A, Bannister GC (2005) The importance of leg length discrepancy after total hip arthroplasty. J Bone and Joint Surgery 87-B(2):155–157

17. Lai KA, Lin CJ, Jou IM, Su FC (2001) Gait analysis after total hip arthroplasty with leg length equalization in women with unilateral congenital complete dislocation of the hip: comparison with untreated patients. J Orthop Res 19:1147–1152

18. McGrory BJ, Morrey BF, Cahalan TD, An K-N, Cabanela ME (1995) Effect of femoral offset on range of motion and abductor muscle strength after total hip arthroplasty. J Bone Joint Surg 77-B(6):865–869

19. Fackler CD, Poss R (1980) Dislocation in total hip arthroplasties. Clin Orthop Relat Res 151:169–178

20. Asayama I, Chamnongkich S, Simpson KJ, Kinsey TL, Mahoney OM (2005) Reconstructed hip joint position and abductor muscle strength after total hip arthroplasty. J Arthroplasty 20(4):414–420

21. Silva M, Lee KH, Heisel C, Dela Rosa MA, Schmalzried TP (2004) The biomechanical results of total hip resurfacing arthroplasty. J Bone and Joint Surg 86-A(1):40–46

22. Loughead JM, Chesney D, Holland JP, McCaskie AW (2005) Comparison of offset in Birmingham hip resurfacing and hybrid total hip arthroplasty. J Bone and Joint Surg 87-B(2):163–166

23. Girard J, Lavigne M, Vendittoli P-A, Roy AG (2006) Biomechanical reconstruction of the hip. J Bone Joint Surg 88-B(6):721–726

24. Ahmad R, Gillespie G, Annamalai S, Barakat MJ, Ahmed SMY, Smith LK, Spencer RF (2009) Leg length and offset following hip resurfacing and hip replacement. Hip Int 19(2):136–140

25. Kwon Y-M, Ostlere SJ, McLardy-Smith P, Athanasou NA, Gill HS, Murray DW (2011) "Asymptomatic" pseudotumors after metal-on-metal hip resurfacing arthroplasty. Prevalence and metal ion study. J Arthroplasty 26(4):511–518

26. Kwon YM, Xia Z, Glyn-Jones S, Beard D, Gill HS, Murray DW (2009) Dose-dependent cytotoxicity of clinically relevant cobalt nanoparticles and ions on macrophages in vitro. Biomed Mater 4(2):025018

27. Langton DJ, Sprowson AP, Joyce TJ, Reed M, Carluke I, Partington P, Nargol AV (2009) Blood metal ion concentrations after hip resurfacing arthroplasty: a comparative study of articular surface replacement and Birmingham hip resurfacing arthroplasties. J Bone Joint Surg 91-B(10):1287–1295

28. Kwon YM, Glyn-Jones S, Simpson DJ, Kamali A, McLardy-Smith P, Gill HS, Murray DW (2010) Analysis of wear retrieved metal-on-metal hip resurfacing implants revised due to pseudotumours. J Bone Joint Surg 92-B(3):356–361

29. Langton DJ, Joyce TJ, Mangat N, Lord J, Van Orsouw M, De Smet K, Nargol AV (2011) Reducing metal ion release following hip resurfacing arthroplasty. Orthop Clin North Am 42 (2):169–180
30. Kwon YM, Mellon SJ, Monk P, Murray DW, Gill HS (2012) In vivo evaluation of edge-loading in metal-on-metal hip resurfacing patients with pseudotumours. Bone Joint Res 1 (4):42–49
31. Mellon SJ, Kwon YM, Glyn-Jones S, Murray DW, Gill HS (2011) The effect of motion patterns on edge-loading of metal-on-metal hip resurfacing. Med Eng Phys 33(10):1212–1220
32. De Haan R, Pattyn C, Gill HS, Murray DW, Campbell PA, De Smet K (2008) Correlation between inclination of the acetabular component and metal ion levels in metal-on-metal hip resurfacing replacement. J Bone Joint Surg 90-B(10):1291–1297
33. De Haan R, Campbell PA, Su EP, De Smet KA (2008) Revision of metal-on-metal resurfacing arthroplasty of the hip: the influence of malpositioning of the components. J Bone Joint Surg 90-B(9):1158–1163
34. Hart AJ, Satchithananda K, Liddle AD, Sabah SA, McRobbie D, Henckel J, Cobb JP, Skinner JA, Mitchell AW (2012) Pseudotumors in association with well-functioning metal-on-metal hip prostheses: a case-control study using three-dimensional computed tomography and magnetic resonance imaging. J Bone Joint Surg 94-A(4):317–325
35. Liu F, Gross TP (2012) A safe zone for acetabular component position in metal-on-metal hip resurfacing arthroplasty: winner of the 2012 HAP PAUL award. J Arthroplasty 28:1224–1230
36. Arndt JM, Wera GD, Goldberg VM (2013) A initial experience with hip resurfacing versus cementless total hip arthroplasty. HSSJ 9:145–149
37. Hess T, Gampe T, Koettgen C, Szawloeski B (2004) Intraoperative navigation for hip resurfacing. Methods and first results. Orthopade 33(10):1183–1193
38. El Hachim M, Penasse M (2014) Our midterm results of the Birmingham hip resurfacing with and without navigation. J Arthroplasty 29:808–812
39. Bailey C, Gul R, Falworth M, Zadow S, Oakeshott R (2009) Component alignment in hip resurfacing using computer navigation. Clin Orthop Relat Res 467:917–922
40. Hodgson A, Helmy N, Masri BA, Greidanus NV, Inkpen KB, Duncan CP, Garbuz DS, Anglin C (2007) Comparative repeatability of guide-pin axis positioning in computer-assisted and manual femoral head resurfacing arthroplasty. Proc Inst Mech Eng H 221(7):713–724
41. Cobb JP, Kannan V, Brust K, Thevendran G (2007) Navigation reduces the learning curve in resurfacing total hip arthroplasty. Clin Orthop Relat Res 463:90–97
42. Romanowski JR, Swank ML (2008) Imageless navigation in hip resurfacing: avoiding component malposition during the surgeon learning curve. J Bone Joint Surg 90-A(Suppl 3):65–70
43. Seyler TM, Lai LP, Sprinkle DI, Ward WG, Jinnah RH (2008) Does computer-assisted surgery improve accuracy and decrease the learning curve in hip resurfacing? A radiographic analysis. J Bone Joint Surg 90-A(Suppl 3):71–80
44. Radermacher K, Portheine F, Anton M, Zimolong A, Kaspers G, Rau G, Staudte HW (1998) Computer assisted orthopaedic surgery with image based individual templates. Clin Orthop Relat Res 354:28–38
45. Birnbaum K, Schkommodau E, Decker N, Prescher A, Klapper U, Radermacher K (2001) Computer-assisted orthopedic surgery with individual templates and comparison to conventional operation method. Spine 26(4):365–370
46. Brown GA, Firoozbakhsh K, DeCoster TA, Reyna JR Jr, Moneim M (2003) Rapid prototyping: the future of trauma surgery? J Bone Joint Surg 85-A(Suppl 4):49–55
47. Owen BD, Christensen GE, Reinhardt JM, Ryken TC (2007) Rapid prototype patient-specific drill template for cervical pedicle screw placement. Comput Aided Surg 12(5):303–308
48. Boonen B, Schotanus MG, Kort NP (2012) Preliminary experience with the patient-specific templating total knee arthroplasty. Acta Orthop 83(4):387–393
49. Ng VY, DeClaire JH, Berend KR, Gulick BC, Lombardi AV Jr (2012) Improved accuracy of alignment with patient-specific positioning guides compared with manual instrumentation in TKA. Clin Orthop Relat Res 470(1):99–107

50. Moopanar TR, Amaranath JE, Sorial RM (2014) Component position alignment with patient-specific jigs in total knee arthroplasty. ANZ J Surg 84(9):628–632
51. Levy JC, Everding NG, Frankle MA, Keppler LJ (2014) Accuracy of patient-specific guided glenoid baseplate positioning for reverse shoulder arthroplasty. J Shoulder Elbow Surg 23 (10):1563–1567
52. Walch G, Vezeridis PS, Boileau P, Deransart P, Chaoui J (2014) Three-dimensional planning and use of patient-specific guides improve glenoid component position: an in vitro study. J Shoulder Elbow Surg 24(20):302–309
53. Du H, Tian XX, Li TS, Yang JS, Li KH, Pei GX, Xie L (2013) Use of patient-specific templates in hip resurfacing arthroplasty: experience from sixteen cases. Int Orthop 37(5): 777–782
54. Zhang YZ, Lu S, Yang Y, Xu YQ, Li YB, Pei GX (2011) Design and primary application of computer-assisted, patient-specific navigational templates in metal-on-metal hip resurfacing arthroplasty. J Arthroplasty 26(7):1083–1087
55. Olsen M, Naudie DD, Edwards RW, Schemitsch EH (2014) Evaluation of a patient specific femoral alignment guide for hip resurfacing. J Arthroplasty 29:590–595
56. Kitada M, Sakai T, Murase T, Hanada T, Nakamura N, Sugano N (2013) Validation of the femoral component placement during hip resurfacing: a comparison between conventional jig, patient-specific template, and CT-based navigation. Int J Med Robotics Comput Assist Surg 9:223–229
57. Sakai T, Hanada T, Murase T, Kitada M, Hamada H, Yoshikawa H, Sugano N (2014) Validation of patient specific guides in total hip arthoplasty. Int J Med Robotocs Comput Assist Surg 10:113–120
58. Kunz M, Rudan JF, Xenoyannis GL, Ellis RE (2010) Computer-assisted hip resurfacing using individualized drill templates. J Arthroplasty 25(4):600–606
59. Kunz M, Rudan JF, Wood GCA, Ellis RE (2011) Registration stability of physical templates in hip surgery. Stud Health Technol Inform 162:283–289
60. Ma B, Ellis RE (2003) Robust registration for computer-integrated orthopedic surgery: laboratory validation and clinical experience. Med Imag Anal 7(3):237–250
61. Hananouchi T, Saito M, Koyama T, Hagio K, Murase T, Sugano N, Yoshikawa H (2008) Tailor-made surgical guide based on rapid prototyping technique for cup insertion in total hip arthroplasty. Int J Med Robotics Comput Assist Surgery 5:164–169
62. Hananouchi T, Saito M, Koyama T, Sugano N, Yoshikawa H (2010) Tailor-based surgical guide reduces incidence of outliers of cup placement. Clin Orthop Relat Res 468:1088–1095
63. Zhang YZ, Chen B, Lu S, Yang Y, Zhao JM, Liu R, Li YB, Pai GX (2011) Preliminary application of computer-assisted patient-specific acetabulum navigational template for total hip arthroplasty in adult single development dysplasia of the hip. Int J Med Robotics Comput Assist Surg 7:469–474
64. Buller L, Smith T, Bryan J, Klika A, Barsoum W, Innotti JP (2013) The use of patient-specific instrumentation improves the accuracy of acetabular component placement. J Arthroplasty 28:631–636
65. Small T, Krebs V, Molloy R, Bryan J, Klika AK, Barsoum WK (2014) Comparison of acetabular shell position using patient specific instrument vs. standard surgical instruments: A randomized clinical trial. J Arthroplasty 29:1030–1037
66. Leeuwen JA, Grøgaard B, Nordsletten L, Röhrl SM (2014) Comparison of planned and achieved implant position in total knee arthroplasty with patient-specific positioning guides. Acta Orthop 11:1–6
67. Cavaignac E, Pailhé R, Laumond G, Murgier J, Reina N, Laffosse JM, Bérard E, Chiron P (2014) Evaluation of the accuracy of patient-specific cutting blocks for total knee arthroplasty: a meta-analysis. Int Orthop (Epub ahead of print)
68. Voleti PB, Hamula MJ, Baldwin KD, Lee GC (2014) Current data do not support routine use of patient-specific instrumentation in total knee arthroplasty. J Arthroplasty 29(9):1709–1712
69. Turmezei TD, Poole KE (2011) Computed tomography of subchrondral bone and osteophytes in hip osteoarthritis: the shape of things to come? Front Endocrinol (Lausanne) 13(2):97

70. Gelse K, Soeder S, Eger W, Dietmar T, Aigner T (2003) Osteohyte development—molecular characterization of different stages. Osteoarthritis cartilage 11(2):141–148
71. Aigner T, Sachsse A, Gebhard PM, Roach HI (2006) Osteoarthritis: pathobiology—targets and ways for therapeutic intervention. Adv Drug Deliv Rev 58(2):128–149
72. Baxter BS, Sorenson JA (1981) Factors affecting the measurement of size and CT number in computed tomography. Invest Radiol 16:337–341
73. Horn BKP (1987) Closed form solution of absolute orientation using unit quaternions. J Opt Soc Amer A 4(44):629–642
74. Kunz M, Balaketheeswaran S, Ellis RE, Rudan JF (2015) The influence of osteophyte depiction in CT for patient-specific guided hip resurfacing procedures. Int J Comput Assist Radiol Surg (Accepted for publication with minor revisions)
75. Cerha O, Kirschner S, Guenther K-P, Luetzner J (2009) Cost analysis for navigation in knee endoprothetics. Orthopaede 38(12):1235–1240 (Article in German)

Printed in the United States
By Bookmasters